Analytical Methods for Geochemical Exploration

Analytical Methods for Geochemical Exploration

J. C. Van Loon and R. R. Barefoot

Departments of Geology and Chemistry
and
The Institute for Environmental Studies
University of Toronto
Toronto, Canada

Academic Press, Inc.
Harcourt Brace Jovanovich, Publishers
San Diego New York Berkeley Boston
London Sydney Tokyo Toronto

ACADEMIC PRESS, INC.
San Diego, California 92101

United Kingdom Edition published by
ACADEMIC PRESS, INC. (LONDON) LTD.
24-28 Oval Road, London NW1 7DX

Library of Congress Cataloging-in-Publication Data

Van Loon, J. C. (Jon Clement), Date
Analytical methods for geochemical exploration / J.C. Van Loon and R.R. Barefoot.
p. cm.
Bibliography: p.
Includes index.
ISBN 0-12-714170-7 (alk. paper)
1. Geochemistry, Analytical. 2. Geochemical prospecting.
I. Barefoot, R. R. II. Title.
QE516.3.V36 1988
622′.13—dc19 87-33382
CIP

PRINTED IN THE UNITED STATES OF AMERICA
88 89 90 91 9 8 7 6 5 4 3 2 1

Contents

5. Field Methods

6. Geochemical Methods for Soils and Related Samples

7. Lithogeochemical Analysis

Preface

Analytical Methods for Geochemical Exploration covers chemical analysis methodology mainly for the determination of base metals and precious metals in geochemical exploration samples. It is meant for the practicing analyst. No principles of geochemical exploration are included since this constitutes specialist material that has been the subject of other books (1, 2). The theory of chemical analytical methods has been kept to a minimum. Complete procedures have been provided in most cases so that no additional sources are necessary to finish the analyses. To avoid misinterpretations, the authors' descriptions of equipment and reagents have been left as complete as possible.

Chemical analysis methodology has been chosen as follows: Preference has been given to procedures that have evidence of thorough testing. At the head of this list are methods that have been tested using standard reference samples or have been a part of interlaboratory comparison studies, or both. Responsibility for the choice of methodology rests totally with the authors. We are painfully aware of the probability that some good methods have been overlooked. For this we apologize. Careful use of the methods printed in this book, however, should result in the acquisition of reliable data.

A wide variety of instrumental techniques is currently used in geochemical exploration. Among these we have found that atomic absorption, plasma source atomic emission, and x-ray fluorescence greatly predominate. To make the book concise and yet as useful to most laboratories as possible, we have limited our coverage primarily to these three techniques. An exception to this rule is Chapter 10, in which plasma source mass spectrometry is introduced. We believe this technique, just now beginning to make an inroad into geochemical work, is of such great promise as to warrant its brief inclusion here. In our view, plasma source mass spectrometry will make possible routine isotopic ratio work. This is an area of great importance to geochemistry. In addition, isotope dilution mass spectrometry, the technique against which much methodology is judged, could become a practical day-to-day method of analysis.

It is difficult to unambiguously structure a book of this type. The analytical method chapters are organized according to sample type and operation except for precious metals determinations and plasma source mass spectrometry. In this way we hoped to keep duplication to a minimum. Other chapters on basic materials and sample preparation emphasize the importance ascribed to the chemical treatments, the choice of reagents, the proper methods of solution storage, the importance of standard reference samples, and so on.

Finally, the authors wish to express deep appreciation to the copyright holders and researchers who gave permission for valuable works to be utilized.

1. A. A. Levinson, "Introduction to Exploration Geochemistry," 2nd ed. Applied Publishing Ltd., Wilmette, Illinois, 1980.
2. A. W. Rose, H. H. Hawkes, and J. S. Webb, "Geochemistry in Mineral Exploration," 2nd ed. Academic Press, New York, 1979.

1 Introduction

The analysis of geological samples presents the ultimate challenge to the analytical chemist. This is true of all aspects of the process, from sampling through sample preparation (physical and chemical) to the final determination. The samples embrace a wide variety of geological and related materials. These include rocks, ores, sediments, soils, vegetation, waters, and vapors.

Geochemical methods of exploration for ore bodies have become popular in the past three decades. They are used not by themselves, but together with geological and geophysical approaches.

There are two types of geochemical survey and they are classified according to the amount of detail employed. These are reconnaissance and detailed surveys. Reconnaissance surveys involve the investigation of a large area with a relatively low density of sample points. Detailed surveys, as the name suggests, involve a relatively high density of sample points which are distributed over a relatively small area. The object of the reconnaissance investigation is to locate possible mineralized locations within a large area. A detailed survey of the outlined location can then be done to locate the ore body more precisely.

I. SAMPLES

A. Rocks

The chemical analysis of rocks for trace and major elements is most commonly undertaken in detailed geochemical surveys. When mineralization occurs there is a release of ore-forming elements into the surrounding rock. This escape of elements occurs by convective transport through fractures and veins and by diffusion through pores. Thus, by sampling this contaminated area in the vicinity of an ore body it will be possible to locate the ore. These anomalous areas are best developed near hydrothermal deposits since fluids have relatively low viscosities.

A wide range of igneous, metamorphic, and sedimentary rocks may be encountered in rock geochemical surveys. Thus, methods of sample preparation must be capable of dissolving silicate material.

Acid mixtures containing hydrofluoric acid may be employed. Usually sulfuric acid is one of the main components of this mixture because its relatively high boiling point allows complete expulsion of traces of fluoride toward the end of the decomposition procedure. More rapid and more complete digestions with these acids can be obtained using pressure. For this purpose Teflon "bombs" are commonly employed. In the present authors' laboratory, thick-walled tightly capped Teflon vessels are used. They are placed in batches of six to eight in a conventional pressure cooker.

If grains of sulfide or oxide minerals are present in the rocks, hydrochloric acid and/or nitric acid may be added to the acid solution. Often a treatment with these acids follows the completion of the initial hydrofluoric acid decomposition step.

Fusions are effective in decomposing silicate rocks. However, contamination introduced by the relatively large amounts of fluxing agent necessary can invalidate the results if trace elements are sought. Blanks must be run with both the fusion technique and the acid approach.

A variety of fluxing agents have been proposed for use with rock samples. These include sodium and potassium pyrosulfate, sodium and potassium carbonate, potassium hydroxide, lithium metaborate, and lithium tetraborate. When oxidizing conditions are useful or essential, sodium peroxide may be added to the fusion mixture.

When the fusion must be dissolved, this is usually accomplished using a dilute mineral acid. X-ray fluorescence methods utilize buttons formed from fused sample using lithium borates. These buttons are cast so that they fit the sample holder exactly.

B. Sediments and Soils

Soils are frequently used as indicators of underlying ore bodies. Ore-forming metals travel outward from the ore body by a variety of mechanisms into the surrounding soil.

Generally in geochemical surveys it is not necessary or desirable to totally decompose soils or sediments. The trace elements of interest are usually adsorbed on particle surfaces, loosely bound in clays, trapped in manganese or iron oxide phases, and associated with organic matter.

Commonly, therefore, strong mineral acid mixtures (e.g., hydrochloric/nitric) are used at elevated temperatures. When manganese or iron oxides are to be attacked a reducing agent can be employed. Organic matter must be oxidized to fully release the associated metals, and for this purpose perchloric acid or hydrogen peroxide can be used.

Recently there has been interest in using "selective" extractants with soils and sediments to develop a better contrast. The purpose is to extract some particular form of the metal with these reagents. In this regard, reducing agents, weak acids, salts of weak acids, and chelating agents have been employed.

C. Vegetation

Twigs and leaves from trees and shrubs are often used as samples for geochemical exploration. Plant roots can extract metals from the soils and these metals are transported to and become fixed in plant tissue, often at elevated levels compared to the soil.

Samples containing a high level of organic matter must be ashed to aid their decomposition. There are two choices, wet and dry ashing.

Wet ashing involves treatment of the samples with an oxidizing mixture of acids. Preeminent among the oxidizing acid mixtures are those involving perchloric acid with nitric acid. For geochemical exploration purposes, however, a 3 : 1 hydrochloric acid : nitric acid mixture is often satisfactory and is faster and easier to employ. Another oxidizing mixture which has gained some favor among geochemical analysts is a solution of hydrogen peroxide and sulfuric acid. However, a procedure involving these two constituents is very labor-intensive, involving dropwise addition of hydrogen peroxide throughout the decomposition.

Dry ashing is generally much less labor-intensive than wet ashing. This approach involves treatment of the sample open to the air in an oven at temperatures above 430°C. Unfortunately, even at this relatively low temperature there will be losses of volatile elements such as arsenic, selenium, and mercury. Many procedures recommend ashing temperatures above 550°C. If such temperatures are used, there is additional danger of loss of zinc, cadmium, and lead.

D. Waters

Water in contact with ore bodies leaches small amounts of the constituent metals. Natural waters have pH values ranging from 6.5 to 8.0. Under these acidity conditions the transition heavy metals are only very slightly soluble. However, if the acidity should increase (e.g., due to oxidizing reactions with sulfide or acid rain), a proportionately larger amount of metal will be mobilized. Natural chelating agents found in some waters may also increase the metal levels found therein.

Gibbs (1) identified five mechanisms by which metal can be transported in waters: in solution, as inorganic and organic complexes, adsorbed, precipitated and coprecipitated in organic solids, and in crystalline sediment material. Whether or not one agrees with such a classification, it is obvious that metals in some forms are more available for reaction in a geochemical environment than others. Thus it is probably useful to distinguish between easily "extractable" (more reactive) metal and "total" metal in waters. Extractable metal in this context is that metal, both bound and unbound, which can be extracted by the organic reagents used according to the proposed procedure. Total metal refers to that metal which can be released by a strong acid (usually hydrochloric–nitric mixtures) digestion. This metal is termed total metal because experience with many waters indicates that over 90% of the metal in the water is released by using such mixtures.

No general accord exists on the method of water pretreatment prior to analysis. Most authors agree on an acidification following filtration. The filtration should be done at the time of sample collection or within a few hours thereafter. It is common to filter waters through a 0.45 μm pore size filter. Metal which passes this pore size is often termed "soluble." The present authors believe that a significant percentage of particulate metal passes the 0.45 μm pore size filter. Work designed to clarify some of the above variables is in progress in this laboratory, embracing a wide range of water sample types.

The levels of metals in water are often below the detection limits of even modern instrumental techniques. Thus, it may be necessary to concentrate the sample. For this purpose an evaporation, chelation followed by solvent extraction, or an ion-exchange chromatographic step can be employed.

E. Vapors

Relatively recently there has been interest in determining substances in the atmosphere or in soil vapors as a means of chemically prospecting for ore bodies. For example, elemental mercury and some mercury compounds have an appreciable vapor pressure at ambient temperatures. Thus, the vapors can be trapped and stripped of mercury and the mercury determined. Vapor methods at this date remain largely in the development stage.

II. ANALYSIS TECHNIQUES

The techniques commonly used for trace element analysis are listed in Table 1.1 (2) together with approximate instrumental detection limits for a selection of elements. The abbreviations are: ASV, anodic stripping voltametry; FL–AAS, flame–atomic absorption spectrometry; ETA–AAS, electrothermal atomization–atomic absorption spectrometry; ICP–AES, inductively coupled plasma–atomic emission spectrometry; NAA, neutron activation analysis.

It is interesting to compare the values in Table 1.1 with values given in Table 1.2 (3) for the average amounts of elements now thought to constitute geochemical samples. When a solution is used for an analysis it must be remembered that at least a factor of 10 dilution, and most often more, will be involved. Thus, when a solution is involved the values in Table 1.2 should be multiplied by at least a factor of 10.

A cursory survey of the literature suggests that between 65 and 75% of all trace element analyses are carried out by the techniques of valence electron atomic spectrometry, with atomic absorption (to this date) being responsible for the largest fraction of this work. Emission spectrometry, with the advent of the inductively coupled plasma (ICP) source, is presently enjoying a renaissance. In the larger

Table 1.1
Detection Limits and Sensitivities of Analysis Techniques (ppb) (2)

Element	ASV detection limit	FL–AAS detection limit	ETA–AAS detection limit	ICP–AES detection limit	NAA sensitivity	X-ray fluorescence detection limit
As	—	100	0.2	20	50	—
Ba	—	200	0.06	5	20	400
Be	—	20	0.001	3	—	1000
Bi	0.05	50	0.3	50	—	2000
Cd	0.005	1	0.003	1	5	2500
Co	—	5	0.6	2	10	200
Cr	—	2	0.4	2	300	700
Cu	0.5	2	0.4	2	2	400
Fe	—	1	0.8	2	2000	300
Hg	1	2200	0.5	50	3	1000
Mn	—	3	0.1	0.5	1	200
Mo	—	10	0.2	5	1000	1500
Ni	—	8	0.1	5	700	200
Pb	0.01	10	0.05	20	500	—
Sb	0.01	30	0.2	200	7	—
Se	—	100	0.1	30	10	—
Sn	0.2	1000	0.4	3	30	—
V	—	20	0.5	2	2	150
Zn	0.4	0.6	0.02	1	100	200

Table 1.2
Average Elemental Abundances (ppb) (3)[a]

Element	Rocks	Soils	Plant ash	Water (fresh)
As	2,000	7,500	—	2
Ba	580,000	300,000	500,000	20
Be	2,000	3,000	700	5
Bi	100	800	700	0.005
Cd	100	300	4,300	0.03
Co	25,000	10,000	5,000	0.1
Cr	100,000	43,000	6,300	1
Cu	50,000	15,000	130,000	3
Fe	4.6%	21,000	1.6%	100
Hg	20	56	10	0.56
Mn	0.1%	320,000	0.7%	15
Mo	1,500	2,500	3,000	2
Ni	75,000	17,000	18,000	2
Pb	10,000	17,000	30,000	3
Sb	100	2,000	1,000	2
Se	100	300	27	0.3
Sn	2,000	10,000	15,000	0.1
V	150,000	57,000	5,000	2
Zn	80,000	36,000	570,000	20

[a]With a few changes.

laboratories this technique will probably replace flame atomic absorption in applications where more than three elements must be determined per sample. Furnace atomic absorption, however, will remain preeminent in the foreseeable future when best detection limits are necessary. However, most trace metal element analysis laboratories do not have access to neutron activation equipment. X-ray fluorescence has some applications for trace element analysis but is better used for the determination of major and minor elements (from 0.01 to 100%). Electrochemical techniques have been used over the years for trace element analysis. Interferences, when complex samples are to be analyzed, are usually too severe for this approach to have general usefulness in biological or environmental trace element analysis.

Inductively coupled plasma emission spectrometry detection limits are in many cases too poor for direct geochemical determinations. Because of the inherent multielement characteristics of ICP emission (particularly with a quantometer), the time per sample analysis is small compared to that with AAS when more than about three elements are to be determined. Thus elemental concentration becomes a practical process when ICP emission is employed. In the case of sediments, soils, and similar samples, detection limits often need not be so good. However, the matrix is very complex and the sample contains high levels of interfering substances. Thus, detection limits for the techniques in these cases will be much poorer than those given in Table 1.1.

Obtaining good detection limits is important, particularly when samples with background levels of the trace elements are to be analyzed. The steady decrease in these background values over the years attests to the fact that improved detection limits are still required if accurate values of trace element levels in geochemical exploration are to be obtained.

The term "trace" in trace element analysis of geochemical samples does not have commonly agreed upon concentration limits. For this monograph trace means any concentration below about 0.01% (i.e., 100 μg/g or 100 μg/ml). Ultratrace is an expression sometimes used for levels below nanograms per milliliter (or nanograms per gram). The term is not employed in this work.

The most commonly used concentration expressions in trace element analysis are summarized in Table 1.3. Sometimes, particularly when referring to detectable quantities by furnace atomic absorption, the absolute value in micrograms, nanograms, or picograms is given.

Table 1.3
Concentration Terms Used in Trace Metal Analysis

	Parts per million (ppm)	Parts per billion (ppb)
mg	mg/liter, mg/kg	—
μg	μg/ml, μg/g	μg/liter, μg/kg
ng	—	ng/ml, ng/g

III. BACKGROUND LEVELS OF THE ELEMENTS

Values of the absolute abundances of elements in the earth's crust (i.e., crustal abundances) are estimates only. The main reasons for the uncertainties are the variable composition of the earth's crust, the incomplete exposure of the crust, and the lack of representative samples for absolute abundance measurements. A number of workers have compiled tables of crustal abundances of some or all of the elements. The results have been summarized in Ref. 4. A discussion of the methods used to obtain the estimates of crustal abundance is found in the same reference. A unit called the "clarke" has been defined as the average percentage (or abundance) of a particular element in a geochemical system. However, clarke values should not be used as a means of assessing enrichment or depletion of elements in rocks.

In order to outline anomalous values it is important that background levels of the elements can be established clearly. Background values of the elements may be very low, particularly in the case of waters and plant material. In this regard, techniques such as x-ray fluorescence and ICP–AES may not have the required sensitivity.

Contamination can present a serious problem in the establishment of background levels. Thus, it is important that blanks be included in each sample run to allow for subtractions of any contaminants.

IV. SEPARATIONS AND CONCENTRATION

Despite recent rapid advances in analytical instrumentation, it is still often necessary to use separation and concentration methods prior to the determinative step. The reasons for doing a separation/concentration step are to bring the concentration of a trace element to a detectable level and/or to separate it from interfering substances (usually high-concentration elements of the sample matrix). Rarely is it necessary to separate the individual trace elements from one another. An extensive section on separation is given in Chapter 6. The reader is referred to a good critical review by Bachmann on separation/concentration (5). Any separational/concentration method is time-consuming and prone to problems with losses and/or contamination. Solvent extraction and ion-exchange chromatography are most commonly used.

A. Blanks

The need for running blanks with each set of determinations cannot be overemphasized. A blank should represent the same quantities of reagents carried through the same procedure as the sample. The purposes of blanks are mainly to establish the detection limits for elements in the procedure and to monitor the procedure with respect to contamination. In geochemical trace element analysis

Table 1.4
Impurities in HCl (ng/g) (6)

Element	American Chemical Society reagent	Commercial high purity	Subboiling distillation
Pb	0.05	<1	0.07
Cd	0.03	0.5	0.02
Cu	4	1	0.1
Ni	6	3	0.2
Cr	2	0.3	0.3

the detectable amount is often set by the concentration of elements in the reagents and in the laboratory environment. To keep blanks low it may be necessary to use specially purified acids and double-distilled or specially deionized water. In the case of the latter the singly distilled water in our laboratory was found to contain 0.1 ppb lead, a level too high for determination of lead in lake waters. Double distillation of the water lowered the lead level by one order of magnitude, making the determination feasible.

Considerable amounts of impurities in the blanks (and hence the sample being analyzed) can arise even from analytical grade reagents. Murphy (6) analyzed various grades of hydrochloric acid for trace elements and some of his results are summarized in Table 1.4.

V. ERROR

Error related to a trace element analysis may occur at all points from sampling through to the determinative step. The magnitude of error encountered in the present authors' experience is sampling > sample preparation > determination.

A discussion of obtaining representative amounts of samples in the field, reducing field samples to sizes needed in the laboratory, and selecting sizes of samples (aliquots) for analyses is found in Chapter 4. A good description of geochemical field sampling has also been written by Maxwell (7).

Sample preparation often includes both physical and chemical preparation steps. Physical sample preparation, including grinding and sieving if necessary, can result in severe contamination if metal implements are employed. Losses of trace metals may result if any fraction of the sample is rejected. Chemical sample preparation is of crucial importance and no unanimity of opinion exists on the proper treatment for each sample type. The topic is covered in Chapter 4. The determinative step, during which the sample is presented to an instrument, is relatively free of interference, yet great care is essential in calibration and the elimination of interferences.

A. Physical Sample Preparation

Rocks or ores require grinding prior to analysis. Serious contamination from the grinding apparatus may occur during this step. Ideally, the sample should be ground between plates manufactured from blocks of the sample being ground. In most cases such an approach is impossible. Thus hardened steels and alumina materials are commonly used to construct grinding apparatus. Agate vessels may also be recommended.

Generally, if a total decomposition must be done, the sample should be ground to pass a 200 mesh sieve. It is important to obtaining a representative sample that *all* the sample be ground to pass through the sieve. Thus regrinding of the material rejected by the sieve is essential.

If a large bulk sample has been submitted, it is important to obtain a representative sample for analysis. This can be done as follows. The sample is dumped onto a clean flat surface. It should then be flattened and divided into four equal parts. The two opposite quarters are rejected and the other two quarters recombined. This procedure is repeated until a sample suitable for taking weighed portions has been obtained. Chapter 4 contains more details about preparation of samples.

B. Accuracy Assessment

At this point, some definitions of terms will be helpful. Accuracy of a determination is a measure of how close the value obtained is to the "true" value. The numerical difference between the values is the error. Precision is a measure of the degree of scatter obtained on replication of the analysis of a sample. A quantitative estimate of precision is expressed in terms of the variance or the standard deviation of a set of results.

In this era of analytical instrument sophistication one might assume that trace metal data in the literature would be on the whole quite reliable. This is far from the case. It is only in the past 5 to 10 years that laboratories in general have employed useful checks on accuracy and precision. This is certainly true for most data obtained prior to the above times.

It is not always necessary to have the highest accuracy in trace metal analysis, but it is crucial to be able to assess the accuracy of an analysis. This is very important in the case of whole rock and ore analyses. An excellent two-volume treatise on accuracy in trace analysis was published in 1976 (8).

Generally four approaches to accuracy assessment are employed in geochemical analysis:

1. Analysis of standard reference samples
2. Recovery studies
3. Use of an absolute method, e.g., isotope dilution
4. Analysis of control and repeat samples

The use of standard reference samples is preferred. These samples have been analyzed carefully in several laboratories, usually by more than one method. However, it is not always possible to obtain such samples for the sample type and elements desired. Fortunately, a number of agencies are greatly expanding their offerings in this area. Chapter 3 contains a listing of some of the most useful standard reference samples known to the authors. The suppliers and the suppliers' addresses are also given.

Standard reference samples are relatively expensive and often in short supply, and hence they may not be available for routine day-by-day control of analytical data. For this purpose "in-house" laboratory control standards may be prepared and employed together with an occasional standard reference sample.

A milestone in the analysis of geological material was reached in 1951 in the publication by Fairbairn (9) of a cooperative study of the precision and accuracy of chemical and spectrochemical analyses of two silicate rocks, a granite G-1 and a diabase W-1, followed in 1960 by a progress report (10). These publications for the first time documented the accuracy and precision obtainable in chemical analyses of silicates. Because a number of different laboratories were involved, this study graphically demonstrated the meaningfulness (or lack thereof) of comparisons of chemical data on silicates in the published literature.

Tables 1.5 and 1.6 are compilations for manganese oxide and total iron, respectively, of data obtained on G-1 and W-1 and published in these reports. Table 1.7 details the values obtained for lead and copper in G-1 to illustrate the variation obtained in the case of trace elements.

Evaluating these data (and the data for other elements contained in the reports), one can gauge the sense of distress which was felt by geologists and analysts

Table 1.5
MnO in W-1 by Different Methods (%)

Colorimetric			Volumetric	
Periodate	Bismuthate	Persulfate	persulfate	Spectrographic
0.13	0.11	0.12	0.13	0.17
0.16	0.17	0.12	0.16	0.15
0.16	0.20	0.18	0.18	0.16
0.16		0.23	0.26	0.17
0.16				0.18
0.17				0.19
0.17				0.21
0.17				
0.17				
0.17				
0.18				
0.18				
0.21				
0.21				
0.22				
0.53				

Table 1.6
Total Fe as Fe_2O_3 in G-1 by Different Methods (%)

$SnCl_2$ reduction	H_2S reduction	Zn amalgam	Ag reductor	SO_2 reduction	Titration using titanous chloride	Spectrographic
1.29	1.86	1.84	1.88	2.00	1.83	1.42
1.91	1.92	1.84	1.90		1.83	1.84
2.04	1.94	1.86	1.92			1.96
2.34	1.99	1.91	1.94			1.99
2.47	2.01	1.91	1.97			2.00
	2.10	2.13				2.04
	2.16	2.20				
	2.26	2.27				
	2.99					

around the world with their publication. These publications were instrumental in a gradual improvement in accuracy of not only geological but also all other types of chemical analyses throughout the world. At present the situation is much improved over that in 1960. However, it is only through vigorous adherence to accuracy assessment programs that improvements will continue.

Table 1.7
Pb and Cu in G-1 (μg/g)

Pb	Cu
Chemical	
14	10
15	16
28	20
28	
46	
47	
48	
51	
Spectrographic	
22	5
22	6
23	8
25	10
26	10
28	11
37	11
47	12
47	13
47	15
48	15
50	16
50	17
52	25

C. Establishment of Precision of Results

The problems giving rise to misleading results in the establishment of precision of results have been outlined by Thompson and Howarth (11). The deficiencies are that

1. Most methods do not allow for changes which occur in the error over the concentration range being studied.
2. Optimistically biased results frequently are obtained.

The two methods discussed below are recommended for overcoming these problems.

When 50 or more duplicated results are available, the variation of standard deviation (s_c) can be expressed as a function of concentration (c) and the standard deviation at zero (s_0).

$$s_c = s_0 + kc$$

The parameters s_0 and k can be used to quantify the precision (p_c) at concentration c by means of the definition

$$p_c = 2s_c/c$$

Thus

$$p_c = 2s_0/c + 2k$$

Thus, the value $2k$ corresponds to the precision which is observed at concentrations well above the detection limit.

In suitable cases, i.e., when there are observations at low concentration, the practical detection limit c_d (when $p_c = 1.0$) can also be estimated from

$$c_d = 2s_0/(1 - 2k)$$

To prevent the acquisition of optimistically biased results from duplicated results the following must be adhered to:

1. Duplicate analyses must be done on splits of all the actual samples or a random selection of these.
2. Each of the two subsamples for duplication must be taken through the *whole* analytical procedure as if it were an independent sample.
3. The position of the second subsample in the sequence must not be systematically related to the first but should be distributed at random in the batch.
4. The data must not be rounded off too severely. At least one significant digit containing uncertainty must be retained.
5. Subzero (or sub-detection limit) measurements, when obtained, must be recorded and used as such and not set to zero or another arbitrary value.
6. Subsamples should be numbered to prevent their identification at the time of analysis.

Method 1 (50 or more duplicated results):

1. From the ($N>50$) pairs of results a_i, b_i ($i = 1, 2, \ldots N$), form lists of the pair means $(a_i + b_i)/2$ and the corresponding absolute differences $|a_i - b_i|$ (do not logtransform the data).
2. Sort the list of means into increasing order and the differences into the corresponding order.
3. Select the first 11 results and calculate the mean of the pair means and the *median* of the differences.
4. Repeat this for successive groups of 11 results and obtain corresponding lists of means and medians. Ignore any terminal group of less than 11 results.
5. Either plot the medians as a function of the means and obtain the intercept and slope of the line graphically by eye, or obtain the same parameters by regression. These parameters correspond respectively to s_0 and k in the first equation.

Method 2 (10 to 50 duplicated results):

1. Specify the precision required in the form $s_c = s_0 + kc$. Either s_c or k could be zero if appropriate.
2. Form two new equations from this:

$$d_{90} = 2.326(s_0 + kc)$$

$$d_{99} = 3.643(s_0 + kc)$$

3. Plot d_{90} and d_{99} over a suitable range of c to form a control chart.
4. As in method 1, obtain the pair means $(a_i + b_i)/2$ and absolute differences $|a_i - b_i|$ of the duplicate results.
5. Plot these points on the control chart.

NOTE: d_{90} and d_{99} are the 90th and 99th percentiles of the absolute difference between duplicates as a function of concentration.

VI. ELEMENTS COVERED

The elements to be covered by procedures in the following chapters are shown in Fig. 1.1.

A. Association of the Elements

Usually the main constituent(s) of an ore is the element(s) which is determined in a geochemical exploration study. For example, copper would usually be the element determined in soils, rocks, and sediments in seeking a copper ore. Such

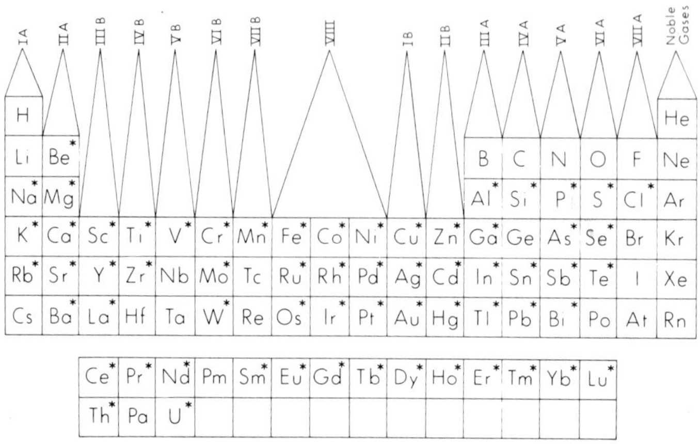

Figure 1.1. Periodic table of the elements. *Elements covered by procedures in this book.

Table 1.8
Some Common Geochemical Associations of Elements[a]

Group	Association
Generally associated elements	K-Rb
	Ca-Sr
	Al-Ga
	Si-Ge
	Zr-Hf
	Nb-Ta
	Rare earths - La - Y
	Pt-Ru-Rh-Pd-Os-Ir
Plutonic rocks	
General association (lithophile elements)	Si-Al-Fe-Mg-Ca-Na-K-Ti-Mn-Zr-Hf-Th-U-B-Be-Li-Sr-Ba-P-V-Cr-Sn-Ga-Nb-Ta-W-the halogens-rare earths
Specific association	
Felsic igneous rocks	Si-K-Na
Alkaline igneous rocks	Al-Na-Zr-Ti-Nb-Ta-F-P-rare earths
Mafic igneous rocks	Fe-Mg-Ti-V
Ultramafic rocks	Mg-Fe-Cr-Ni-Co
Some pegmatitic differentiates	Li-Be-B-Rb-Cs-rare earths-Nb-Ta-U-Th
Some contact metasomatic deposits	Mo-W-Sn
Potash feldspars	K-Ba-Pb
Many other potash materials	K-Na-Rb-Cs-Ti
Ferromagnesian minerals	Fe-Mg-Mn-Cu-Zn-Co-Ni
Sedimentary rocks	
Fe oxides	Fe-As-Co-Ni-Se
Mn oxides	Mn-As-Ba-Co-Mo-Ni-V-Zn
Phosphorite	P-Ag-Mo-Pb-F-U
Black shales	Al-Ag-As-Au-Bi-Cd-Mo-Ni-Pb-Sb-V-Zn

Source: Refs. 12–14.

[a]For additional association in ore bodies see Table 1.9.

Table 1.9
Associated Elements (Pathfinders) Useful in Ore Typing

Type of deposit	Major components	Associated elements
Magmatic deposits		
Chromite ores (Bushveld)	Cr	Ni, Fe, Mg
Layered magnetite (Bushveld)	Fe	V, Ti, P
Immiscible Cu-Ni-sulfide (Sudbury)	Cu, Ni, S	Pt, Co, As, Au
Pt-Ni-Cu in layered intrusion (Bushveld)	Pt, Ni, Cu	Cr, Co, S
Immiscible Fe-Ti-oxide (Allard Lake)	Fe, Ti	P
Nb-Ta carbonatite (Oka)	Nb, Ta	Na, Zr, P
Rare-metal pegmatite	Be, Li, Cs, Rb	B, U, Th rare earths
Hydrothermal deposits		
Porphyry copper (Bingham)	Cu, S	Mo, Au, Ag, Re, As, Pb, Zn, K
Porphyry molybdenum (Climax)	Mo, S	W, Sn, F, Cu
Skarn-magnetite (Iron Springs)	Fe	Cu, Co, S
Skarn-Cu (Yerington)	Cu, Fe, S	Au, Ag
Skarn-Pb-Zn (Hanover)	Pb, Zn, S	Cu, Co
Skarn-W-Mo-Sn (Bishop)	W, Mo, Sn	F, S, Cu, Be, Bi
Base-metal veins	Pb, Zn, Cu, S	Ag, Au, As, Sb, Mn
Sn-W greisens	Sn, W	Cu, Mo, Bi, Li, Rb, Si, Cs, Re, F, B
Sn-sulfide vein	Sn, S	Cu, Pb, Zn, Ag, Sb
Co-Ni-Ag vein (cobalt)	Co, Ni, Ag, S	As, Sb, Bi, U
"Epithermal" precious metal	Au, Ag	Sb, As, Hg, Te, Se, S, U
Mercury	Hg, S	Sb, As
Uranium vein	U	Mo, Pb, F
Copper in basalt (Lake Superior type)	Cu	Ag, As, S
Volcanogenic massive-sulfide Cu	Cu, S	Zn, Au
Volcanogenic massive-sulfide Zn-Cu-Pb	Zn, Pb, Cu, S	Ag, Ba, Au, As
Au-As-rich Fe formation	Au, As, S	Sb
Mississippi Valley Pb-Zn	Zn, Pb, S	Ba, F, Cd, Cu, Ni, Co, Hg
Mississippi Valley fluoride	F	Ba, Pb, Zn
Sandstone-type U	U	Se, Mo, V, Cu, Pb
Red-bed Cu	Cu, S	Ag, Pb
Calcrete U	U	V
Sedimentary types		
Copper shale (Kupferschiefer)	Cu, S	Ag, Zn, Pb, Co, Ni, Cd, Hg
Copper sandstone	Cu, S	Ag, Co, Ni

Source: Refs. 14 and 15.

an element is termed an indicator element. Sometimes, however, the main ore-bearing element is immobile or yields confusing data. In these cases a pathfinder element should be considered.

The reasons for choosing a pathfinder element would be as follows:

1. It may be easier to detect.
2. A high-mobility element is more easily used than an immobile element.
3. The pathfinder element sometimes produces clearer patterns of anomalous values.

Common geochemical associations are given in Table 1.8, Ref. 3. A list of pathfinder elements is given in Table 1.9, Ref. 3.

REFERENCES

1. R. J. Gibbs, *Science* **180,** 71 (1973).
2. G. H. Morrison, *CRC Crit. Rev. Anal. Chem.* **8,** 287 (1979).
3. A. W. Rose, H. E. Hawkes, and J. S. Webb, "Geochemistry in Mineral Exploration," 2nd ed. Academic Press, London, 1979.
4. G. J. S. Govett, "Handbook of Exploration Geochemistry." Elsevier, Amsterdam, 1983.
5. K. Bachmann, *CRC Crt. Rev. Anal. Chem.* **12,** 1 (1981).
6. T. J. Murphy, *NBS Spec. Publ. (U.S.)* **422,** 509 (1976).
7. J. A. Maxwell, *NBS Spec. Publ. (U.S.)* **422,** 285 (1976).
8. P. D. LaFleur, ed., "Accuracy in Trace Analysis: Sampling, Sample Handling, Analysis," Vols. 1 and 2, NBS Spec. Publ. No. 422. U.S. Govt. Printing Office, Washington, D.C., 1976.
9. W. H. Fairbairn, *Geol. Surv. Bull. (U.S.)* **980,** 21 (1951).
10. R. E. Stevens, *Geol. Surv. Bull. (U.S.)* **1113** (1960).
11. M. Thompson and R. J. Howarth, *J. Geochem. Explor.* **9,** 23 (1978).
12. V. M. Goldschmidt, "Geochemistry." (Clarendon), London and New York, 1954.
13. K. B. Krauskopf, "Economic Geology—50th Anniversary Volume." Econ. Geol. Publ. Co., Lancaster, Pennsylvania, 1955.
14. R. W. Boyle, *Geol. Surv. Pap. (Geol. Surv. Can.)* **74–45** (1974).
15. A. A. Beus and S. V. Grigorian, "Geochemical Exploration Methods for Mineral Deposits." Applied Publishing Co., Wilmette, Illinois, 1977.

2

Principles of Determinative Methods

I. ATOMIC ABSORPTION SPECTROMETRY

A. Theoretical Considerations

Atomic absorption spectrometry is an analytical technique for the determination of elements based on the absorption of radiant energy by free atoms in their ground state. A physical representation of an atom is as a central nucleus surrounded by a number of electrons in orbitals. Each element has a characteristic number and arrangement of electrons. An atom is said to be in its most stable or ground state when the electrons are in their lowest energy levels. (Under normal laboratory conditions atoms are bound to one another in the form of chemical compounds.) When free atoms in the ground state are subjected to a source of radiant energy of appropriate discrete wavelength, energy can be absorbed by the atom population. This results in promotion of electrons to higher energy levels, as in Fig. 2.1.

In returning to the ground state, the atoms emit energy in one or several stages. Atoms of different elements will absorb energy at wavelengths which are characteristic of the elements. An absorption spectrum of an element is generally composed of only a few lines. This is in contrast to emission produced thermally, which results in a relatively complex spectrum. Atomic absorption spectra thus have an advantage of very few line overlaps as compared with emission spectra.

The essential components of a system for atomic absorption analysis are shown in Fig. 2.2. A source of radiant energy, A, such as a hollow-cathode lamp illuminates a cloud of atoms, B. The free atoms are usually formed thermally in a flame, furnace, etc. Radiation which traverses B is focused on the entrance slit of a monochromator, C. The monochromator is set at a wavelength corresponding to one of the emission lines of A. The radiation passed by C is focused on the detector, D.

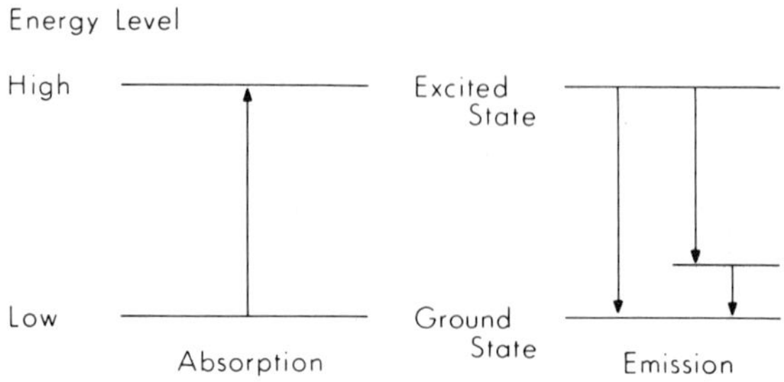

Figure 2.1. Absorption and emission of radiation.

If the source emits the line spectrum of element X, and atoms of element X are present in B, some of the incident radiation I_0 is absorbed. As a result, the transmitted radiation I received by D is reduced in intensity compared to I_0. This is an indication of the presence of X in the sample. The output of D can then be related quantitatively to the proportion of X in the sample as follows.

In Fig. 2.2 the I_0 being emitted by the source is attenuated by absorption in an atom cloud, and a reduced intensity I is detected. Transmittance T is defined as the ratio of the final to the initial intensity:

$$T = I/I_0$$
$$\%T = I/I_0 \times 100 = \text{percent transmission}$$

A term used more conveniently in quantitative work is the absorbance A. It is defined as follows:

$$A = \log(I_0/I) = \log(1/T)$$

Absorbance is related to concentration by means of the Beer–Lambert law:

$$A = abc$$

where a is the absorption coefficient, specific for each atomic species, b the length of the absorption path, and c the concentration of absorbing atoms in the path.

In quantitative analysis, a set of experimental conditions is selected. Then, with a and b constant, concentration should be related to absorbance A in a linear manner. The relationship between A and c is established by recording the absorbances of a series of standard solutions. The instrument is usually set for zero absorbance with distilled water as the analyte. The absorbance of a reagent blank,

Figure 2.2. Atomic absorption process.

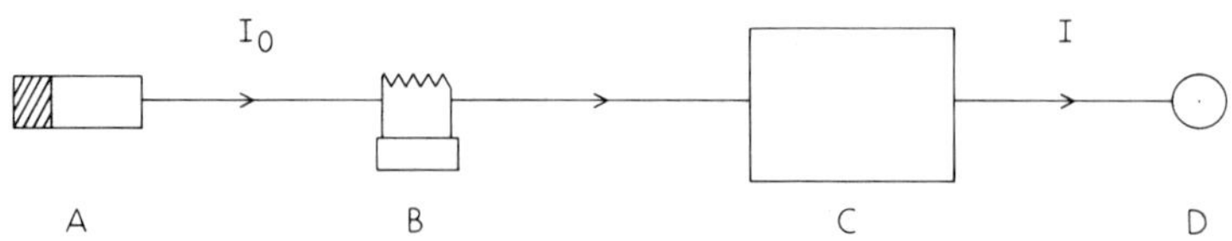

if different from zero, is subtracted from the absorbances of the standards. The net absorbance values of the standards are plotted against the corresponding concentrations to yield a calibration line or curve. This is illustrated in Fig. 2.6. Linearity in atomic absorption analysis can be expected over a range of 0.4–0.5 absorbance unit. Over wider ranges, curvature of a calibration line, usually toward the concentration axis, is encountered. Concentrations are calculated by means of interpolations. Modern instruments can be calibrated to provide direct readings of concentrations, so that the operator does not need to construct a calibration curve.

B. Instrumentation for Atomic Absorption

The basic components of an atomic absorption system are shown in Fig. 2.2. In single-beam instruments, all of the radiation from the source lamp is directed through the cloud of atoms generated in the flame or electrothermal atomizer. Double-beam instruments divide the radiation by means of a beam chopper and mirrors. In this design, a constant portion of the incident radiation, called the reference beam, is passed around the atomizer. The remaining portion, or sample beam, is directed through the atomizer. Then the beams are combined again and focused on the inlet slit of the monochromator. The output signals of the detector for sample and reference beams are amplified separately and compared in a bridge circuit. The net signal is transmitted to the readout device.

In some instruments the source lamp intensity is modulated at a particular frequency such as 50 Hz, and the amplifier is tuned to the same frequency. Another means of source modulation is to place a rotating chopper between the source and the flame. In this manner, the radiation such as flame emission which is not excluded by the relatively narrow bandpass of the instrument is rejected and the system responds only to radiation from the lamp. Readings are taken with and without samples. The net absorbance values are proportional to the quantities of the element in the sample.

1. Atomizers

As noted earlier, the function of the atomizer component is to provide an environment for the production of atoms in the ground state for periods long enough for quantitative measurements to be made. Flames provide a convenient and economical method of formation of atoms. The two oxidant/fuel mixtures most often used in atomic absorption are air/acetylene and nitrous oxide/acetylene. In premix burners, the approximate temperature of the former is 2400 K and of the latter is 3200 K. The proportion of fuel may be varied to yield lean to rich flames depending on the element being analyzed. The temperature will vary slightly depending on the fuel-to-oxidant ratio.

About 30 elements can be determined satisfactorily in air/acetylene flames. Elements which form highly refractory oxides, such that the dissociation energies of the M–O bond are larger than about 5 eV, require the hotter nitrous oxide/acetylene flame. Examples are aluminum, titanium, and zirconium. The hotter flame is also used to minimize some interferences, as will be explained later. Propane and hydrogen have also been used as fuels; but, because of the low temperatures of such flames, there are only limited applications.

A number of elements are listed in Table 2.1 with the wavelengths of their resonance lines and their detection limits.

Figure 2.3 illustrates the structure of a typical flame. Best sensitivities usually occur slightly above the inner cone, in the interconal zone about 0.3–0.5 mm above the burner slot.

2. Nebulizers

The purpose of the nebulizer is to convert a liquid sample into a mist or aerosol. The mist is mixed with oxidant and fuel gases in a spray chamber. It is then carried to the burner in the gas flow. The operation can be summarized as follows. The gas flows to the burner are adjusted to the desired ratio of oxidant to fuel, and the burner is lit. A flexible capillary tube is placed in the sample solution. Sample

Table 2.1
Wavelengths of Resonance Lines and Detection Limits of Elements in Flame Atomizers

Element	Wavelength (nm)	Detection limit[d] (μg/ml)	Element	Wavelength (nm)	Detection limit[d] (μg/ml)
Ag	328.1	0.0002	Na	589.0	0.0002
Al[a]	309.3	0.02	Ni	232.2	0.005
As	193.7	0.2	Os[a]	290.9	0.08
Au	242.8	0.01	Pb[c]	283.3	0.01
B[a]	249.8	0.7	Pd	247.6	0.02
Ba[ab]	553.6	0.008	Pt	265.9	0.05
Be[e]	234.9	0.001	Rh	243.5	0.004
Bi	223.1	0.025	Ru	349.9	0.07
Ca	422.7	0.0005	Sb	217.9	0.04
Cd[c]	228.8	0.001	Se[c]	196.0	0.20
Co	240.7	0.01	Si[a]	251.6	0.02
Cr	357.9	0.003	Su[ac]	224.6	0.07
Cu	324.8	0.002	Sr[b]	460.7	0.002
Fe	248.3	0.003	Te[c]	214.3	0.03
Hg[c]	253.6	0.25	Ti[a]	364.3	0.04
Ir	264.0	0.6	Tl[c]	276.8	0.01
K	766.5	0.002	V[a]	318.4	0.04
Li	670.8	0.0003	Zn	213.9	0.001
Mg	285.2	0.001			
Mn	279.5	0.002			
Mo[a]	313.3	0.02			

[a] Nitrous oxide/acetylene flame.
[b] 1000 μg/ml potassium added to suppress ionization.
[c] Electrodeless discharge lamp.
[d] Detection limit = concentration × standard deviation/mean.

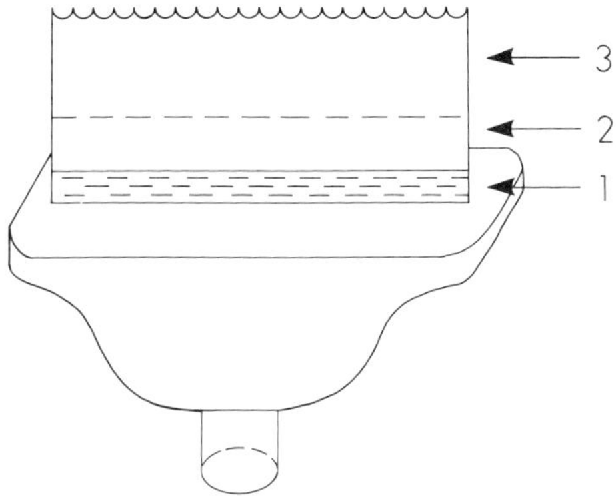

Figure 2.3. Zones of a flame from a premix burner: 1, primary combustion zone; 2, interconal zone; 3, combustion zone.

liquid is drawn up the capillary tube and into the mixing chamber by the reduced pressure at the end of the inlet tube caused by the concentric flow of gas around the tip (venturi effect). Shearing of the liquid at the tube orifice produces droplets. In some designs, an impact bead is situated close to the orifice to generate additional small droplets. An alternative design uses a flow spoiler to aid in removing larger droplets from the gas stream. In passing through the spray chamber, droplets larger than 5–10 μm are retained as condensate and flow to waste. Only about 10% of the sample in the form of small droplets reaches the burner. The response of the system is fast, and steady readings are usually achieved within a few seconds.

The performance of a particular design of nebulizer/burner system can be optimized by adjusting the gas flow rates according to the manufacturer's recommendations. Liquids of different viscosities are drawn through the narrow capillary tube at different rates. Therefore, the viscosities of samples and standards must be similar in order to maintain equal rates of sample delivery to the flame. Samples containing insoluble matter must be filtered so that solids will not clog the nebulizer. Corrosion of nebulizers by acid solutions can be a problem. For most laboratories, an acid-resistant nebulizer (e.g., Pt/Ir or PTFE) is a good investment. Some organic solvents may attack gaskets and other sealants used in the system. The user should determine whether a particular solvent will be compatible with the equipment. Attention must be given to the liquid trap in the waste line. A liquid head of about 5 cm should always be present to prevent combustible gas from escaping to the atmosphere.

Other nebulizer designs included ultrasonic and cross-flow nebulization. They have not found wide acceptance in atomic absorption work.

3. Burners

Most of the burners used for atomic absorption are of the premix type as shown in Fig. 2.3. Fuel and oxidant gases are mixed with sample droplets and then proceed to the combustion area. Burner heads usually contain a single slot 5–10 cm long.

Although a long path length is desirable in order to increase the absorbance of sample, and hence sensitivity, practical considerations of flame stability limit the length. Nitrous oxide/acetylene burners must have a slot not exceeding 5 cm in length. Because of the high flame temperatures, they should be constructed of heat-resistant alloys. If samples have large dissolved solids contents, a burner with a wider slot should be used to avoid clogging. A three-slot burner head, called a Boling burner, has the advantage that its flame fills the horizontal aperture of the optical system. This can result in more absorption for a given concentration of analyte.

When the sample in the form of a liquid enters the flame, the solvent evaporates, or burns if it is combustible, leaving fine particles of compounds. The compounds begin to vaporize as they pass into hotter regions of the flame, and then they dissociate into atoms. The flame temperature must be high enough to provide efficient atomization. As explained previously, refractory compounds require hot flames generated by nitrous oxide/acetylene. Atoms of elements which are easily ionized can lose one or more electrons, particularly in hot flames. This will result in ionization interference.

4. Electrothermal Atomizers

Electrothermal atomizers constitute a class of flameless atomizers for atomic absorption work. They are electrically heated devices which are capable of achieving the high temperatures required to produce analyte atoms. Most commercial instruments make provision for mounting an electrothermal unit in place of a nebulizer/burner assembly so that the atomizers may be interchanged as they are needed.

Electrothermal atomizers possess some advantages and disadvantages in comparison with flame atomizers. An increase in sensitivity of two or three orders of magnitude can be achieved for many elements, as illustrated in Table 2.2. Since

Table 2.2
Detection Limits for Analyte in Low-Salt[a] Coastal Waters, Perkin-Elmer 604 and HGA 2100

Element	Detection limit (pg)	Element	Detection limit (pg)	Element	Detection limit (pg)
Ag	0.5	Ir	1000	Se	200
As	25	Mn	0.8	Sn	200
Au	25	Mo	15	Te	100
Bi	30	Ni	30	Ti	500
Cd	0.5	Pb	5	V	200
Co	10	Pd	30	Zn	0.2
Cu	5	Rh	50		
Fe	10	Sb	10		

[a]Interference-free matrix.

sample volumes of less than 50 μl are normally used, this is an advantage when the total volume of a sample is limited. Electrothermal atomizers can be used for some types of solid samples. The major disadvantages of electrothermal devices are the severe background and matrix interferences which are characteristic of these atomizers. Very rapid heating of samples, particularly those containing refractory elements such as vanadium, is necessary in order to obtain sharp absorption peaks. Since the atoms are present for such a short duration, the detector/recorder system must respond rapidly. Finally, a longer time (3–5 min) is required for each determination than is needed for an analysis with a flame atomizer.

Atomizer Design Two main designs of atomizer have been available commercially. These are tubular graphite furnaces and carbon rods. Other atomizers such as crucible furnaces, quartz tubes, carbon filaments, wire loops, and metal strips have been described, but none has been generally accepted.

L'vov (1,2) designed the first tubular furnace specifically for atomic absorption work. It consisted of a small graphite tube heated to a constant high temperature. Samples were introduced through a hole in the bottom of the furnace by means of a separately heated graphite electrode. Radiation from a spectral source passed through the furnace along the central longitudinal axis. An inert gas protected the graphite against oxidation. Samples were volatilized into the hot graphite tube by heating the electrode.

The tubular furnace atomizers sold by most manufacturers are derived from the design of Massmann (3). A small-diameter graphite tube is temperature programmed so that samples deposited on the inside wall of the tube can be dried, ashed, and then atomized at a high temperature. The Woodriff furnace (4) is a larger unit designed for operation at constant temperature. It has not been as popular as the Massmann design and has not been produced commercially.

Early carbon rod atomizers consisted of a carbon rod clamped at both ends in an electrical contact. The sample was placed in an indentation in the center of the rod. In later models, a graphite cylinder was inserted between two short pieces of carbon rod at the center of the atomizer. In both designs, a gas sheath was blown around the atomizer components, but precise control of the gas flow in the area of the sample was not possible. Carbon rod atomizers require less power than graphite furnace devices. The detection limits for many elements are about the same in a rod atomizer as in a furnace.

Operation The power supply for the furnace is a very important part of the equipment. The operator must select the temperatures and times for each step of the heating cycle. Modern units are more versatile than the older models in allowing the operator to designate rates of change in temperatures between steps and in extending the number of steps. Normally, there are three programmable cycles. In the first, the sample is dried. The second stage is carried out at a higher temperature with the purpose of removing organic matter and evaporating low-boiling components of the sample matrix. The term charring or ashing is applied to this step. The operator should use the highest char temperature which does not result in appreciable loss of analyte. The optimum temperature is found by preparing an ash/atomization curve for each element being determined in a particular sample. Directions are given in the manufacturers' manuals and in other monographs (5,6). Operating manuals also contain information on maximum char temperatures for the elements, but these may not apply to the sample in question. Removal of as much as possible of the potential interfering substances at this stage is important. Matrix modifiers can be added to aid in the volatilization of some salts and to increase the char temperatures of elements which evaporate easily. More details will be provided later. The third step is the atomization of the analyte. Temperatures ranging to 3000°C are necessary, depending on the element. Of equal importance is rapid rise of temperature so that the analyte evaporates and atomizes within as short a time as possible. Capacitive heating to bring the furnace very rapidly to the required temperature followed by resistive heating to maintain the temperature has been used by some workers (7). Interferences and background were greatly reduced. At the end of the cycle, the furnace is cooled to room temperature by the water circulation around the shell. Some samples may leave residues. Another high-temperature heating period is applied to volatilize such material. However, any loose ash must be removed with a small clean brush.

In older equipment, signals are recorded on a strip chart, and peak heights are taken as measures of concentration. Newer instruments are capable of measuring either peak heights or peak areas. The latter are preferable when rates of atomization are low or variable due to sample matrix. Recorders and electronic integrators must have short response times. For example, a recorder with a full scale response of ½ second or less is required. Examination of the peak shape is very useful in the development of new methods and in the diagnosis of problems. Cathode ray tube (CRT) displays are commonly used for this purpose.

Furnace Tubes and Rods Graphite furnace tubes and rods have been studied extensively because the properties of the graphite and the changes in properties which occur during a series of analyses have a direct influence on the quality of the results. The need for an inert atmosphere to prevent oxidation of the furnace at high temperature was noted previously. Changes in the atomizer surface usually result in decreases in peak heights for a given concentration of element. Also, the electrical resistance of the tube changes, and thus its thermal properties change. The porosity of graphite is an important factor. Changes in porosity as the tube ages influence the sensitivities of analyte elements. The usual method of stabilizing a tube and of making it less porous is to coat it with a layer of pyrolytic graphite.

Coated tubes are available from equipment manufacturers. A coating can be applied while the tube is mounted in the furnace by introducing some hydrocarbon gas such as methane mixed with argon when the tube is hot. However, the coating may not be as uniform as the coatings on purchased tubes. The use of coated tubes leads to increases in sensitivities for most elements and to decreases in interferences. Pyrolytic tubes are less inclined to form stable carbides with elements such as vanadium, titanium, and tungsten.

Metals such as tantalum and molybdenum have been applied as coatings to improve atomizer performance. The effects of metallic coatings are variable, depending on the type of sample. The reader should consult the literature for information about particular samples.

A graphite platform situated in the central part of the furnace beneath the sample port is used to improve atomization (8). The samples are atomized from the platform rather than from the wall of the furnace. During the sharp rise in temperature in the atomization step, the temperature of the platform lags behind the wall temperatures because the platform is heated mainly by radiant energy. This means that the sample is atomized into a hot, almost constant-temperature environment when maximum furnace temperature has been achieved. Reduced matrix effects and sharper absorption peaks have been reported as advantages of platforms.

Samples Liquids constitute most of the samples used in electrothermal analyzers. Samples are dispensed from micropipets inserted through the sample port of a tubular furnace or into the depression in a carbon rod. Small sample volumes provide the advantage of concentrating the analyte on a small area of the atomizer. Deposition of sample over a large area of the atomizer is not desirable because uneven heating along the graphite element will alter the rate of atomization of the total sample. Large volumes of liquid can be added in increments of 10–20 μl. After each addition, the solvent is evaporated. If the sample contains organic materials and/or significant concentrations of salts, an ashing step should be included after each addition. When large numbers of samples must be analyzed, an autosampler is a good investment. The precision of volumetric measurements is usually better with an autosampler than with manual operations.

Prevention of contamination is very important in electrothermal work because of the sensitivity of the method. Precautions must be taken to minimize the introduction of contaminants from reagents, equipment, and the working environment. Volatile halides of elements may be lost during the charring step of the heating cycle. Oxy acids, especially nitric acid, are preferable in solutions of samples and standards.

5. Radiation Sources

Hollow-Cathode Lamps The most widely used radiation sources in atomic absorption work are hollow-cathode spectral lamps. The purpose of a single-element lamp is the production of resonance radiation of narrow linewidth, usually < 0.001 nm, of the element used as the cathode. Most elements have spectra consisting of several resonance lines as well as other lines. The most intense resonance line is used and is listed in Table 2.1. Resonance radiation results from the process of excitation of atoms in the ground state and the subsequent transition from the excited to the ground state.

A lamp, as shown in Fig. 2.4, is filled with an inert gas such as argon. A small current ionizes some of the gas. Interactions with the gas ions remove some atoms from the cathode. With excitation, these atoms provide the resonance radiation.

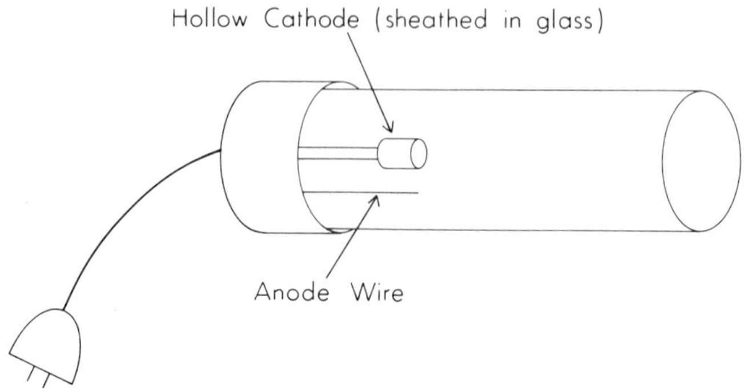

Figure 2.4. Hollow-cathode lamp.

The lifetime of a lamp will depend on the element used as the cathode, hours of use, strength of the current, and filler gas leakage. However, good-quality lamps will last for several years. Each lamp should be operated for a brief time (about 30 min) each month even if it has not been required for analytical work. A weak lamp can sometimes be regenerated by reversing the leads to the electrodes and operating the lamp at a few milliamperes for about 10 min.

Some multielement lamps are marketed on the basis of saving the cost of several single-element lamps. Dual-element lamps of closely related elements such as magnesium/calcium and sodium/potassium have performed well in the authors' laboratory. However, most multielement lamps are not recommended because of their overall poor performance.

Lamps should be operated at as low a current as possible for achieving the sensitivity required. The manufacturer's recommendation for maximum current should be noted. Large currents will diminish the lifetime of the lamp and may also broaden the resonance line. Nonlinear calibration curves and decreased sensitivities are results of line broadening.

Electrodeless Discharge Lamps Electrodeless discharge lamps are alternative radiation sources which produce more intense spectra than the corresponding hollow-cathode lamps. They are recommended for elements such as arsenic, selenium, tellurium, tin, and lead. A separate power supply is required. The construction of a lamp is illustrated in Fig. 2.5.

A small amount of metal or metal salt, such as an iodide, and an inert gas are sealed within a silica bulb. The bulb is mounted in the cavity of a radio-frequency (rf) coil. Radiation from the lamp is transmitted through a quartz window. The lamp has about the same overall size as a hollow-cathode lamp and is positioned in the lamp compartment of the instrument in the same way. Stable lamp output is usually obtained after about 15 min. The manufacturer's recommendations for operating the lamp should be followed.

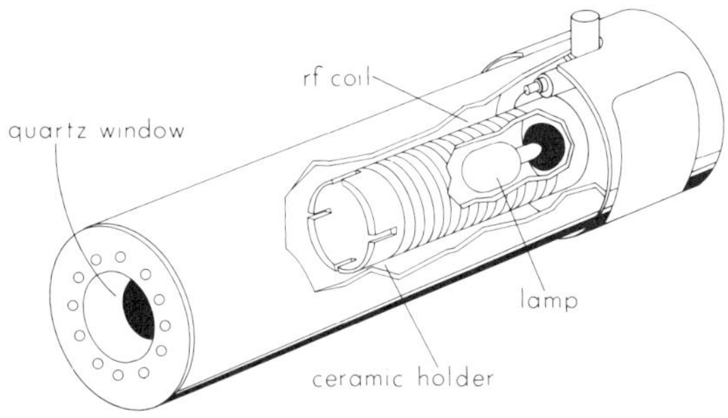

Figure 2.5. Electrodeless discharge lamp.

6. Interferences

As in all analytical methods, interferences are important considerations because they decrease accuracy and increase detection limits. Knowledge of the types of interferences which may arise in atomic absorption and how they may be overcome is essential. Some metals such as copper and manganese do not have serious problems in flame atomic absorption even in the analysis of a range of sample types. The various kinds of interferences are discussed in the following sections.

Atomic Spectral Interferences A spectral interference can occur when the absorption profile of a second element overlaps the absorption line of the analyte. The result is a positive error because the detector signal is the sum of the two absorbances. Because of the narrow widths of atomic emission lines, this type of interference is rarely encountered. Tabular information on spectral interferences discovered by a number of workers has appeared in other publications (5,9).

Nonspecific Interferences The main sources of these interferences are molecular absorption and light scattering in the atomizer. The former occurs when a molecular species has an absorption profile that overlaps the line of the element being measured. The latter occurs when solid particles deflect some of the source radiation away from the beam received by the monochromator and detector. Samples which contain large proportions of refractory elements frequently produce light scattering.

Both interferences are more serious in the wavelength region below 250 nm; flames are less subject to these problems than are electrothermal analyzers. Background correction is used to overcome the interferences, as described later.

Chemical Interferences These interferences occur in both flame and electrothermal analyzers. They are caused by chemical interactions in the sample which alter the proportion of the analyte which is atomized or the rate at which analyte atomization takes place. Pure standard solutions may not be subject to the same chemical interactions, so erroneous results are possible in direct comparisons of samples and standards. There are several ways to overcome chemical interferences. The first is to alter the physical parameters of the atomizer. The interference of phosphate with the determination of calcium in flame atomizers is much less in a hot nitrous oxide/acetylene flame than in an air/acetylene flame. In electrothermal analyzers, adjustments of drying and charring times and temperatures to volatilize as much as possible of the interfering matrix is a good procedure.

Another method is matrix matching. The compositions of standard solutions are altered to correspond as closely as possible to those of the samples. Matrix matching may not be possible if the overall composition of the samples is not known. If the composition is complex, matching is not practical. Standard additions of measured volumes of standard solution to sample aliquots are an alternative. This method of calibration is explained in more detail later. In electrothermal work, the chemical form of the element in the standard additions should be the same as that in the sample.

A third approach involves modification of the sample matrix by the addition of another chemical. Some examples will be described. In the determination of calcium in the presence of phosphate by means of flame atomization, releasing agents are added to prevent the formation of stable calcium phosphates. The releasing agent can be an organic complexing agent such as EDTA, which reacts with calcium (and other cations), or lanthanum and strontium salts to react with the phosphate. In electrothermal work, some elements volatilize at low temperatures, and interferences cannot be eliminated without loss of analyte. Ediger (10) has shown that the addition of a nickel salt to samples containing arsenic, selenium, and tellurium will stabilize these elements so that higher charring temperatures can be used. In the presence of nickel, both selenium and tellurium can be heated to 1200°C without significant losses. Ammonium phosphate will stabilize cadmium, and potassium dichromate and sodium sulfide will stabilize mercury.

Many studies of cation/cation interferences have been reported. The subject is complex. Depression and enhancement of signals are observed in many samples. A summary of the extensive literature on interferences in electrothermal atomization has been prepared by Slavin and Manning (11). Matrix matching and modification do not overcome all chemical interferences. Separation of the elements of interest from the bulk of the sample may be the only practical way to deal with severe problems.

Ionization Interferences Atoms of some elements can become ionized in flames or electrothermal atomizers. As a result, the atom population diminishes and sensitivity declines. The extent of the interference is dependent on the concentrations of the easily ionized elements. Sodium, potassium, rubidium, and cesium ionize readily in air/acetylene flames. The magnitude of the problem increases from sodium to cesium. At higher temperatures in nitrous oxide/acetylene flames, calcium, strontium, and the rare earths are ionized. Ionization effects can be controlled with additions of an easily ionized element to standards in order to compensate for the presence of a particular element in the sample. For example, if elements are being measured in the presence of a major amount of sodium, the standards should also contain a large amount (1000 ppm) of sodium.

Physical Interferences The viscosity and surface tension of a solution will influence its rate of flow into the nebulizer. This means that the atom populations in the atomizer will differ for the same concentration of elements in solutions of different viscosities. Aqueous solutions sometimes contain large concentrations of mineral acids. Standard solutions should be made up with approximately the same acid concentrations.

In electrothermal work, liquids are introduced into the atomizer by means of pipets, so the rule of matching the physical properties of samples and standards applies here also. The degree of spreading of liquids along the surfaces of electrothermal atomizers depends on the composition of the liquid. Organic solvents tend to spread over larger areas than aqueous solutions. Since the surface temperature of the atomizer varies along the length of the tube or rod, rates of atomization will vary also. Therefore, sample spreading can lead to errors in analysis if standards do not behave in the same manner.

7. Background Correction

An important method of overcoming nonspecific absorption is background correction. Although the problem is more severe with electrothermal atomizers, nonspecific interferences also occur with flame atomizers. They are of particular concern when the instrument is being operated near the detection limit. Of the four methods discussed here, simultaneous correction by means of a continuum emission source is the most popular.

Continuum Method A deuterium arc lamp can be used over the spectral region 190–300 nm to measure continuous background radiation or scatter in the wavelength range of the resonance line. For longer wavelengths, either a tungsten–iodine lamp or a mercury–xenon lamp is necessary. Both single- and double-beam instruments are equipped or can be modified so that signals from the line source and the continuum source are measured alternately. The difference signal or background-corrected absorbance is recorded. Some sensitivity may be lost if the slit width is too narrow. The absorbance due to the continuum radiation then becomes a significant part of the total.

In setting up a background corrector, the manufacturer's instructions should be followed in order to have the two beams (a) optically coincident, (b) filling the same fraction of the optical aperture, and (c) of approximately equivalent signal intensities. Most correctors will compensate for signals up to a maximum of 1 absorbance unit. If this value is exceeded, the sample composition must be changed in some way to reduce the background.

Nonabsorbing-Line Method In this method, the background signal is measured at a wavelength not more than 20 nm from the resonance line. The background absorbance is then subtracted from the absorbance of the sample. Thus, each sample must be analyzed at the two chosen wavelengths unless a dual-channel instrument is being used. The line chosen must be a nonabsorbing line of the element being measured. Such a line may be present in the spectrum of the same hollow-cathode lamp, or another lamp may be required. This method is not used for electrothermal atomizers. Background signals in electrothermal atomizers are less reproducible than in flames and they vary more from one determination to the next.

Zeeman Effect Method The Zeeman effect is used in another means of simultaneous background correction, but the applications have been limited mainly to instruments equipped with electrothermal atomizers (12). The resonance line is split into its Zeeman components, designated as π and $\pm\sigma$, as a result of the application of a magnetic field around the atomizer. Polarizers are placed in the optical path so that the absorption signal and background are measured on the central component. The background is measured on the $\pm$ components, and it is subtracted from the central component.

Line Self-Reversal Method When a hollow-cathode lamp is operated in the normal mode with a low current, radiation is absorbed by the analyte element and the background. If the current in the lamp is increased excessively, the intensity at the center of the line decreases greatly, and atomic absorbance by the analyte is also reduced. This is known as self-reversal. However, the background absorbs almost the same proportion of radiation when the lamp is in either the self-reversal or the normal mode. In the Smith–Hieftje system (13), the lamp current is varied so that high current pulses occur for very brief periods of time. The difference between the detector signals produced during normal operation and during self-reversal is the corrected absorbance. Some advantages of the system are that it corrects at all wavelengths and that it cannot be misaligned. A disadvantage is a reduction in sensitivity of about 40% on average. There is also some question about the adverse effect of such operation on lamp lifetime.

8. Calibration

Direct Comparison with Standards About three to five standards of increasing concentration and a blank are prepared in order to cover the expected range of the element to be determined. Because of the relatively short linear working range in atomic absorption (up to 2.5 orders of magnitude), adjustments in analyte concentrations may be necessary to obtain a linear calibration graph. As explained earlier, the overall compositions of standard and sample solutions should be matched as closely as possible. If additions are required to overcome interferences, they should be included in all solutions. Measurements in the range of 0.1 to 0.6 absorbance unit are preferred because the best precision is obtained in this region. A number of measurements of each standard are recorded, and the averages are used to draw a graph of concentration versus absorbance. The best accuracy is available when the sample concentrations occur in a linear part of the calibration curve. Extrapolations are not recommended, but they may be necessary if the amount of sample is limited.

Modern instruments contain microprocessor-controlled functions which allow for direct readout of concentrations. In calibration work, linear regression and curve-fitting programs are carried out automatically. Instruments with provision for multiple-point calibrations are preferred over those with single-point calibrations.

During analyses of a series of samples, standards should be checked to determine whether a significant change in response has occurred. An updated calibration curve can be applied if necessary. When sufficient readings are taken during the run, statistical calculations are possible. Some modern instruments have programs for such calculations. Experience has shown that multiple sample readings will improve results only if they are accompanied by multiple standard readings.

Standard Additions An alternative to direct comparisons of samples and standards is the addition of increments of standard solutions to aliquots of the sample. This method is time-consuming. It is used when interference problems occur (except for nonspecific interferences). The approach is particularly valuable in electrothermal work. The usual procedure is to prepare three aliquots. One of the aliquots is spiked with a quantity of standard so that the resulting analyte concentration is approximately twice that of the original sample. A second aliquot is spiked with sufficient standard to yield three times the initial analyte concentration. These solutions and the original sample aliquot are diluted to volume, and multiple readings are recorded. A graph of absorbance versus concentration is prepared as shown in Fig. 2.6.

The line is extrapolated to intersect the concentration axis. The value of the intercept is the concentration of the analyte. For the method to be successful, the

Figure 2.6. Calibration graph for method of standard additions.

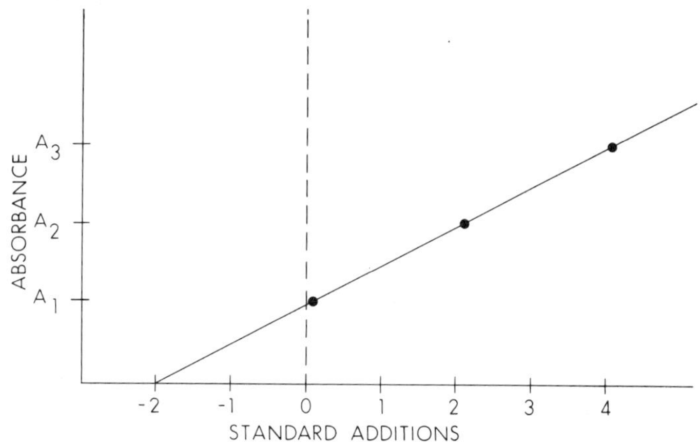

calibration graph must be linear within the range of concentrations as in Fig. 2.6. Care must be taken in work with electrothermal analyzers as the linear range is often very short. Background correction and additions to suppress interferences should also be used.

Use of Standard Reference Materials A number of types of geological samples have been certified as standards which contain stated concentrations of various elements. If a group of standards that covers concentration ranges of one or more elements of interest is available, it can be used for the preparation of calibration curves. The samples should have roughly the same composition as the standards. In this manner, a sample of complex composition is matched with standards, and contributions from the matrix are compensated. The subject is discussed in Chapters 1 and 3. Table 3.2 lists many geological standards. The limited supply and the costs of reference materials must be considered.

High-Precision Measurements The purpose of this procedure is to obtain data of higher precision than is available by means of the other calibration methods. An instrument equipped for scale expansion is required. Two standards are needed. They must bracket the concentration of the analyte within about $\pm 5\%$. The lower standard is set at a low absorbance value. The scale is expanded in order to set the upper standard near full scale. A graph is prepared, and the two standard points are joined by a straight line. Sample concentrations are obtained from the graph. All precautions to minimize interferences and to correct for background apply.

9. Special Procedures

Procedures based on flame and electrothermal atomizations are satisfactory for determinations of many elements of geological interest. Some elements have poor sensitivities in flames, as shown in Table 2.1. Interferences may also severely limit the determination of some elements with resonance lines below 300 nm. Two procedures which can decrease interferences and improve sensitivities of a few elements are outlined here. These are based on (a) the formation of volatile hydrides and (b) the evolution of mercury as vapor.

Hydride Formation The chalcocide elements, that is, arsenic, antimony, selenium, tellurium, and bismuth, together with tin, lead, and germanium can be evolved from their solutions as volatile, stable hydrides. Sodium borohydride is the preferred strong reductant used to form the hydrides. The hydrides must be decomposed to liberate the atoms for measurements. Atomization can be accomplished in a flame or in a heated tube.

Sensitivities can be improved further by injecting the hydride as a pulse into the atomizer. Electrodeless discharge lamps are better than hollow-cathode lamps for these analyses because they can provide about 10 times the intensity of radiation of the latter.

Hydride generation apparatus is available as an accessory from a number of manufacturers of atomic absorption equipment. The essential components are a hydride generation vessel and a gas transfer system for injecting the hydride into the atomizer. Various kinds of equipment including continuous-flow injection have been described. The atomizers used most often are air/hydrogen burners and low-temperature quartz tube furnaces. The latter are preferred because of the lower background and the longer residence times of atoms in the optical path.

The hydride generation method is subject to many interferences. Interferences depend on the method of reduction, the sample matrix, and the type of atomizer. Most workers have used the sodium borohydride method of reduction because of the rapidity of hydride formation. In any of the reduction methods, interferences occur during both the generation of hydride in the solution and the release of the hydride from the sample. The main reason for solution interference is that the formation of the desired hydride is slowed down by competing reactions involving other sample components and perhaps by coprecipitation or formation of insoluble compounds in some samples. The valence state of the analyte and the general solution environment also influence rate of hydride formation.

A study of the effects of 48 elements on the determination of As, Sb, Se, Te, Bi, Ge, and Sn was carried out by Smith (14). The following elements always interfered: Cu, Ag, Au, Pt, Rh, Ru, Ni, and Co. Alkali and alkaline earth elements and B, Al, Ga, Tr, Zr, Hf, Hg, La, Mn, V, and Y did not interfere. There are some interelement interferences among the hydride-forming elements. Usually these are in the range of 10–50% suppression of the signals at levels of 1000-fold more than the analyte. Inorganic acids used in sample preparation have a suppressing effect. Nitric acid causes severe suppression of selenium.

Gas phase interferences are related to the mechanism of hydride atomization. For example, a quartz tube analyzer was subject to less interference than an argon–hydrogen flame or a graphite furnace (15).

Many workers have tried to reduce the interferences due to sample matrix by increasing acid and sodium borohydride concentrations and by adding complexing agents of various kinds prior to hydride formation. Coprecipitation of interfering elements with hydrated oxides such as $Fe(OH)_3$ and separations on ion-exchange resins have been tried. However, standard additions are recommended as the best method for dealing with interferences. Standard spikes and reagent blanks should be carried through as much of the procedure as possible.

Mercury Cold Vapor Absorption Mercury is seldom measured in flame atomizers because of the relatively poor sensitivity obtained and the small concentrations of mercury usually found in samples. On the other hand, the high vapor pressure of mercury metal and some mercury compounds and the ease of reduction of many mercury compounds place severe limitations on the use of electrothermal atomizers. An alternative method involves quantitative reduction of mercury to the metallic form and evolution of the metal as vapor into a cell situated in the optical beam of a spectrometer. The relatively high vapor pressure of mercury at room temperatures means that mercury can be stripped easily from a reducing solution by means of air or an inert gas. The concentration of mercury in the stream is then measured at 254 nm.

Samples containing mercury compounds should be dried near room temperature to avoid losses, unless mercury is present as stable compounds. Solid samples should be stored in a refrigerator or kept frozen. Liquid samples are filtered and acidified with enough nitric acid to make a 0.5 *M* solution. The sample should also contain 0.05% potassium dichromate as a preservative. An alternative is the addition of gold at 10 μg/ml to acidified samples. Decompositions of organic samples containing mercury must always take place in the presence of an oxidizing agent such as nitric acid, or sulfuric acid and potassium permanganate, in order to retain the mercury. In pyrolytic decompositions, gold metal can be used to amalgamate mercury. The amalgam is then heated to evolve mercury vapor, which is swept into the atomizer by a flow of carrier gas. In analytical work, amalgamation is used also to concentrate mercury for subsequent measurement. Mercury compounds in the gas phase must be decomposed prior to entering the atomizer. A pyrolyzer in the sample line serves the purpose. The mercury can then be amalgamated on gold foil.

Apparatus for the generation of mercury vapor from aqueous samples, e.g., Fig. 2.7, is available from laboratory supply companies, or it can be assembled from components to suit the requirements of the analytical program. It usually consists of a test tube or flask for the reduction and sparging of mercury, a drying tube, a measuring cell, and gas-handling lines (16). A UV spectrometer or a good-quality mercury vapor meter can be substituted for an atomic absorption spectrometer. To obtain the best sensitivity, dead volume in the system must be min-

Figure 2.7. Apparatus for mercury determination: A, compressed air or nitrogen; B, needle valve; C, flow meter; D, two-way stopcock; E, ground-glass joint; F, sample container; G, diffuser; H, drying tube; J, gas connection to spectrometer.

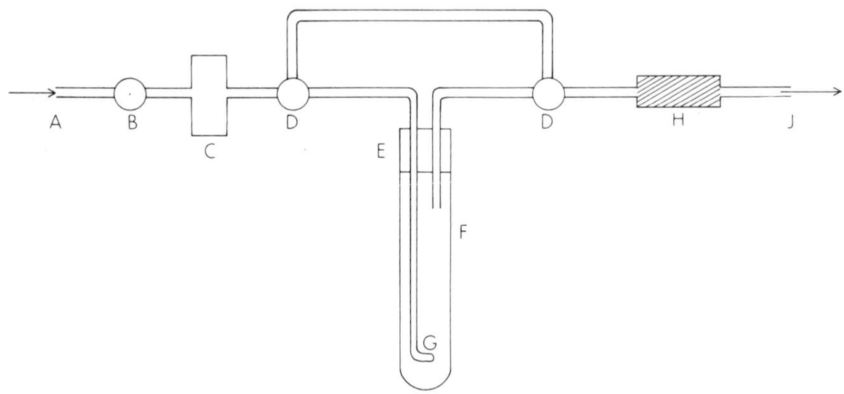

imal. The peak concentrations are measured as the mercury passes through the cell.

Gases such as chlorine and organic vapors which absorb in the UV spectral region around 254 nm will cause positive errors. A number of anions and cations interfere by suppressing the reduction of mercury compounds to the metal and the evolution of mercury vapor. Bromide, nitrate, phosphate, perchlorate, sulfide, and sulfate are anionic interferences. Interfering cations include gold, platinum, palladium, silver, copper, and possibly selenium and tellurium. Matrix matching is important in compensating for interferences.

10. Selection of Instruments

A list of commercial instruments together with brief descriptions is published in "Annual Reports on Analytical Atomic Spectroscopy" by the Royal Society of Chemistry, London. The list covers single- and double-beam instruments of major manufacturers in North America, the Western European area, Australia, and Japan. A major factor in the choice of an instrument will be the type and volume of work being undertaken in a laboratory. A flame instrument is best for high-accuracy analyses of major and minor components of samples. Either a single- or a double-beam instrument is satisfactory. Although trace element analyses can be carried out satisfactorily with a good-quality flame instrument, an electrothermal atomizer is recommended for this type of work. The detection limits for many elements are two to three orders of magnitude better with electrothermal atomizers. Some form of background correction is essential for flameless atomizers. Another factor to be considered is the work load. Provided that there is satisfactory sensitivity for the elements of interest, a flame atomizer is capable of handling about three times as many samples per hour as a typical flameless atomizer.

An air/acetylene burner is satisfactory for many elements, but a nitrous oxide/acetylene burner must be used for refractory elements. Tubular furnace electrothermal atomizers are recommended rather than graphite rods.

An autosampler is a good investment for large work loads with either flame or flameless atomizers. Autosamplers simplify the preparation and programming of standard additions and matrix modifiers.

Some instruments are equipped with multilamp turrets. While a lamp is being used, several other lamps can be warming up. Thus, rapid lamp changes are possible when several elements must be measured. Some new instruments are highly automated. Analytical conditions for a series of elements including sequences of samples and standards can be selected by the operator in advance. In fact, analytical programs can be stored in the computer memory for retrieval when required. While digital meters, printers, and strip chart recorders were used extensively in the past, a considerable variety of automated data acquisition equipment is available for new instruments. The extent of the investment in such equipment will depend on the needs of the laboratory.

II. INDUCTIVELY COUPLED PLASMA–ATOMIC EMISSION SPECTROMETRY

A. Introduction

Emission spectrometry employing arc and spark sources has been used for many years as a qualitative and quantitative tool in analytical chemistry laboratories. Simultaneous multielement analysis for 70 or more elements is possible with this equipment. Types of samples include metals, alloys, rocks, minerals, sediments, and solutions. The sensitivity available for many elements has made emission spectroscopy a popular choice for trace element analyses. A number of publications contain detailed descriptions of emission spectrometry and its applications in geochemical work (17–20).

Over the years, improvements have been made in emission sources, optical components, detectors, and data processors. The recent introduction of the inductively coupled plasma (ICP) as a new kind of emission source for a familiar instrument marked another stage in this development. In contrast to earlier sources, which were suitable for solid samples, ICP sources are designed almost solely for liquid samples. The ICP sampling system can be adapted for the introduction of vapors and gases, and research is continuing of methods of introducing solids.

A schematic diagram of an ICP quantometer is shown in Fig. 2.8.

Figure 2.8. Schematic diagram of an ICP quantometer.

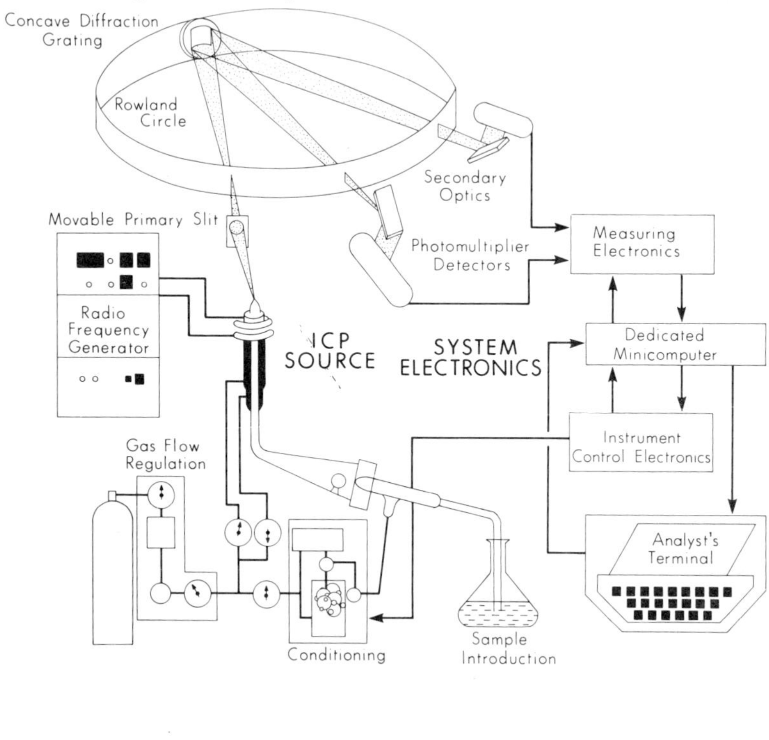

B. Instrument Components

1. Plasma Torch

In emission spectrometry, the purpose of the source is to provide sufficient energy to atomize samples and then to produce excited-state atoms. When the excited atoms return to their ground states, they emit their characteristic line spectra. Plasma torches are capable of reaching very high temperatures (>8000 K). A plasma is a gas containing a significant fraction of its atoms and molecules in an ionized state. Plasmas can interact with magnetic fields. The inductive coupling of a time-varying magnetic field is the basis of the operation of an ICP torch. A typical plasma torch is illustrated in Fig. 2.9.

The plasma is usually formed from argon gas. The argon is split into three metered streams, which enter the torch at the points indicated in Fig. 2.9. The sample in the form of an aerosol, vapor, or particles enters with the argon along the central channel at 1. The main flow of argon to sustain the plasma enters at 2. Because the hottest area of the plasma can reach temperatures as high as 10,000 K, a tangential flow of argon, entering at 3, cools the outer walls of the torch. This flow also centers the plasma radially. To establish the argon plasma, some ions must be formed initially in the region of the induction coil space. This is accomplished by means of a Tesla coil. The high-frequency currents in the induction coil generate oscillating magnetic fields. The lines of force from these

Figure 2.9. Plasma torch; 1, sample aerosol and argon; 2, argon gas inlet; 3, argon coolant inlet; 4, induction coils; 5, plasma.

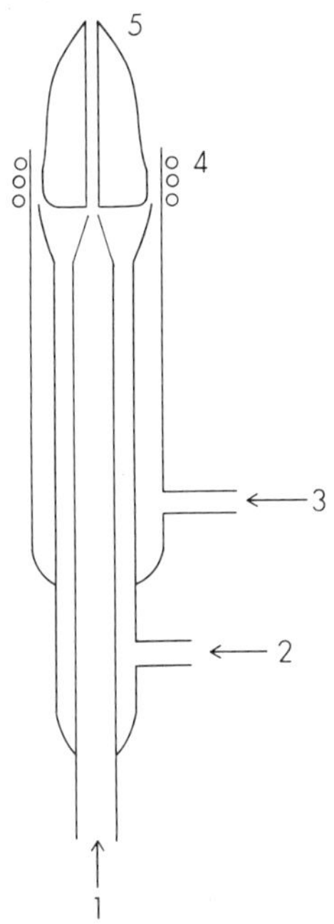

fields are axially oriented inside the tube and follow elliptical closed paths outside the coil. The argon ions flow in closed annular paths inside the quartz tube in response to the magnetic field. As the ions are accelerated, resistance to their flow produces ohmic heating, and this results in additional ionization of argon. The overall result is the continuous generation of a plasma.

The sample flows into and through the plasma along a narrow axial channel where the temperature is in the range of 6000–7000 K. Interactions with the plasma and the high temperature of the plasma combine to vaporize the solvent and atomize the analyte. The emission from the torch is viewed at a point yielding the largest signal-to-noise ratios for the analyte atoms. This region is about 1 to 3 cm above the induction coil. Although the best viewing height varies according to the atomic species, a compromise height is selected for multielement analyses in order to fix the optical configuration. Calibration curves relating emission intensities to concentrations are usually linear over a range of better than five orders of magnitude. Thus, determinations of major, minor, and trace components are often possible without sample dilution. ICP detection limits are generally better than those in flame atomic absorption, arc, and spark but are not as good as those in furnace atomic absorption. They range over 0.1 to 10 parts per billion (ppb).

2. Nebulizer

As noted previously, liquid and gaseous samples are employed in ICP sources. Volatile metal hydrides (arsenic, selenium, etc.) are easily introduced into the plasma by means of argon carrier gas. Liquid samples must be introduced as fine droplets. As in atomic absorption work, pneumatic and ultrasonic nebulizers have been adapted for this purpose. The two main types of pneumatic nebulizers are concentric and cross-flow designs. A peristaltic pump is required to pump sample liquids to the nebulizer. In the concentric design, a flow of argon gas is directed through a small annular space around the tip of a capillary tube which carries the liquid sampled. A venturi effect is produced at the tip, and liquid is drawn into the flowing argon. Shear forces cause the disintegration of the liquid into droplets. The droplets are carried through a spray chamber equipped with a drain and then to the plasma torch. Some spray chambers contain an impact bead to provide additional small droplets.

Cross-flow nebulizers consist of two capillary tubes, one for sample and one for gas, situated at 90° to one another. The relative positions of the tubes can be adjusted to provide maximum shearing of the liquid and thus aerosol formation. As in the concentric design, the aerosol passes through a spray chamber, where large droplets and excess liquid are separated. Liquid samples must be free of suspended solids because solids can block the capillary tubes. Also, samples containing large concentrations of dissolved salts may plug the nebulizer because evaporation of liquid around the nozzle results in the deposition of solids. Even though blockages do not occur, carryover of sample in the succeeding analyses may be a problem. The salt effect can be reduced by saturating the argon with water before it enters the nebulizer. In addition, some commercial equipment contains a nebulizer tip washer. This device automatically injects a drop of water into the argon stream to rinse the nebulizer tip at the conclusion of each analytical cycle. Solutions which contain up to 10% dissolved organic matter and as much as 30% inorganic salts have been nebulized satisfactorily. However, the maximum salt concentration that can be tolerated will depend on how closely the solution composition approaches the saturation point. As in atomic absorption work, water rinses must alternate with samples. The compositions of samples and standards should be matched as closely as possible to ensure that changes in physical properties affecting liquid uptake rates will not lead to errors. If matching is not possible, then an internal standard method of compensation is available and will be described later. Other designs and types of nebulizers have been used. Ultrasonic nebulizers require desolvation of the aerosol/argon stream because the larger proportion of liquid will cause instability of the plasma and may extinguish it entirely. Babington-type nebulizers (21,22) have been used by some workers because of their greater tolerance of dissolved solids in the sample solutions.

3. Spectrometer

An outstanding feature of ICP–AES instruments is their ability to measure many elements in a sample in either a simultaneous or a rapid sequential manner. Sequential instruments are favored for research and for programs where samples vary greatly in composition. Simultaneous instruments provide faster analyses and are particularly suited to more routine work.

4. Simultaneous Analysis

The simultaneous measurement of 50 or more elements in a sample is possible with a polychromator. One such design is shown in Fig. 2.8.

Radiation from the ICP is focused on the entrance slit of the optical system. A concave diffraction grating and exit slits are arranged on a Rowland circle. The positions of the exit slits are chosen so that they isolate specific wavelengths of the diffracted radiation. Thus, an emission line of a particular element is focused on a designated slit. Provided that there is no overlapping of lines, the detector which serves the slit will respond only to the concentration of that element as a function of radiation intensity. The signals from the photomultipliers are processed in a dedicated minicomputer. Data from analyses of standards and for background intensities are stored in the computer. The computer can print out background-corrected concentration values for the elements in the analytical program. One limitation on the application of this design of instrument is that the number of elements together with their spectral lines must be specified when the instrument is purchased, so that the optical array can be fixed. It is very difficult to add other elements or to change the line selection once the instrument has been manufactured and installed.

Another type of polychromator of rather novel design employs an echelle grating. The grating is capable of high resolution over a broad spectral range (180–800 nm). An optical arrangement for this grating is shown in Fig. 2.10.

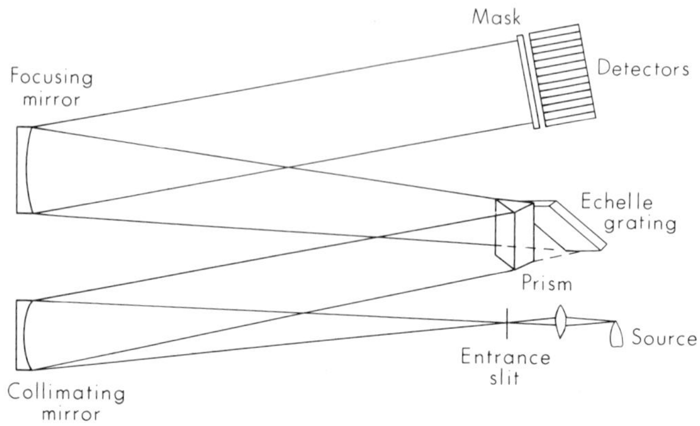

Figure 2.10. Polychromator with an echelle grating.

The radiation from the source passes through a prism placed in the optical path. The prism acts as an order disperser in a direction perpendicular to the diffraction grating. Radiation reflected from the grating forms a two-dimensional spectrum. An array of detectors is mounted at the focal plane. Each detector is positioned so that it monitors a specific emission line of an element. This is accomplished with the help of a mask with apertures that limit the transmitted radiation. Again, the wavelengths to be monitored must be selected beforehand. If the operator needs to change the number of elements to be determined or to use alternative emission lines, the mask can be replaced by a new one containing the required apertures. Replacement masks or cassettes are supplied by the equipment manufacturer.

5. Sequential Analysis

Sequential spectrometers use scanning monochromators with computer interfacing for drive control. Emission lines to be monitored are chosen, and the monochromator is centered on each line in turn. Scanning rates and integration times for each analytical line can be varied. Improvements in wavelength scan mechanisms ensure good accuracy at high speeds. The flexibility of the scanning monochromator design is an advantage when samples of different elemental compositions must be analyzed. The operator is not limited by a prior selection of emission lines.

C. Interferences

Some of the interelement effects encountered in atomic absorption work are either absent or greatly reduced in an ICP source. The reason is that the high atomization temperatures in a chemically inert atmosphere and the relatively long residence time of the analyte in the plasma result in almost complete dissociation and atomization of the analyte. Refractory elements such as tungsten and vanadium are analyzed under the same conditions as easily atomized elements such as copper. Ionization interferences are much less severe than in flames. The interferences can be divided into two main categories: (1) spectral and (2) nonspectral. These types will be described briefly in the following sections and some ways of overcoming or correcting for them will be discussed.

1. Spectral Interferences

Spectral interferences are of major concern in emission work. One such interference, called line coincidence, is the overlap of an emission line of the analyte by a line of another sample component. The reason for the overlaps is that the emission spectra of many elements are complex and yield many lines. The magnitude of the problem depends on the resolving power of the spectrometer. Echelle spectrometers give especially good resolution. Line coincidence will result in positive errors unless the operator recognizes the problem. Reference books which list many spectral interferences are available (23,24).

A number of methods of overcoming interferences caused by line coincidence are available. The first is the use of reference data in the selection of an analyte line free of interference from lines of other elements in the sample (25). However, the sensitivity of such lines may not be sufficient for trace element determinations.

The second approach is to measure the extent of interferences quantitatively as a function of the concentration of the interfering element. If element A is being determined and element B interferes, an interference coefficient K_{AB} is required in order to subtract the contribution of B from the total line intensity. The coefficient is calculated by using a pure solution of B and measuring the line intensity in the absence of A. Thus

$$C_A \text{ (corrected)} = C_A \text{ (total)} - K_{AB} C_B \text{ (actual)}$$

where C stands for concentration. The concentrations of B must be known in the samples being analyzed. If more than one element interferes, then appropriate corrections must be applied for each element.

A third method is the use of multielement standards in the preparation of calibration curves. Their purpose is compensation of the interfering elements within the concentration range of the series of standards. Again, the operator must know the concentrations of the interfering elements in the samples.

Spectral interferences also arise from background radiation generated in the plasma by the reagents and gases used in the work. For example, weak band emissions from OH, NO, and CN contribute to the background. The background spectrum of a sample may change in comparison with that of a calibration standard. Elements such as aluminum, magnesium, and calcium cause background shifts. Organic materials in the sample can also change the background. Stray light within the spectrometer will increase the background.

When instrumental conditions for an analysis are being established, examinations of background contributions to the chosen emission lines should be made. Corrections may not be required for all lines. The emission line of the analyte is scanned by measuring intensities at small increments over the range of ±0.04 nm of the wavelength. The results are plotted to yield a wavelength intensity profile. A standard, a blank, and a sample of any suspected interfering element are treated similarly. If these data show that background correction is necessary, intensity readings are taken at 0.02–0.03 nm above and below the wavelength of the line. The average value is subtracted from the total intensity of the line. The close proximity of an emission line from another element in the sample may allow only one background reading to be used for correction. Details of instrument operation for wavelength intensity profiling are given in manuals provided by the instrument manufacturers. In a multichannel instrument, the primary slit is moved slightly backward and forward over the region of the line while intensity readings are recorded. Wavelength-scanning instruments can be programmed for this task. Once the background corrections have been determined, they can be applied automatically in modern instruments.

Interference effects which have been tabulated quantitatively in the literature apply only to a particular instrument and sample composition. However, they indicate the magnitude of the effects which may occur in work of a similar kind.

2. Nonspectral Interferences

These interferences arise from physical and chemical processes that affect the transport, atomization, and excitation of the analyte. They are classified as nonspecific and element-specific.

Nonspecific interferences are related to the physical properties of the analyte. In solution work, physical properties of the solution affect the rate of introduction of the sample into the nebulizer and hence the plasma. The subject has been discussed in the section on atomic absorption. Matrix matching of samples and standards particularly with respect to total dissolved solids and acid concentration is recommended as a means of overcoming these interferences.

Element-specific interferences originate in the plasma and involve chemical and atomization effects. Signals can be enhanced or depressed. Although solutions containing elements such as calcium and phosphate, calcium and aluminum, and combinations of alkali metals have been studied by a number of researchers, conflicting results have been reported. These interferences are complex in nature and involve volatilization of analyte, temperature changes in the plasma, and ionization suppression (26). Generally, interferences can be reduced by optimizing the operation conditions of the ICP, e.g., changing rf power, argon flow rate in the sample line, and observation height in the plasma.

D. Calibration

1. Direct Comparisons with Standards

The preparation of single-element standards is described in Chapter 3. For emission work, the quality of solution standards is very important. Only the highest purity chemicals, reagents, and water should be used in order to minimize the number and amounts of impurities in the standards.

Multielement standards are very useful in emission spectroscopy. They are readily prepared by mixing appropriate volumes of the single-element standards. A set of standards will allow multielement calibrations to be made over a range of concentrations. The stability of multielement standards is an important consideration. Anions, such as sulfate and chromate, which form precipitates with some metal ions should be excluded. If silver is a component of the standard, then chloride should be absent. Thallium can precipitate as a barium salt from solutions containing both elements. In strongly acidic solutions, combinations of niobium, tantalum, molybdenum, and tungsten may be unstable (27). Therefore, caution is advised when standards containing these elements are prepared.

Before a series of measurements is begun, the analytical response to the emission line of an element must be optimized. In some spectrometers, this is accomplished by means of slight movements of the primary slit so that the line profile is centered on the exit slit. In sequential analyzers, the software program contains the necessary operating details so that the optical system is centered on each emission line in turn.

The ratio of the intensities of analyte line and background depends mainly on (a) power of the source, (b) argon flow rate in the sample line, and (c) observation height above the induction coil. These three parameters must be controlled precisely in order that good analytical results may be obtained. Compromise conditions for plasma operation for a multielement ICP spectrometer have been recommended (28): observation height about 15 mm above the coil, aerosol argon flow rate 0.35–0.85 liter/min, and power 1.2–1.3 kW for a 1½-turn load coil or 1.0–1.1 kW for a 3–4-turn coil.

The preparation of calibration curves will involve background corrections and corrections for line overlaps as explained earlier. Most ICP calibration curves are linear. Some workers have reported nonlinearities in curves covering extended concentration ranges. The data systems of most modern instruments include curve-fitting capabilities, and slight deviations from linearity can be accommodated. At least five standards should be used, and more are desirable.

2. Internal Reference Element

The performance of the nebulizer is dependent on the physical properties of the solution. Standards and samples should be matched as closely as possible in their concentrations of acids and dissolved salts. If matching is not feasible, an alternative procedure is to use an internal reference element in standards and samples. The reference element should be one that is not expected to be present in the samples. Measured concentrations of the reference are added to all solutions. The reference element is analyzed together with the other elements, and its value in the samples is compared with the expected concentration. If the two are different, a factor is calculated which will convert the concentration of the reference to the correct result. The factor, which compensates for differences in transport of samples and standards to the plasma, is applied to the results for all elements in the sample.

3. Standard Reference Materials

If a series of standard reference materials is available, it can be used for calibration purposes provided that the sample compositions fall within the same range. Individual standard reference materials can be included in analyses of samples of similar compositions to check on the accuracy of the results. Significant differences between the specified values and the results of analysis of a standard may help to identify problems in the method. A list of geological standard reference materials is contained in Table 3.3.

E. Selection of Instruments

Commercially available plasma spectrometers are listed and described briefly in the "Annual Reports on Analytical Atomic Spectroscopy" of the Royal Society of Chemistry. One of the main advantages of ICP-AES instruments over atomic absorption units is the capability of performing rapid multielement analyses. Atomic absorption is essentially limited to the determination of one element per sample. Emission spectrometers are designed for either simultaneous or sequential determinations of elements. The main advantage of a simultaneous direct reading instrument is a high throughput of samples. This system is recommended when many elements must be measured in large numbers of samples as in routine laboratory operation. The main disadvantage of such an instrument is that a selection of lines must be made at the time of purchase so that the optical channels corresponding to each element can be set. If different or additional lines are required later, the channels can be altered only with difficulty. However, the echelle design of direct reading spectrometer is more flexible. The wavelength selection can be changed by altering the positions of the exit slits serving the detector array.

Wavelength-scanning instruments offer greater flexibility than polychromators for making and revising line selections to suit the application. Computerized, slew-scanning monochromators are available from several manufacturers. This design is the best choice for a laboratory where a limited number of elements must be determined in a variety of samples. In emission work, the resolving power of a particular instrument is an important consideration. This is usually quoted as dispersion (nanometers per millimeter). The lower the number, the better the capability of isolating a line. The optical characteristics of the spectrometer must receive careful consideration. The capability of line scanning and automatic background correction are also important. All modern instruments are equipped with computer-controlled readout systems.

III. X-RAY FLUORESCENCE ANALYSIS

A. Introduction

X-rays constitute a form of radiation defined by the wavelength range of 0.01–10 nm in the electromagnetic spectrum. There are two forms of x-rays, namely continuous radiation and characteristic radiation. The former is the result of a high-energy electron beam interacting with the clouds of electrons surrounding atomic nuclei. Characteristic radiation results from the ejection of inner-orbital electrons by high-energy photons, electrons, protons, or other charged particles. This is followed by transitions of other orbital electrons from high to low energy states. The transitions give rise to series of x-ray lines. In practice, a sample (usually a solid) is bombarded by high-energy particles. The x-rays emitted by the elements composing the sample have intensities and wavelengths which are characteristic of the atoms of the various elements. Thus, x-ray emission spectrometry is a technique that provides both qualitative and quantitative analytical data. The spectrometers are capable of measuring both wavelengths and intensities of x-rays emitted by the sample. The identity of an element emitting a particular line is established by measuring the wavelength of that line and comparing the wavelength with those in tables of standards. The intensity of an x-ray line of a particular element is converted to a unit of concentration by means of a previously prepared calibration curve.

Elements yield more than one series of x-ray emission lines. The relative intensities of the lines in each series are dependent on the probabilities of electron transitions. Commercial spectrometers are designed to detect and measure the intensities of many of the *K* and *L* series lines and a few of the *M* series lines of the heavy elements. Because *K* lines are the most intense, they are most often used in analytical work. *L* series lines are used in analyses of heavy elements, as some of the *K* series lines are more difficult to generate. Longer x-ray wavelengths of the lighter (or lower atomic number) elements, i.e., elements below titanium in the periodic classification, are absorbed strongly by air. To overcome this problem, air is replaced by helium, or the gas in the instrument is evacuated. Even in vacuum spectrometers, sensitivities of x-ray lines diminish drastically below atomic number 12 (magnesium). X-ray fluorescence is not used for analyses of elements below atomic number 9 (fluorine).

In the early work on x-ray analyses, primary x-rays were used to excite the characteristic secondary x-rays from the sample. Hence, the technique is described as x-ray fluorescence spectrometry.

In analyses of geological samples, x-ray fluorescence (XRF) is used for measurements of major elements, as well as many minor and trace elements. Modern instruments are capable of high-speed analyses with good sensitivity to the low parts per million range. In comparison with the other instrumental methods described in this book, XRF has an advantage in the time required for sample preparation if loose or pressed powders are suitable as samples. Some additional comparisons will be made later.

Detailed discussion of the theoretical basis of x-ray spectrometry will not be part of this book. The description which follows deals with practical aspects. For additional information, the reader should consult the textbooks listed in Refs. 29 and 30.

B. Instruments

X-ray fluorescence spectrometers are of two types. In the wavelength-dispersive design, a single crystal diffracts the x-rays emitted from the sample in order to isolate narrow wavelength bands. In energy-dispersive spectrometers, the polychromatic beam of radiation from the sample falls on a detector, which isolates narrow energy bands. Figures 2.11 and 2.12 are schematic diagrams illustrating the major components of typical spectrometers.

In x-ray-dispersive instruments, an intense beam of radiation is directed at the surface of the specimen, which is almost always a solid in geological work. Preparations of samples are described in detail in Chapter 4. Part of the characteristic radiation generated is collected. This is accomplished by means of a collimator or a slit which allows the radiation from the sample compartment to fall onto the surface of an analyzing crystal. Bragg's law governs the diffraction of individual x-rays according to the equation $n\lambda = 2d \sin \theta$, where λ is the wavelength of the radiation, n the diffraction order, d the interatomic layer spacing in the crystal, and θ the angle of diffraction. The diffracted beams are collimated so that the diffracted radiation falls on the surface of a photon detector. The detectors are gas flow or scintillation counters whose outputs are voltage pulses. The pulses are conditioned, integrated, and displayed as measures of line intensities.

Wavelength-dispersive instruments are single-channel or multichannel (simultaneous) instruments. The former design is equipped with a number of inter-

Figure 2.11. Wavelength-dispersive instrument: 1, x-ray tube; 2, sample; 3, collimator; 4, diffracting crystal; 5a, gas proportional counter; 5b, scintillation counter.

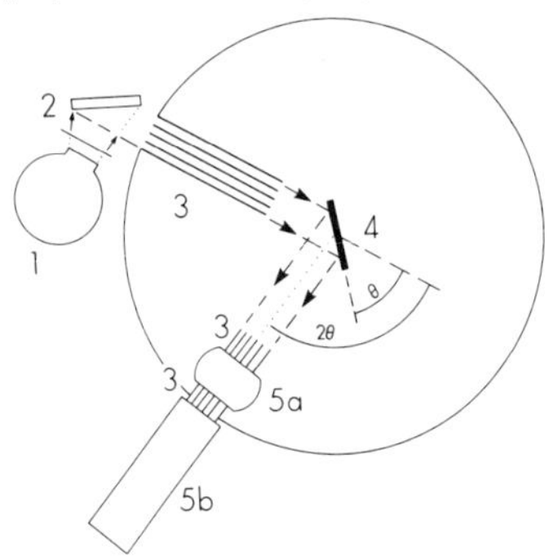

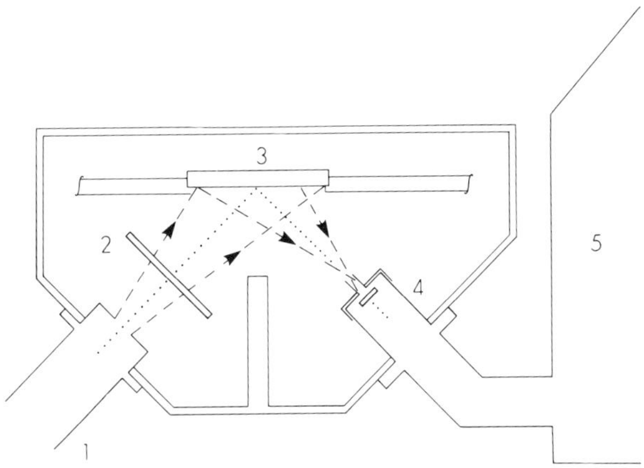

Figure 2.12. Energy-dispersive instrument: 1, x-ray tube; 2, filter; 3, sample; 4, Si(Li) detector; 5, liquid nitrogen cryostat.

changeable crystals so that an appropriate range of wavelengths can be selected for the analysis at hand. A multichannel instrument is designed so that a number of single-channel spectrometers can be arranged radially about an x-ray tube and sample. Each of the subassemblies has its own collimator, analyzing crystal, and detector. In this manner, up to 28 elements in the sample can be measured simultaneously. Most of the channels can be preset for specific analyte lines. In addition, one or more channels can be equipped as for scanning and thus are suitable for qualitative work. The scanning channels can also be used for measurements of lines of elements which are required less frequently than the elements with their own preset channels. During an analysis, each channel accumulates counts for a specific time or until a standard channel accumulates a designated count (one of the channels is used as the standard channel). When the counting period has been concluded, the count in each channel is either printed or used by a computer in a program which yields the quantitative results.

A single-channel instrument is also called a sequential x-ray spectrometer. With automation, a number of elements in the same sample can be measured according to programs provided by the manufacturer and/or the operator.

Energy-dispersive spectrometers (Fig. 2.12) consist of an excitation source, a spectrometer, and a detector. All of the x-rays from the sample are directed into the detector at once; the detector generates a pulse of electric current having a height proportional to the energy of each x-ray photon. The most popular detector is a lithium-drifted silicon Si(Li) unit. The detector must be maintained at liquid nitrogen temperature. The detector together with the associated electronics provides an x-ray energy spectrum. The instruments include a minicomputer to help in peak identifications and quantitative analyses.

C. Excitation Sources

1. X-Ray Tubes

Although different types of high-energy particles can excite characteristic x-rays from a sample, x-ray tubes are used in most commercial instruments. However, the tubes used in wavelength-dispersive (WD) spectrometers are operated at much higher power settings than those in energy-dispersive (ED) instruments. A major requirement is a stable, efficient power supply. Sources for WD instruments should be capable of delivering 2–4 kW of power, whereas 500 W is usually sufficient for ED instruments.

The high-power, thick-anode x-ray tubes are operated to excite simultaneously most of the elements in the sample. The high power will also result in high background for some elements, but operation at high power settings is necessary to overcome losses of x-rays in the collimators and crystal. The tubes are designed to work over a range of voltages and currents, but the total power on a tube must not exceed the manufacturer's recommended limit. The upper limit is usually in the range of 2–4 kW. The optimum excitation voltages for most elements in a sample are greater than 50 kV in order to excite the *K* lines. Adjustment of the tube voltages over a range of 60–100 kV is desirable so that the best sensitivity for a broad range of elements can be obtained. The most popular anode metals are W, Cr, Mo, Rh, Ag, and Au. A choice of anode can be made on the basis that the characteristic lines of the target do not interfere with the lines of the elements in the sample. In addition, it may be possible to choose an x-ray tube in order to enhance the sensitivities of element lines of slightly longer wavelength or lower energy than the anode lines.

In ED instruments a low-powered x-ray tube is required as an excitation source, as mentioned previously. In addition, the tubes are pulsed in some modern instruments. The primary x-rays are filtered to minimize background, and then they are directed onto the sample. In an alternative arrangement, the x-ray tube excites secondary x-rays from a pure element sample which is located between the tube and the sample for analysis. The secondary x-rays irradiate the sample.

2. Radioisotopes

The power supplies for x-ray tubes are bulky because the tubes consume a large amount of power. In contrast, radioisotope sources are compact and have no requirements for electrical power. Thus, radioisotope sources have been used especially in portable equipment (see Chapter 5). Radioactive sources have frequently been combined with NaI(Tl) or Si(Li) detectors to take advantage of their high detection efficiencies. However, radioisotope sources provide smaller output intensities and less flexibility in selection of excitation spectra than do x-ray tubes. X-ray tubes are preferred, therefore, for more complex analytical problems encountered in geochemical projects.

The radioisotope source is encapsulated and then shielded so that the detector is not exposed to direct radiation from the source. In addition, the entire chamber

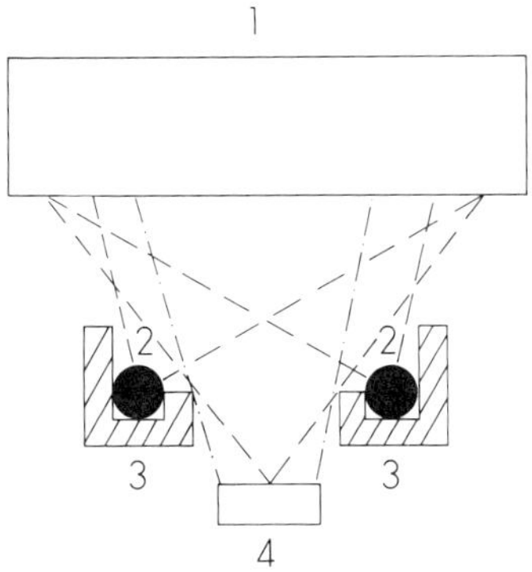

Figure 2.13. Geometry for radioisotope source: 1, sample; 2, radioisotope source; 3, shielding; 4, detector.

containing the source, sample, and detector must be shielded to protect personnel. Some of the properties that are important in choosing a source are the types of emitted radiation, the total energy of the source, the activity of the source, and the half-life. A number of configurations of source, sample, and detector have been used. Figure 2.13 illustrates a configuration of an annular source which is useful for a small-area detector such as Si(Li). For most applications, the activities of the source range from 0.5 to 100 millicuries (mCi). Table 2.3 contains some data pertaining to the commonly used sources.

D. Collimators

In wavelength-dispersive spectrometers, the radiation from the x-ray tube covers a large area of the surface of the sample. Collimators placed in the beam between the sample and the flat crystal (i.e., the primary collimator) and between the crystal and the detector (i.e., the secondary collimator) provide parallel beams. In flat-crystal spectrometers, Soller-type collimators are used. These collimators are made of a series of thin, parallel metal plates. They limit the divergence of the x-ray

Table 2.3
Radioisotopes as Sources

Isotope	Half-life (years)	Typical activity (mCi)	Applications
^{55}Fe	2.7	5–100	Excitation of light-element *K* lines Na–Ti
^{109}Cd	1.3	0.5–100	Excitation of medium *K* lines Cr–Nb and *K* lines of heavy elements (Pb, Hg, Pt, Au)
^{241}Am	458	1–50	Excitation of *K* lines of medium and heavy elements
^{3}H	12.3	3000–5000	Broadband excitation
^{147}Pm	2.6	500	Broadband excitation

beams, since only those beams parallel to the plates of the collimator arrive at the crystal and the detector. The primary collimator is fixed, while the secondary collimator moves with the detector. One is usually coarse (short and/or wide-spaced plates) and the other fine (long and/or close-spaced plates). In a combination detector, a coarse collimator is placed between the forward flow-proportional counter and the scintillation counter. A fine collimator will produce narrower peaks, lower background, higher resolution, and lower analyte-line intensity. Collimators are chosen to yield a resolution adequate for the measurements.

E. Analyzer Crystals

The crystal is a thin section of a crystalline material which has been cut or cleaved parallel to a set of crystallographic planes. The crystal can be flat or curved. As noted earlier, parallel-plate collimators are used with flat crystals. With a curved crystal, the collimators are replaced by slits. The function of the crystal is the dispersion of the secondary x-ray beam into a wavelength spectrum. The crystal rotates from a position parallel to the collimated secondary beam through progressively larger angles. As it rotates, it passes through the Bragg angle for each wavelength up to a limit and diffracts each wavelength in turn.

In most crystal spectrometers, the maximum attainable diffraction angle is about 70–75°. According to the Bragg equation, the maximum wavelength which can be measured with a particular crystal is $1.8d$, where d is the interplanar spacing of the crystal. One crystal cannot cover the full wavelength range since there would be poor angular dispersion at the shorter wavelengths. For this reason, several analyzing crystals are usually available for the spectrometer. The crystals for various parts of the wavelength range can be interchanged readily. Crystals with large interplanar spacing are required for analyses of elements of low atomic number such as sodium and magnesium. For elements of larger atomic number, a crystal with a smaller d value is necessary. A list of some of the crystals used frequently is shown in Table 2.4.

Table 2.4
Some Analyzing Crystals for XRF Spectrometers

Crystal	Name	$2d$ (nm)	Elements: K_α lines	Elements: L_α lines
Topaz	Aluminum fluosilicate	0.2712	Yb to V	U to Ce
LiF (220)	Lithium fluoride	0.2848	Er to Ti	U to La
LiF (200)	Lithium fluoride	0.4028	Ce to K	U to In
PE	Pentaerythritol	0.8742	Zr to Al	Sn to Rb
ADP	Ammonium dihydrogen phosphate	1.065	Al to Na	Pa to As
TlAP	Thallium acid phthalate	2.590	Si to F	
KAP	Potassium acid phthalate	2.663	Si to F	

F. Detectors

In wavelength-dispersive spectrometers, gas flow proportional and NaI(Tl) scintillation counters are used to convert the diffracted characteristic x-rays to voltage pulses. The pulses are integrated and displayed in some manner as measures of the x-ray line intensities. In energy-dispersive spectrometers, the detector is generally made of lithium-drifted silicon, Si(Li), which is a proportional counter of high resolution.

1. Gas Flow Proportional Counter

These counters are designed for the detection of low-energy (long-wavelength) x-rays. The gas is usually composed of 10% methane in argon. A constant flow of gas through the counter is necessary because it can diffuse through the thin window material. In the counter, incoming signals are amplified by selecting a voltage (1000–3000 V) for the insulated center wire. The mean pulse amplitude output is directly proportional to the energy of the x-ray photon which causes the signal. Thus, pulse height selection can be used to eliminate other signals. The dead time of the counter is small, e.g., 0.5 μs. Dead time is the time interval after the initiation of a pulse of normal amplitude during which the detector is insensitive and does not respond to a second x-ray photon.

These counters are used for measurements of K_α radiation of elements below arsenic in the periodic classification. The response of the counter for elements of larger atomic number decreases to the point that sensitivity is too low except for major concentrations of elements.

2. Scintillation Counter

The scintillation counter commonly used in x-ray work is made of an NaI(Tl) crystal optically coupled to a photomultiplier tube. The window is made of beryllium about 0.1 mm thick. When an x-ray photon enters the crystal, the energy is absorbed and excited states are formed. The decay from excited to normal state results in the emission of a scintillation or flash of light. The intensity of a scintillation is proportional to the x-ray energy. The light falls on the photocathode of the photomultiplier, and the number of electrons emitted from the cathode is proportional to the light intensity. Thus, the overall output of the counter is proportional to the energy of the incident x-ray photon. As with the gas flow proportional counter, pulse height selection is used to eliminate background signals. The dead time of the counter is about 0.2 μs. This counter is used primarily for the measurement of emissions from the K lines of elements heavier than iron in the periodic classification (wavelengths shorter than 0.21 nm).

In some instruments, either a gas flow proportional or a scintillation counter can be selected as the detector. In others, the two counters can be used in tandem and operated simultaneously. In this arrangement, the proportional counter is mounted first, and the scintillation counter detects x-rays which pass through the proportional counter. The thin window and good pulse height resolution make a gas flow proportional counter very effective for low-energy (long-wavelength) x-rays. The scintillation counter has poorer pulse height resolution, but it is efficient at shorter wavelengths.

3. Energy-Dispersive Detector

As stated previously, the detector in an energy-dispersive spectrometer receives all the x-rays from the sample at once. The commonly used detector is a lithium-drifted silicon crystal, designated Si(Li). It is a high-resolution proportional detector. A typical resolution figure for this detector is 160–180 eV. It can be considered as a solid-state equivalent of the gas flow detector, where there is unity gas gain. In the detector, x-ray photons are stopped and absorbed. This leads to the creation of a cloud of ionization in the form of electron–hole pairs. The number of electron–hole pairs which corresponds to the total electric charge released by the photon is proportional to the energy of the photon. A high voltage which is applied across the detector sweeps the charge from the detector. The charge is collected by a preamplifier, and a voltage pulse is produced. The pulse is proportional to the x-ray photon energy. In energy-dispersive spectrometers, a range of photon energies reaches the detector, and a proportional range of voltage pulses is generated. The pulses are sorted by a multichannel analyzer to form an energy spectrum of 400 or more channels. This is displayed visually. Other accessories allow spectra to be stored, spectra to be overlaid for comparison purposes, and background to be subtracted. In addition, a minicomputer component of the spectrometer allows quantitative analyses to be carried out, peaks to be identified, etc.

The Si(Li) detector and some elements of the preamplifier are contained in a light-tight vacuum cryostat and operated at 77 K (liquid nitrogen). X-rays enter the detector through a thin beryllium window.

The Si(Li) detector operates well over the range of 1–40 keV. However, the *K* lines of elements above cesium (atomic number 55) fall in the energy range of 30–140 keV. For this range, a high-purity germanium detector is much more efficient.

G. Quantitative Analysis

In x-ray fluorescence work, quantitative work of good accuracy depends on two main considerations. The first is the availability of calibration standards of known composition which are similar to the unknowns in both chemical composition and physical characteristics. The second consideration is to overcome or minimize matrix effects. A number of standardization methods will be discussed as they apply to geochemical work.

1. General Method of Calibration Standards

The method involves comparisons of measurements of the intensities of selected lines of elements in samples with the same set of lines in a series of standards. In this manner, a relationship between x-ray intensity and chemical composition is established for each element of interest. The data from the standards are used to establish calibration or working curves. As stated above, the standards must be similar to the samples. The following characteristics are important: (a) The physical forms are the same, e.g., loose powder briquet, bead, thin film. (b) The concentrations of the analyte(s) are in the same ranges. (c) The physical characteristics are closely matched with respect to particle size, particle size range, surface, and packing density.

The calibration curve is prepared by plotting intensity of the measured analyte line versus concentration. Usually, the background intensity measured at a wavelength close to the line is subtracted from the total intensity so that the curve passes through the origin. The calibration curve need not be linear, but a linear curve is the most useful. For a linear calibration, only two standards are necessary to define the curve each time it is used.

Some problems may arise in this work. The working standards must contain concentrations of the analyte(s) over the desired range, and also the concentrations of the potential absorbers and enhancers must be in the same ranges as in the samples. A second consideration is that the useful working range of the calibration curve may be short if the total intensity changes only slightly with changes in analyte concentration. Finally, if the deviation of data points is not acceptable, other problems in the method may be present.

2. Thin-Film Calibration

When the thickness of a sample is reduced sufficiently, matrix effects are overcome. In a very thin sample, each atom absorbs and emits independently of other atoms. That is, each atom interacts with the primary beam and fluoresces independently of the others. Thus, in very thin films of the same thickness, the intensities of the analyte lines are directly proportional to analyte concentrations. A thin specimen is defined as satisfying the relationship $m(\mu/\rho)' < 1$, where m is mass per unit area (grams per centimeter squared) and $(\mu/\rho)'$ is the sum of mass absorption coefficients of the sample for primary and analyte line x-rays. Some major elements in rock samples have been measured in this way (31).

3. Dilution Methods

Variations in the total composition of a sample constitute the major cause of interelement interactions. A way to overcome matrix differences between samples and between samples and standards is to add a diluent. This can be accomplished in two ways. In the first, a small proportion of a heavy metal absorber is added to both samples and standards. Examples of such diluents are La_2O_3 and $BaSO_4$. The mass absorption coefficient of a sample is raised to a high enough level that variations in the absorption coefficient due to changing concentrations of matrix elements are reduced. Thus, the matrix elements have much less influence on the intensity of the analyte line. The method is called a high-absorption method. The diluent must be mixed thoroughly with the sample in powder and briquet forms. It can also be used in a fused disk procedure.

An alternative is the addition of a relatively large amount of a diluent which has a very low absorption coefficient. The effect of matrix elements on the intensity of the analyte line becomes insignificant if enough diluent is added. The disadvantage of a low-absorption method is that the intensity of the analyte line may be diminished too much, and the final intensity may not be sufficient. If a moderate dilution ratio is used (10 : 1 diluent : sample) interelement effects usually persist. Usually, a ratio of 100 : 1 is needed, and this limits the work to major elements.

4. Compensation for Matrix Effects

Matrix interferences can be handled by means of compensation methods. Two such approaches rely on information provided by intensities of scattered lines from the sample and by internal standards.

Background in x-ray spectra originates from the scattering of primary radiation and from instrument noise. The incoherent or Compton scattering is dependent on the sample matrix. In addition, there is an inverse relationship between the mass absorption coefficient and the Compton peak intensity. Feather and Willis (32) used these relationships to develop a simple method for background and matrix corrections of spectral peaks in trace element analyses of geological materials. They obtained relative accuracies of 2–5%. This method of matrix compensation is useful because no special sample treatment or large numbers of calibration standards are needed.

Kikkert (33) described a simple method of predicting background intensities from measured peak and Compton intensities. In the same paper, he presented a simple, accurate method of matrix correction based on Compton scattering and included corrections for significant absorption edges.

Methods based on internal standards tend to be time-consuming and often impractical. The objective is to add an element (or more than one element) to a sample so that the added element (internal standard) is affected in the same way as the analyte element(s) by the matrix. The ratio of line intensities of analyte element to internal standard will then be independent of the matrix. A separate internal standard may have to be chosen for each element to be determined.

A special application is the addition of known increments of the analyte element to the sample. This is known as the method of standard additions or spiking. It is described in Section I,B,8 of this chapter in connection with atomic absorption. Spiking has been useful sometimes in analyses of trace elements in rocks.

5. Mathematical Corrections for Matrix Effects

A great deal of work has been done on devising absolute methods of relating intensities of x-ray lines to chemical composition. Early work was concerned with the development of empirical relationships between characteristic x-ray wavelength intensities and chemical compositions. The limitations on the use of the relationships included the availability of computers with sufficient core space.

The "fundamental parameters" method (34) attempted to obtain an absolute intensity/concentration algorithm from measured primary spectra. In this work, standards would not be needed. However, a powerful computer is required for the calculations. In addition, there are uncertainties in data on mass absorption coefficients and fluorescent yields available in the literature. Also, as Jenkins (35) has pointed out, most workers have used the spectral distribution data reported in other publications rather than generate their own data.

Another method is called the "effective wavelength" method and was incorporated in the program CORSET by Stephenson (36). In this work, the primary spectral source distribution was replaced by an effective wavelength. The method has been of particular value in analyses of samples of complex compositions such as rocks and minerals. It is quickly and easily applied.

"Alpha" correction models are illustrated by the following three equations.

$$W_i/R_i = K_i + \Sigma\, \alpha_{ij} W_j \qquad \text{Lachance–Traill}$$

$$W_i/R_i = K_i + \Sigma\, \alpha_{ij} W_j + \Sigma\, \beta_{ik} \frac{W_k}{i + W_i} \qquad \text{Rasberry–Heinrich}$$

$$W_i/R_i = K_i + \Sigma\, \alpha_{ij} W_j + \Sigma\, \gamma_{ij} W_j^2 \qquad \text{Claisse–Quintin}$$

In the equations listed above W_i is the concentration of element i, R_i the characteristic line intensity of element i, K_i the slope of the calibration curve of element i, α_{ij} the correction for absorption effect of element j on element i, and β_{ij} the corresponding correction for enhancement.

The Claisse–Quintin model is a shortened form of the original. It contains additional higher order terms to account for the polychromatic x-ray beam. The terms can be calculated when the spectral distribution of the primary beam is known.

In the Rasberry–Heinrich equation, the α coefficients are used when absorption is the predominant effect of element j on element i. When the predominant effect of element k on element i is enhancement, the β coefficients are used.

Empirical determinations of the α (or β) coefficients are made by measuring x-ray intensities of a set consisting of many standards which correspond in composition with the samples. Regression analyses, graphic methods, or algorithms are then applied to calculate the values of the coefficients.

A summary of the relationships among the various models is found in Ref. 30, pp. 481 *et seq*. This reference is also recommended for detailed discussions and explanations of the various correction programs listed in this section.

H. Comparisons of Energy-Dispersive and Wavelength-Dispersive Performances

Analyses of silicate rocks have been carried out by means of ED–XRF and the results compared with WD–XRF data from the literature by Potts *et al.* (37). Both major and trace elements were determined. The comparisons included limits of determination, analysis times, instrumental precision, and routine performance. The limits of determination of the lighter major elements by means of ED–XRF were inferior to the corresponding limits of WD–XRF. However, the analytical precisions at the levels normally encountered in silicate rocks were almost the same. In analyses of some heavier trace elements, limits of determination, instrumental precision, and routine analytical performance were equivalent to results obtained by WD–XRF. The authors stated that the overall accuracy of either technique can be influenced significantly by uncertainties in the recommended values for international rock reference materials that are used for standardization and calibration.

I. Comparison of Wavelength-Dispersive–XRF and ICP–AES Analyses

Hannaker *et al.* (38) published results of analyses of major and minor constituents of 54 Indian rocks. They were particularly interested in the analytical capability of ICP–AES for routine analysis and compared ICP–AES data with data obtained by means of WD–XRF. The samples for ICP–AES work were dissolved, whereas fused disks were used for the x-ray work. The ICP–AES work technique was evaluated by analyzing each rock four times in nine batches over a period of 9 days. The agreement between the two techniques was reported to be good. However, the authors expressed concern about the accuracy of the determinations of zirconium and chromium by means of ICP–AES. There was difficulty in dissolving minerals containing these elements in acids. Fusion or pressure decomposition techniques rather than acid digestion would yield better accuracy. The results obtained for 20 major and trace elements showed that ICP–AES provided a very reproducible means of analysis of components of rocks.

REFERENCES

1. B. V. L'vov, *Spectrochim. Acta* **17,** 76 (1961).
2. B. V. L'vov, *Spectrochim. Acta, Part B* **24B,** 53 (1969).
3. H. Massmann, *Spectrochim. Acta, Part B* **28B,** 215 (1968).
4. R. Woodriff and G. Ramelow, *Spectrochim. Acta, Part B* **28B,** 665 (1968).
5. W. J. Price, "Spectrochemical Analysis by Atomic Absorption." Heyden, London, 1979.
6. C. W. Fuller, "Electrothermal Atomization for Atomic Absorption Spectrometry." Chemical Society, London, 1977.
7. C. L. Chakrabarti, H. A. Hamed, C. C. Wan, W. C. Li, P. C. Bertels, D. C. Gregoire, and S. Lee, *Anal. Chem.* **52,** 167 (1980).
8. B. V. L'vov, *Spectrochim. Acta, Part B* **33B,** 153 (1978).
9. J. C. Van Loon, "Analytical Atomic Absorption Spectroscopy: Selected Methods." Academic Press, New York, 1980.
10. R. D. Ediger, *At. Absorpt. Newsl.* **14,** 127 (1975).
11. W. Slavin and D. C. Manning, *Prog. Anal. At. Spectrosc.* **5,** 243 (1982).
12. H. Koizumi, K. Yasuda, and M. Katayama, *Anal. Chem.* **44,** 1106 (1977).
13. S. Smith, R. G. Schleicher, and G. M. Hieftje, *Pittsburgh Conf. Anal. Chem. Appl. Spectrosc., 33rd, 1982,* Pap. No. 442 (1982).
14. A. E. Smith, *Analyst (London)* **100,** 300 (1975).
15. F. D. Pierce and H. R. Brown, *Anal. Chem.* **49,** 1417 (1977).
16. W. R. Hatch and W. L. Ott, *Anal. Chem.* **40,** 2085 (1968).
17. B. F. Scribner and M. Margoshes, *in* "Treatise on Analytical Chemistry" (I. M. Kolthoff and P. J. Elving, eds.), Part 1, Vol. 6, p. 3347. Wiley (Interscience), New York, 1965.

18. M. Slavin, "Emission Spectrochemical Analysis." Wiley (Interscience), New York, 1971.
19. H. Bennett, *in* "Physiochemical Methods of Mineral Analysis" (A. W. Nichol, ed.), p. 156. Plenum, New York, 1975.
20. R. E. Stanton, "Analytical Methods for Use in Geochemical Exploration." Wiley, New York, 1976.
21. P. W. J. M. Boumans and F. J. DeBoer, *Spectrochim. Acta, Part B* **31B,** 355 (1976).
22. P. J. McKinnon, K. C. Giess, and T. V. Knight, *in* "Developments in Atomic Plasma Spectrochemical Analysis" (R. M. Barnes, ed.), p. 287. Heyden, London, 1981.
23. P. W. J. M. Boumans "Line Coincidence Tables for Inductively Coupled Plasma Atomic Emission Spectrometry." Pergamon, Oxford, 1980.
24. M. L. Parsons, A. Forster, and D. Anderson, "An Atlas of Spectral Interferences in ICP Spectroscopy." Plenum, New York, 1980.
25. G. R. Harrison, "MIT Wavelength Tables with Intensities in Arc, Spark or Discharge Tube." Wiley, New York, 1939.
26. G. R. Kornblum and L. de Galan, *Spectrochim. Acta, Part B* **32B,** 455 (1977).
27. N. R. McQuaker, P. D. Kluckner, and G. N. Chang, *Anal. Chem.* **51,** 888 (1979).
28. S. S. Berman and J. W. McLaren, *Appl. Spectrosc.* **32,** 372 (1978).
29. E. P. Bertin, "Introduction to X-Ray Spectrometric Analysis." Plenum, New York, 1978.
30. R. Jenkins, R. W. Gould, and D. Gedke, "Quantitative X-Ray Spectrometry." Dekker, New York, 1981.
31. K. Govindaraju and R. Montanari, *X-Ray Spectrom.* **7,** 148 (1978).
32. C. E. Feather and J. P. Willis, *X-Ray Spectrom.* **5,** 41 (1976).
33. J. Kikkert, *Spectrochim. Acta, Part B* **38B,** 809 (1983).
34. J. W. Criss and L. S. Birks, *Anal. Chem.* **40,** 1080 (1968).
35. R. Jenkins, *Pure Appl. Chem.* **49,** 1583 (1977).
36. D. A. Stephenson, *Anal. Chem.* **43,** 310 (1971).
37. P. J. Potts, P. C. Webb, and J. S. Watson, *Analyst (London)* **110,** 507 (1985).
38. P. Hannaker, M. Jaukka, and S. K. Senn, *Chem. Geol.* **42,** 319 (1984).

3
Basic Materials

The preparation of solutions of samples for analysis and of standards requires great care. In work involving measurements of traces of elements, very dilute solutions are used frequently as working standards. Contributions of the element in question from reagents, water, glassware, and storage containers have a direct relationship to the accuracy of the result. The working environment must be considered also. Airborne particulate matter and vapors (e.g., mercury) can contaminate samples and standards. The effects of minor contamination may not be perceived in concentrated standard solutions, but dilute standards can be altered significantly. While blanks are necessary in all of the work, they may not account for contamination from all sources. The following sections contain information on ways to minimize contamination and to preserve standard solutions.

I. WATER

Although water is not usually regarded as a chemical in analytical work, nevertheless it constitutes the bulk of the solutions. Thus, water purification is important. Raw water sources contain sodium, calcium, magnesium, iron, and silicon as the main dissolved elements. Copper, zinc, and aluminum may also be present in significant amounts. The concentrations of these elements can be used as a means of checking the water quality. However, purity is frequently specified in terms of total solids content and electrical resistivity. Good-quality water contains less than 0.1 ppm solids and has a resistivity of 2–10 megohms/cm.

Some professional organizations have established standards for water quality. There are several different types or grades of water depending on the requirements of the laboratory.

Type I—reagent grade water. This is the highest purity water. It is used for trace metal analysis and for other work where the water must be free of inorganic impurities. The specific resistance at 25°C is in the range 10–17 megohms/cm. The contents of organic impurities, microorganisms, and suspended solids are very low.

Type II—analytical grade water. This water is less pure than type I and is suitable for many analytical procedures except the determinations of trace elements. The specific resistance should be above 1.0 megohms/cm at 25°C.

Types III and IV—laboratory grade water. Water of this grade is useful for glassware rinsing; the resistance is at least 0.2 megohms/cm at 25°C.

Specifications for water published by the American Society for Testing and Materials (ASTM) and the College of American Pathologists (CAP) are shown in summary form in Table 3.1. American Chemical Society (ACS) specifications are similar to CAP type II. Specifications published by the National Committee for Clinical Laboratory Standards (NCCLS) are similar to the CAP values.

Type II water can be obtained by distillation from glass or quartz. The resistivity should be in the range of 2–10 megohms/cm at 25°C, and dissolved solids should be less than 0.1 ppm. The cost of a commercial still which will produce 3 liters of water per hour is about $2000 (U.S.).

Type I water can be obtained by means of two distillation stages in quartz apparatus. Another method of obtaining this grade is to purify or polish any of the other types of water by means of a mixed bed ion-exchange resin. The resins are the strong acid and strong base varieties. When the cation and anion resins are present in the H^+ and OH^- forms, respectively, other ions such as Na^+, Ca^{2+}, and SO_4^{2-} in the water are exchanged for H^+ and OH^- ions, which are released from the resin. Packaged resins for water purification can be obtained from laboratory equipment companies. For example, a unit would cost about $400 (U.S.).

Table 3.1
Laboratory Reagent Grade Water Specifications

	CAP type			ASTM type			
	I	II	III	I	II	III	IV
Specific resistance (megohms/cm) (25°C)	10	2.0	0.1	16.6	1.0	1.0	0.2
Silicate (mg/liter)	0.05	0.1	1.0	—	—	—	—
Heavy metals (mg/liter)	0.01	0.01	0.01	—	—	—	—
Potassium permanganate Reduction (min)	60	60	60	60	60	10	10
Sodium (mg/liter)	0.1	0.1	0.1	—	—	—	—
Hardness	Neg.	Neg.	Neg.	—	—	—	—
Ammonia	0.1	0.1	0.1	—	—	—	—
Bacterial growth	10	10^4	—	—	—	—	—
pH	—	—	5.0	—	—	6.2	5.0
	—	—	8.0	—	—	7.5	8.0
CO_2 (mg/liter)	3	3	3	—	—	—	—

While the mixed bed resins can treat any raw water and purify it to various degrees, the lifetime of a resin charge will depend on the purity of the feedwater. The most economical way to use the ion-exchange cartridges is to purify distilled water or water from a reverse osmosis unit. If raw water is being purified by ion exchange, then a prefilter is recommended in order to remove suspended material that would foul the resins. The performance of the ion-exchange bed is monitored by measuring the resistivity of the effluent. When the resistivity falls below 10 megohms/cm, the bed should be renewed.

Reverse osmosis is another method for the purification of raw water. In this process, water is forced through a membrane which contains very small pores. The membrane will remove most particles, bacteria, dissolved inorganics, and organics of molecular weight larger than 300. Strongly ionized polyvalent ions are removed at efficiences greater than 99%, while monovalent ions like sodium are removed at about 90% efficiency. A filter should be placed before the reverse osmosis membrane to prevent fouling and to extend the life of the membrane. Different feedwaters may require different types of membranes, and the manufacturer of the equipment should be consulted. The product of reverse osmosis is equivalent to water of types III and IV. The water quality is not as good as that of the distillate from a good still. However, the water from a reverse osmosis unit is satisfactory for further purification by means of ion exchange.

Organic contaminants in the raw water can be destroyed by means of alkaline potassium permanganate added to the boiler of the still. An alternative is the adsorption of organic matter on a bed of activated charcoal. If ion exchange is being used for purification of raw water, the charcoal bed can be installed ahead of the resin bed. The ion-exchange beads are sources of minute quantities of organic compounds in the processed water. In geochemical work, such contamination is not significant.

Storage of purified water even in carefully cleaned containers of plastic or glass over a period of weeks will result in contamination from the walls of the containers. Purification of water as needed is the only way to ensure that it will be of the highest quality.

II. REAGENTS

Very pure metals (99.9% purity or greater) are the preferred starting material for the preparation of standard stock solutions. Suitable chemicals are listed by a number of scientific supply companies, e.g., Spex Industries, Inc., J. T. Baker Chemical Co., and Fisher Scientific Co. When suitable metallic forms of elements are not available, metal salts or oxides of reagent grade quality or better can be used. Standard solutions for electrothermal atomic absorption and inductively coupled plasma–atomic emission spectrometry should be prepared from ultra-high-purity chemicals.

Acids, bases, and other chemicals used in preparation of standards as well as for analyses of samples should also be of good quality. Reagent grade chemicals are suitable for determinations of major and minor elements. Trace element analyses will require high-purity chemicals in order that blanks can be controlled at low values for the elements in question. High-purity chemicals are available from scientific supply companies; examples are Utrex and Instra-Analyzed (J. T. Baker Chemical Co.) and Suprapur, Aristar, Analar (BDH). These are up to 10 times more expensive than reagent grade chemicals. The product specifications should be consulted so that the most suitable grade of chemical can be selected. If there is no specification covering particular elements in a product, then its suitability must be determined in the laboratory.

Subboiling stills are used for the production of small volumes of acid of the highest purity (1–3). Quartz stills have been used for hydrocholoric, nitric, sulfuric, and perchloric acids and Teflon and polypropylene stills for hydrofluoric acid. Because the distilled acids dissolve impurities from even the cleanest containers, the products should be distilled as needed. Water produced in subboiling stills is not superior to distilled, deionized water (1).

III. CONTAINERS

Glass and several types of plastic containers have been used successfully as storage containers for standard solutions, samples, and chemical reagents. Of these, high-density linear polyethylene containers are recommended. Studies have shown that dilute acidified aqueous solutions can be stored for up to 1 year without detectable changes in concentrations (4). A study was made of contamination of pure water and 10% nitric acid stored in high-density polyethylene bottles (5). The bottles were cleaned with 25% nitric acid, filled with the samples, and stored in a freezer for up to 4 years. Concentrations of contaminents in micrograms per liter were Fe, 0.4; Cr, 0.6; Ni, 1.0; Tl, 0.6; Co, 1.5; Mn, 0.14; Ag, 0.2; Cu, 1.5; Cd, 0.2; Pb, 2.4; and Zn, 2.4. Up to 1.0 μg/liter of Pb and Zn came from the nitric acid sample. Polyethylene containers are cheap and nonbreakable and are minimal sources of metal contamination. Teflon containers are strong and chemically inert but are too expensive for all but special applications. Some workers have noted small metal particles embedded in the walls of Teflon containers (1). Glass containers are inferior to plastic because of adsorption of dissolved materials on the walls and ion-exchange reactions between glass and solutes. However, glass containers are preferable for storage of mercury solutions. Amber glass containers are recommended for silver and gold solutions.

All containers must be washed thoroughly to remove foreign material such as particles, and residues of processing chemicals. A sequence of a detergent wash, a rinse with 1 : 1 hydrochloric or 1 : 1 nitric acid, and several rinses with distilled water is sufficient for most work with new containers. Containers to be used for standard solutions of trace elements and other laboratoryware to be used in trace element analyses should be soaked in 1 : 1 nitric acid for 2 or 3 days. Several methods of cleaning polyethylene containers for storage of water samples have been compared (6). A 48-h soak in 10% nitric acid was recommended. Teflon surfaces can be cleaned with aqua regia at about 50°C for several hours. Directions for cleaning Teflon containers for storage of high-purity acids at the U.S. National Bureau of Standards have been published (7,8). Do not use chromic acid solutions to clean glass. All laboratoryware (beakers, flasks, pipets, etc.) used in trace element analyses should be designated for this purpose and not mixed with equipment used in general work.

Containers of solutions and samples must be protected as much as possible from contaminants in the laboratory air and from evaporation losses. Some plastic materials allow liquids to permeate slowly through the container walls. The polyethylene containers described earlier minimize evaporation losses.

In the laboratory, contamination of samples and standards can be caused by particulate matter and vapors in the atmosphere as well as from reagents and laboratoryware. Trace metal analyses are best carried out in a specially designed or "clean" laboratory. A class 100 laboratory has been built at the National Bureau of Standards, Gaithersburg, Maryland; for such work (9). The designation "class 100" means that there are fewer than 100 particles larger than 0.5 μm per cubic foot of air in the laboratory. This is achieved mainly by filtering all incoming air through special high-efficiency filters as well as by eliminating other sources of contamination such as exposed metal surfaces. The laboratory is kept at a positive pressure, and class 100 laminar-flow fume hoods are used for sample dissolutions and evaporations.

However, most chemists, including ourselves, do not have access to such a laboratory, so some measures must be taken to improve a conventional laboratory. All windows can be sealed, and all incoming air can be filtered. Because trace metal contamination can arise from exposed metal surfaces, all such surfaces can be painted with an acid-resistant nonmetallic paint. Dusty operations such as grinding and sieving should be carried out in other areas. In the laboratory, all surfaces should be cleaned regularly with either a vacuum cleaner or damp cloths.

IV. STABILITY OF STANDARD SOLUTIONS

No storage conditions can guarantee the stability of standard solutions for an indefinite period. Stock solutions containing 1000 μg/ml of elements stored in clean containers will be stable for up to 1 year under normal laboratory conditions. Diluted standards having concentrations of less than about 1 ppm may be stable for a very short period. Solutions containing 1–10 μg/ml must not be stored for more than 48 h in Pyrex containers. Losses of up to 21% were recorded (10). Many workers recommend that these standards be prepared daily. Intermediate concentrations are stable for weeks or months, but they should be checked on a regular schedule or at least each month.

V. PREPARATION OF STANDARD SOLUTIONS

Directions for the preparation of aqueous standard solutions of many elements of geological importance are given in Table 3.2. A list of all elemental standards has been published (11). For reasons of convenience and stability, concentrated solutions of 1000 μ/ml (or 1000 ppm) are recommended. Working standards are obtained after appropriate dilutions. All metals and compounds should be dried and then cooled in a desiccator prior to weighing. Class A volumetric flasks and pipets should be employed. Diluted standards must have acid concentrations of about 1% for stability. Exceptions are elements such as sodium, potassium, and calcium, which do not hydrolyze in solution. Solutions containing several elements as combined standards can be prepared from the stock solutions. Precipitation will occur if solutions of incompatible constituents are combined. For example, a silver standard must not be combined with another standard containing a halide. Nitric acid is preferred for solutions to be used in electrothermal work. For additional information on multielement standards for atomic emission work, see Chapter 2.

Scientific supply houses are alternative sources of standard solutions. Specially prepared standards for electrothermal and atomic emission analyses are also available. The authors experienced some problems with commercial standard solutions (e.g., gold) in earlier times. However, no problems have been encountered in our laboratories during the past 3 years.

Nonaqueous standard solutions of elements are necessary for direct analyses of organic liquids. The most suitable solvents for atomic absorption spectroscopy are aliphatic esters and ketones (e.g., methyl isobutyl ketone) and alkanes of about C_{10} chain length. The solvent compositions of samples and standards should be matched as closely as possible since solvent properties are very important. Introduction of nonaqueous solvents in inductively coupled plasmas has limitations, and the reader should consult original publications for details.

Table 3.2
Preparation of Standard Solutions

Element	Procedure (1000 μg/ml in a final volume of 1 liter)
Aluminum	Dissolve 1.000 g of metal in a minimum volume of 2 *M* HCl. Dilute to volume.
Antimony	Dissolve 1.000 g of metal in 10 ml of HNO_3 and 5 ml of HCl. Dilute to volume.
Arsenic	Dissolve 1.320 g of As_2O_3 in 3 ml of 8 *M* HCl. Dilute to volume.
Barium	Dissolve 1.516 g of $BaCl_2$ in water and 5 ml of HNO_3. Dilute to volume.
Beryllium	Dissolve 19.65 g of $BeSO_4 \cdot 4H_2O$ in water and 5 ml of HNO_3. Dilute to volume.
Bismuth	Dissolve 1.000 g of metal in 8 ml of HNO_3. Boil gently. Dilute to volume.
Boron	Dissolve 5.720 g of H_3BO_3 in water. Dilute to volume.
Cadium	Dissolve 1.000 g in 10 ml of 0.2 *M* HCl. Dilute to volume.
Calcium	Transfer 2.497 g of $CaCO_3$ to a liter volumetric flask. Add 100 ml of water. Add slowly 30 ml of 1 : 3 HCl. Swirl the contents. Dilute to volume.

(continues)

Table 3.2 (*continued*)

Element	Procedure (1000 μg/ml in a final volume of 1 liter)
Cerium	Dissolve 4.515 g of $(NH_4)_4Ce(SO_4)_4 \cdot 2H_2O$ in 500 ml of water and 30 ml of H_2SO_4. Cool and dilute to volume.
Chromium	Dissolve 2.829 g of $K_2Cr_2O_7$ in water and dilute to volume.
Cobalt	Dissolve 1.000 g of metal in 50 ml of 6 *M* HCl. Dilute to volume.
Copper	Dissolve 1.000 g of metal in 50 ml of 5 *M* HNO_3. Dilute to volume.
Gold	Dissolve 1.000 g of metal in 10 ml of hot HNO_3 with HCl added dropwise. Boil and dilute to volume.
Iridium	Dissolve 2.465 g of Na_2IrCl_6 in water. Dilute to volume.
Iron	Dissolve 1.000 g of metal in 20 ml of 5 *M* HCl and 5 ml of HNO_3. Dilute to volume.
Lanthanum	Dissolve 1.172 g of La_2O_3 in 50 ml of 5 *M* HCl. Dilute to volume.
Lead	Dissolve 1.000 g of metal in 10 ml of HNO_3. Dilute to volume.
Lithium	Dissolve 5.323 g of Li_2CO_3 in 300 ml of water and 15 ml of HNO_3. Swirl to release CO_2. Dilute to volume.
Magnesium	Dissolve 1.000 g of metal in 50 ml of 1 *M* HCl. Dilute to volume.
Manganese	Dissolve 1.000 g of metal in 10 ml of HCl and 1 ml of HNO_3. Dilute to volume.
Mercury	Dissolve 1.000 g in 10 ml of 5 *M* HNO_3. Dilute to volume.
Molybdenum	Dissolve 2.042 g of $(NH_4)_2MoO_4$ in water. Dilute to volume.
Nickel	Dissolve 1.000 g in 10 ml of hot HNO_3. Dilute to volume.
Osmium	Dissolve 1.336 g of OsO_4 in water. Dilute to volume. Prepare this only as required because it is not stable.
Palladium	Dissolve 1.000 g of metal in 10 ml of hot HNO_3 with HCl added dropwise. Dilute to volume.
Platinum	Dissolve 1.000 g of metal in 40 ml of hot aqua regia. Dilute to volume.
Potassium	Dissolve 2.814 g of KNO_3 in water. Dilute to volume.
Rhodium	Dissolve 3.858 g of $(NH_4)_3RhCl_6 \cdot 1.5H_2O$ in water. Dilute to volume.
Ruthenium	Dissolve 1.317 g of RuO_2 in 15 ml of HNO_3. Dilute to volume.
Selenium	Dissolve 1.000 g of metal in 5 ml of HNO_3. Dilute to volume.
Silicon	Fuse 1.070 g of SiO_2 with 2.30 g of Na_2CO_3 in a platinum crucible. Dissolve the cool melt in warm water. Cool. Dilute to volume.
Silver	Dissolve 1.000 g of metal in 10 ml of HNO_3. Dilute to volume. Store in amber glass in a dark cupboard.
Sodium	Dissolve 2.542 g of NaCl in water. Dilute to volume.
Strontium	Dissolve 1.685 g of $SrCO_3$ in 300 ml of water and 10 ml of HNO_3. Swirl. Dilute to volume.
Tellurium	Dissolve 1.251 g of TeO_2 in 10 ml of HCl. Dilute to volume.
Thorium	Dissolve 2.379 g of $Th(NO_3)_4 \cdot 4H_2O$ in water. Add 5 ml of HNO_3. Dilute to volume.
Tin	Dissolve 1.000 g of metal in 15 ml of warm HCl. Dilute to volume.
Titanium	Dissolve 1.000 g of metal in 10 ml of H_2SO_4 with dropwise addition of HNO_3. Dilute to volume.
Tungsten	Dissolve 1.794 g of $Na_2WO_4 \cdot 2H_2O$ in water. Dilute to volume.
Uranium	Dissolve 2.109 g of $UO_2(NO_3)_2 \cdot 6H_2O$ in water. Dilute to volume.
Vanadium	Dissolve 2.296 g of NH_4VO_3 in 100 ml of water and 10 ml of HNO_3. Dilute to volume.
Zinc	Dissolve 1.000 g of metal in 10 ml of HNO_3. Dilute to volume.
Rare earths	Dissolve the weight of compound listed below in 50 ml of 2 *M* HCl. Dilute to volume.

Element	Form	Weight (g)
Dysprosium	Dy_2O_3	1.148
Erbium	Er_2O_3	1.144
Europium	Eu_2O_3	1.158
Gadolinium	Gd_2O_3	1.152
Holmium	Ho_2O_3	1.145
Lutetium	$LuCl_3$	1.608
Neodymium	$NdCl_3$	1.737
Praseodymium	Pr_2O_3	1.170
Samarium	Sm_2O_3	1.160
Terbium	$TbCl_3$	1.669
Thulium	Tm_2O_3	1.142
Ytterbium	$YbCl_3$	1.615

VI. STANDARD REFERENCE MATERIALS

Standard reference materials (SRMs) are substances which are prepared in large batches for distribution by a number of different organizations. They are also called standard samples or reference samples. The most important feature of an SRM is the statement of its composition in terms of major, minor, and trace elements. The number of elements and their levels vary with the SRM. An understanding of the reliability of the analytical results used for establishing the composition of each SRM is important. The value for a particular element in an SRM will be based on a number of analyses carried out in one organization by several people or the combined results of analyses from different laboratories. The analytical methods may be the same or different, e.g., chemical, atomic absorption, or x-ray. A great deal of time and effort is required in obtaining and compiling the data.

In analytical work, SRMs provide a means of (a) checking the accuracies of new or existing analytical methods, (b) calibrating instruments, and (c) calibrating secondary standards for use in a laboratory. The costs and limited supplies of SRMs reduce the possibility of their use in routine quality control work. Most of the analytical methods in this book include data obtained from analyses of selected SRMs.

A number of SRMs of interest in geochemical analyses are listed in Table 3.3. Table 3.4 contains the addresses of the organizations that have issued the materials. The list is not comprehensive in that new standards may appear each year and the supplies of older standards may become exhausted. However, the list represents a collection of the useful SRMs at the time of writing. It is impossible to include compositions of the SRMs in a textbook of this size and scope. The reader is advised to consult the data provided by the issuing organizations. An excellent

Table 3.3
Standard Samples of Geological Materials

Type	Sample No.	Source[a]
Andesite	AGV-1	U.S. Geological Survey (1)
	JA-1	Geological Survey of Japan (2)
	GSD-2	Institute of Geophysical and Geochemical Exploration (China) (20)
Anorthosite	AN-G	(8)
Ash, fly	NBS-1633a	U.S. National Bureau of Standards (4)
Basalt	BCR-1, BHVO-1	(1)
	BIR-1	(1)
	BE-N	(8)
	BM	Zentrales Geologisches Institut (East Germany) (5)
	BR	Centre de Recherches Pétrographiques et Géochimiques (France) (6)
	JB-1, JB-2	(2)
	NBS-688	(4)
Bauxite	BX-N	(3)
	NBS-69b	(4)
	NBS-696	(4)
	NBS-697	(4)
Biotite	Mica Fe	(6)
Clay, kaolinite	KK	Ustav Nerostnych Surovin (Czechoslovakia) (7)
Clay, flint	NBS-97a	(4)
Clay, plastic	NBS-98a	(4)
Diabase	DNC-1,	(1)
	W-2	(1)
Diorite	DR-N	(3)
Dolerite	S-18	National Institute for Metallurgy (South Africa) (9)
Dunite	DTS-1	(1)
	SARM-6	(9)
Feldspar, potash	BCS-376	British Chemical Standards (U.K.) (10)
	FK-N	(3)
	NBS-70a	(4)
Feldspar, soda	BCS-375	(10)
	NBS-99a	(4)
Gabbro	MRG-1	Canadian Certified Reference Materials Project (11)
	SGD-1A	Institut Geokhimii, Irkutsk (USSR) (12)
Glass	NBS-91	(4)
	VS-N	(3)
Glauconite	GL-O	(3)
Granite	GA, GH, GS-N	(6)
	G-2	(1)
	GM	(5)
	MA-N	Groupe Internationale de Travail/International Working Group (8)
	SARM-1	(9)
Albitized	SG-1A	(12)
grandiorite	GSP-1	(1)
	JG-1	(2)

Type	Sample No.	Source[a]
Iron ore	ES-681-1	(14)
Jasperoid	GXR-1	(10)
Kyanite	DT-N	(3)
Larvikite	ASK-1	Analytisk Sporelement Komite (Norway) (13)
Latite, quartz	QLO-1	(1)
Lujavrite	SARM-3	(9)
Mill head	GXR-4	U.S. Geological Survey and Association of Exploration Geochemists (14)
Marine mud	MAG-1	(1)
Marl	MO8-1	Institut de Recherches de la Sidérurgie (France) (15)
Norite	SARM-4	(9)
Peridotite	PCC-1	(1)
Phlogopite	Mica Mg	(6)
Pyroxenite	SARM-5	(9)
Rhyolite	RGM-1	(1)
Sand, glass	NBS-81a; 165a	(4)
	SS	(7)
Schist	ASK-2	(13)
	SDC-1	(1)
Sediment	GSD-1 to 12	(20)
	GXR-3	(14)
	MAG-1	(1)
	NBS-1645; 1646	(4)
	PSJ-1	(18)
	BCSS-1; MESS-1	(21)
	SL-1	(16)
Serpentine	SW	(5)
	UB-N	(3)
Shale	SCo-1; SGR-1	(1)
	TS	(5)
Silica	BCS-267; 313	(10)
	NBS-81a; 165a	(4)
	SS	(7)
Sillimanite	BCS-309	(10)
Slag	BCS-367	(10)
	ES-878-1	(15)
	SLg-1	(11)
Slate	TB	(5)
Soil	GXR-2; 5; 6	(14)
	SO-1; 2; 3; 4	(11)
	SOIL-5	International Atomic Energy Agency (16)
Syenite	SARM-2	(9)
	NS-1	Leningrad State University (USSR) (17)
	STM-1	(1)
	Sy-2; 3	(11)

(continues)

Table 3.3 (*continued*)

Type	Sample No.	Source[a]
Trap	ST-1A	(12)
Platinum metals	SARM-7	(9)
	PTC-1 (flotation sulfide concentrate)	(11)
	PTM-1 (copper–nickel slag)	(11)
	PTA (alluvial black sand)	(11)
Leaves,		
Citrus	NBS-1572	(4)
Tomato	NBS-1573	(4)
Pepperbush	No. 1	(18)
Pine needles	NBS-1575a	(4)
Water	NBS-1643a	(4)
	Trace Metals,	
	WPI-III	(19)
	USGS 65	(1)
	USGS 75	(1)

[a]Numbers in parentheses refer to the organizations listed in Table 3.4. Names of the organizations are given at first occurrence.

Table 3.4
Addresses of Organizations Supplying Standard Reference Materials

(1) Geological Survey
U.S. Department of the Interior
Reston, VA 22092
U.S.A.

(2) Geological Survey of Japan
1-3 Higashi 1-Chome
Yatebemachi
Tsukuba-Geen, Ibaragi
Japan

(3) Association Nationale de la Recherche Technique
B.P. No. 20
54501 Vandoeuvre-Nancy CEDEX
France

(4) National Bureau of Standards
Office of Standard Reference Materials
Washington, DC 20234
U.S.A.

(5) Zentrales Geologisches Institut
Invalidenstrasse 44
DDR-104 Berlin
Deutsche Demokratische Republic

(6) Centre de Recherches Pétrographiques et Géochimiques
Address as in (3)

(7) Ustav Nerostnych Surovin
Institute of Mineral Raw Materials
28403 Kutna Hora-Sedlec
Czechoslovakia

(8) Groupe Internationale de Travail/International Working Group
B.P. No. 20,
54501 Vandoeuvre-Nancy CEDEX
France

(9) National Institute for Metallurgy
Private Bag X3015
Randburg 2125
South Africa

(10) Bureau of Analyzed Samples Ltd.
Newham Hall, Newby
Middlesbrough
Cleveland, TS8 9EA
England

(11) Canada Centre for Mineral and Energy Technology
Energy, Mines and Resources Canada
555 Booth Street
Ottawa, Ontario K1A OG1
Canada

(12) Institut Geokhimii
P.B. 701
Irkutsk-33
USSR

(13) Analytisk Sporelement Komite
Rogaland District High School
Studiesenteret Ullhandhaug
N-4001 Stavanger
Norway

(14) U.S. Geological Survey
Box 25046, MS973
Federal Center
Denver, CO 80225
U.S.A.

(15) Institut de Recherches de la Sidérurgie
Station d'essais
57210 Maizières-lès-Metz
France

(16) International Atomic Energy Agency
Analytical Quality Control Services
Laboratory Seibersdorf
P.O. Box 100
A-1400 Vienna
Austria

(*continues*)

Table 3.4 (*continued*)

(17)	Department of Mineralogy Leningrad State University Leningrad V-164 USSR
(18)	National Institute for Environmental Studies P.O. Yatabe Ibaraki 305 Japan
(19)	Environmental Protection Agency Environmental Monitoring and Support Laboratory Cincinnati, OH 45268 U.S.A.
(20)	Institute of Geophysical and Geochemical Exploration Langzfang Hebei 102801 People's Republic of China
(21)	National Research Council Montreal Road Ottawa, Ontario K1A OR9 Canada

publication on this subject is "Studies in 'Standard Samples' of Silicate Rocks and Minerals 1969–1982" by S. Abbey (12). The author has collected and evaluated data for a large number of standard samples. In addition, he has provided tables of concentration ladders for major, minor, and trace elements of the available samples. A study of this report is highly recommended. Another valuable source of information is "1984 Compilation of Working Values and Sample Descriptions for 170 International Reference Materials for Mainly Silicate Rocks and Minerals" by K. Govindaraju (13). The journal *Geostandards Newsletter* is a good source of information on standards, as is "Annual Reports on Analytical Atomic Spectroscopy" (14).

Secondary standard samples are materials which have been analyzed by organizations such as Mintek and designated as suitable standards for their own internal use. They are prepared in large enough quantities to be used frequently over a period of months or years, provided that the compositions are stable. Their preparation, assuming that they are solids, includes grinding and mixing to obtain a uniform product. As noted previously, their compositions can be compared with those of the more expensive reference materials. Secondary standards are necessities if there are no relevant standard reference materials available.

VII. SAFETY CONSIDERATIONS

Safety must always be a major consideration in laboratory work and design. All laboratory personnel should have a knowledge of the hazards associated with chemicals, compressed gases, and the instruments which they use. Laboratory design must be such that the exposure of personnel to hazardous operations is minimized. Safety equipment should be available and safety rules must be established for the laboratory. Many organizations have safety manuals. The following sections constitute a brief survey of some important points. Additional information can be found in Refs. 15–17, in instrument manuals supplied by manufacturers, and in pertinent government regulations. A circular has been published on the dangers and handling of hazardous chemicals in the geological laboratory (18).

It is important that all laboratory personnel realize the importance of each individual's responsibility for safety. Managers, supervisors, and teachers must place emphasis on safety programs in their organizations.

A. Laboratory

1. Fume Hoods

A laboratory in which toxic, flammable, and corrosive chemicals are used in sample preparation must have adequate ventilation to protect the health of the staff and to prevent damage to equipment and the building from fires and corrosion. Fume hoods must have face velocities above 0.5 m/s, and air flows should be checked regularly. All chemical operations such as digestions with strong acids must be carried out in fume hoods. Materials of construction of hoods and ducting must be able to withstand corrosive atmospheres. Hoods in which perchloric acid is used for evaporations should be separated from other hoods and be washed down frequently to remove any condensate.

2. Eyewash Fountain

Splashes of chemicals in the eyes cause serious injuries. Prompt and prolonged irrigation of the eyes with water is an essential first-aid measure. An eyewash fountain should be situated in the area where chemicals are used and preferably near a main door.

3. Safety Shower

In an accident which results in chemicals being sprayed over a person, or in clothing catching fire, access to a safety shower is important. Most new laboratories include a shower in the work area. In older structures, a shower may be located adjacent to the main laboratory or located centrally for several rooms.

4. First-Aid Box

Treatment of minor injuries will require band aids, cotton bandage, surgical tape, and disinfectant. A notebook should be available so that records can be kept of minor injuries and treatments. Phone numbers of an ambulance service, fire department, and poison control center should be displayed prominently.

Because strong acids are used frequently in many laboratories for sample preparations, the treatment of burns to the skin from contact with acids is mentioned here. The burned area must be washed with copious volumes of cold water as soon as possible to remove the residual acid. Then the area can be bathed with dilute sodium bicarbonate solution. In addition, burns from hydrofluoric acid should be treated with Hyamine solution after a thorough rinsing with water.

5. Fire Extinguishers

General-purpose extinguishers charged with carbon dioxide or dry chemical must be available in the working areas. Personnel must be instructed in the use of extinguishers to deal with small fires. When any extinguisher has been used, it must be refilled even though a considerable proportion of the charge remains. Extinguishers are checked on a regular schedule.

B. Personal Items of Safety Equipment

In most laboratories, workers must wear eye protection. Eye glasses with safety-type lenses provide satisfactory protection for regular laboratory workers. Plastic, disposable glasses are adequate for visitors. Face shields and goggles should also be available. In some laboratories, workers must wear face shields when they are diluting or transferring concentrated acids and bases. Shields provide extra protection from splashes of liquids. Contact lenses should not be worn in a chemical laboratory.

Laboratory coats, rubber and plastic aprons, overalls, and coveralls help in protecting workers from injuries and contamination from chemicals. Hand protection is provided by rubber and plastic gloves. However, organic solvents can permeate most types of rubber and plastic, so these gloves offer only limited protection. Asbestos or heavy leather gloves are recommended for work involving fluxing and ashing. Safety shoes are a good investment for persons whose work includes moving large, heavy objects. Personnel working in an area where the noise level is above 85 dB must wear protective earmuffs. Filter dust masks provide respiratory protection for persons who work with crushing and grinding equipment.

C. Chemicals (19,20)

1. Acids

Inorganic acids, namely hydrochloric, hydrofluoric, nitric, sulfuric, and perchloric acids, are used to dissolve samples for analysis. Many procedures call for combinations of acids. Carelessness in handling, diluting, and heating acids is the cause of many accidents.

Concentrated nitric, hydrochloric, and hydrofluoric acids evolve corrosive, toxic fumes at room temperature. Dilutions and transfers of the concentrated acids should be carried out in fume hoods. Dilution of sulfuric acid liberates a large amount of heat. Dilute solutions of acids should always be prepared by adding concentrated acid slowly with stirring to the dilution water. Water should never be added directly to hot, concentrated acids.

Nitric and perchloric acids are strong oxidants, and concentrated sulfuric acid is an oxidant at high temperatures. The hot, concentrated acids must never be added directly to easily oxidizable organic matter or to strongly reducing substances since explosions and fires may result. Aqua regia is an oxidizing mixed acid prepared from one part of nitric acid to three parts of hydrochloric acid. As the acid generates fumes, it should be prepared only when required and then stored in a fume hood.

Special precautions are necessary when perchloric acid is used. As stated previously, perchloric acid is a strong oxidant. Hot, concentrated acid must never be added to organic matter or to strongly reducing substances. Organic and some inorganic perchlorates are unstable substances. Fires and explosions are caused by decomposition of perchlorates. In procedures requiring the destruction of organic matter, a preliminary treatment with nitric acid will destroy easily oxidizable organic substances. In solutions containing organic materials, perchloric acid, either alone or with other acids, must be added to the cold solutions. Subsequent heating will concentrate the acid and will allow the oxidation and decomposition of the organic material to proceed slowly. Evaporations of perchloric acid must be carried out in special fume hoods.

Spills of perchloric acid must be cleaned up immediately. Dilute the acid with water, and neutralize with sodium carbonate. Any cloths or mops used in the cleanup are rinsed thoroughly with water several times. The acid should never be used where there are porous, wooden surfaces. Contaminated wood may ignite spontaneously years after it has soaked up perchloric acid.

Fume hoods for perchloric acid work must be constructed from materials that will not form organic perchlorates. This applies to ducting and all other materials exposed to perchloric acid vapors. Stainless steel is often used. Inert silica cements are recommended. The hoods must be washed down with water at the end of each day.

2. Bases

Sodium hydroxide and potassium hydroxide are available in the form of pellets, flakes, and solutions. They are corrosive as either solids or liquids. The solids are kept in closed containers to minimize reactions with moisture and carbon dioxide in the air. The liquids also absorb carbon dioxide. Solutions are stored in plastic containers.

Concentrated ammonium hydroxide evolves ammonia fumes at room temperature. The base is not as corrosive as sodium hydroxide.

3. Solvents

Organic solvents are used in some procedures for the concentration of trace elements from aqueous solutions. Care must be taken since many solvents and their vapors are both flammable and toxic. Ether vapors are notorious fire hazards. In addition, ethers will form unstable organic peroxides during storage. Skin contact with any solvent should be avoided.

4. Other Chemicals (19,20)

A detailed description of the hazardous properties of chemicals is beyond the scope of this book. Workers should exercise caution when dealing with cyanides, chromates, compounds of arsenic, mercury, lead, and cadmium, and mercury metal. Acidification of solutions of cyanides, sulfides, and sulfites will liberate toxic fumes. Because of the vapor pressure of metallic mercury, any spills must be cleaned up promptly.

5. Disposal of Chemical Wastes (21)

Large volumes of concentrated acids and bases must never be flushed down a drain. Neutralization and dilution are recommended. Organic solvents not miscible with water are collected for disposal by a licensed company. The same procedure should be followed for poisonous wastes such as chromates. Water-miscible organic solvents such as alcohols and acetone can be diluted with water and small volumes disposed of as domestic waste. However, any local regulations pertaining to the disposal of wastes must be followed.

6. Storage of Chemicals

Winchester bottles or other large containers of concentrated acid are stored on lead trays in vented cupboards. Organic solvents should be stored in a ventilated, cool area away from heat and sources of ignition. Special metal cabinets are available for solvent storage. All bottles and other containers should be identified by labels.

D. Gas Cylinders

Gas cylinders should be moved on a suitable hand truck. The protective cap must always be in place when a cylinder is moved so that the cylinder valve is not damaged. Never attempt to move a cylinder with a pressure regulator attached. No part of a cylinder should be heated to a temperature higher than about 50°C. In laboratories, cylinders are secured firmly to benches or walls. Small cylinders can be supported in ring stands. Pressure regulators are required for all high-pressure cylinders. The choice of regulator depends on the type of gas, cylinder pressure, and delivery pressure. The Compressed Gas Association has specified various cylinder valve outlet connection types for several families of gases. The purpose is to prevent the interchange of regulating equipment between gases that are not compatible. Equipment for oxygen should never be intechanged for use with other gases. All gas connections must be checked for leaks. Leaks of flammable gases such as acetylene and hydrogen are fire and explosion hazards.

E. Instruments

1. Atomic Absorption and Emission Spectrometers

Fume extractors are essential in order to vent all combustion products and waste gases. Venting of fumes applies to electrothermal analyzers as well, even though sample volumes are small. Many elemental vapors such as mercury, lead, and cadmium are toxic. Compressed air for air/acetylene burners is frequently distributed throughout laboratories from a central compressor. The air must be filtered to remove oil droplets and particulate matter. Acetylene must not be drawn from a cylinder when the pressure has fallen below 520 kPa since acetone vapor can enter the burner. A flame arrestor on the acetylene line is recommended.

Manufacturers' instructions should be followed for lighting and extinguishing flame atomizers. Nitrous oxide/acetylene flames require a high-temperature burner. The maintenance of waste liquid traps for flame atomizers is essential in order to prevent gas leaks from the nebulizer.

2. Inductively Coupled Plasma–Atomic Emission Spectrometers

Adequate ventilation of the hot gas and fumes from the plasma compartment is very important for the same reasons as for atomic absorption instruments. Commercial units have a safety interlock on the plasma compartment so that the door cannot be opened when the plasma has been lit. The window should be made of UV-absorbing glass. If the plasma is not shielded, workers should wear protective eyeglasses. Shade 12 glass is recommended.

Since most equipment includes a nebulizer for the introduction of liquid samples, the provision of a liquid waste trap and reservoir is important. The high voltages from the generator are hazardous, and the manufacturer's instructions must be followed for safe operation of the equipment.

3. Inductively Coupled Plasma–Mass Spectrometers

The safety precautions noted for emission spectrometers apply to these instruments also.

4. X-Ray Spectrometers

Three principal hazards are associated with x-ray equipment: (a) radiation from x-ray tubes and radioisotopes, (b) electric shock from x-ray generators, and (c) laboratory contamination from escape and dispersion of radioactive materials.

Commercial x-ray spectrometers are designed so that the shielding of personnel from x-rays is adequate when the tubes are operated at maximum power. When a new unit is being evaluated, the buyer should ask for information from the manufacturer about radiation levels and verify that safety standards which apply in the area will be satisfied. The parts of the instrument most subject to leakage of x-rays are the interface between the x-ray tube head and the sample compartment; the sample compartment itself, especially if the shutter assembly is defective; and the neighborhood of the analyzing crystal, since some of the diffracted and scattered x-rays do not enter the collimator.

The most serious injuries are the result of direct exposure to high-intensity x-ray beams. Even momentary exposures are dangerous, as will be explained later.

Although radiation from an isotope source is of much lower intensity than that from a high-power tube, shielding is very important and the source must be shielded when the equipment is not being used. Care must be taken to avoid damaging or rupturing the container protecting the source.

The fundamental unit for measurement of the quantity of ionizing radiation is the roentgen, R. It is defined as the radiation flux which produces 2.08×10^9 ion pairs in 1 cm^3 of dry air at 0°C and 1 atmosphere pressure. The energy equivalent is 84 ergs.

The unit of biological exposure or dose is the roentgen absorbed dose, rad. This is defined as the amount of ionizing radiation which deposits 100 ergs per gram in an absorber.

The unit of biological radiation damage is the roentgen equivalent man, abbreviated as rem. It is the absorbed dose in rads multiplied by a factor which applies to the equivalent absorption in living tissue. Thus, rem = dose in rad × RBE, where RBE is the relative biological effectiveness of the radiation.

In practice, for the radiation from x-ray spectrometers operating up to 100 kV, dose in rem = dose in rad = dose in roentgen; i.e., the three units are taken to be equivalent. Measurements of x-ray intensities are stated as roentgens per hour or milliroentgens per hour. The accumulated dose is calculated from the product of radiation intensity and exposure time. The normal background radiation level is about 500 milliroentgens (mR) per year.

The maximum permissible whole-body dose rate is 5 rem per year, or 3 rem per quarter-year. The maximum permissible accumulated lifetime whole-body dose is $5(y-18)$R, where y is the age in years and must be >18. If a normal adult receives radiation at a dose rate no larger than the maxima stated above, the body can repair minor damage resulting from the radiation. However, the objective is to reduce the occupational exposure to as low a dose rate as possible. High-power x-ray tubes can generate a dose rate of 100 R per second. A dose of 300 rem will redden the skin, and a dose of 500 rem will burn the skin. Thus, very short exposures to high-intensity x-ray beams result in serious injuries.

The x-ray instrument should be located in an area of the laboratory where only those assigned to x-ray work can enter on a regular basis while the equipment is operating. A new installation must be surveyed for radiation levels in the area surrounding the equipment. Then surveys should be conducted every 3 to 6 months. If any major change is made in the configuration of the components of the instrument, a new survey should be carried out before operations are resumed on a regular basis. In most localities, records of surveys must be available for inspection, and the work area must have signs posted.

The survey meter used for the measurements should be calibrated to respond to x-rays with energies of 100 keV or less, depending on the strength of the source. Readings should be taken as close as possible to the instrument, with particular attention being given to the sample compartment and the analyzing crystal as mentioned earlier. The current acceptable level of radiation from an x-ray spectrometer is 0.5 mR per hour measured at a distance of 5 cm from the surfaces of the equipment. All personnel who work with x-ray equipment must carry passive detectors so that cumulative records of exposures can be maintained. The detectors can be film badges or pocket dosimeters. The films are changed on a regular basis.

Finally, a responsible person must be familiar with all regulations pertaining to the operation of x-ray equipment in the particular part of the country. Addition information can be found in Refs. 22 and 23.

F. Electrical Circuits

All electrical equipment must conform with safety standards (24). Grounding of equipment is essential. Connections of equipment to outlet boxes should be made through three-prong plugs. Using metal water pipes as grounds is no longer recommended. A grounding rod should be used; the rod is connected to the water pipes and metal frame of the building.

REFERENCES

1. R. W. Dabeka, A. Mykytiuk, S. S. Berman, and D. S. Russell, *Anal. Chem.* **48,** 1203 (1976).
2. E. C. Kuehner, R. Alvarez, P. J. Paulsen, and T. J. Murphy, *Anal. Chem.* **44,** 2050 (1972).
3. J. R. Moody and E. S. Beary, *Talanta* **29,** 1003 (1982).
4. R. E. Thiers, *Methods Biochem. Anal.* **5,** 273 (1957).
5. J. W. Marchant and B. C. Klopper, *J. Geochem. Explor.* **9,** 103 (1978).
6. D. P. H. Laxen and R. M. Harrison, *Anal. Chem.* **53,** 345 (1981).
7. J. R. Moody and R. M. Lindstrom, *Anal. Chem.* **49,** 2264 (1977).
8. E. J. Marenthal and D. A. Becker, *NBS Tech. Note (U.S.)* **929** (1976).
9. P. D. Lafleur, ed., "Accuracy in Trace Analysis: Sampling, Sample Handling and Analysis," Vols. 1 and 2, NBS Spec. Publ. No. 422. U.S. Govt. Printing Office, Washington, D.C., 1976.
10. Y. Samiullah, *J. Geochem. Explor.* **23,** 193 (1985).
11. J. A. Dean and T. C. Rains, *NBS Spec. Publ. (U.S.)* **492** (1977).
12. S. Abbey, *Geol. Surv. Pap. (Geol. Surv. Can.)* **83–15** (1983).
13. K. Govindaraju, *Geostand. Newsl.* **8,** Spec. Issue, July (1984).
14. "Annual Reports on Analytical Atomic Spectroscopy." Royal Society of Chemistry, London, 1987.
15. I. M. Kolthoff, P. J. Elving, and F. H. Stross, "Treatise on Analytical Chemistry," Part III, Vol. 1. Wiley, New York, 1967.
16. N. V. Steere, ed., "CRC Handbook of Laboratory Safety," 2nd ed. CRC Press, Boca Raton, Florida, 1971.
17. Manufacturing Chemists Association, "Guide for Safety in the Chemical Laboratory," 2nd ed. Van Nostrand-Reinhold, New York, 1972.
18. A. Riedmiller, P. L. Hauff, and R. W. Matthias, *Geol. Surv. Circ. (U.S.)* **924** (1984).
19. N. I. Sax, ed., "Dangerous Properties of Industrial Materials," 6th ed. Van Nostrand-Reinhold, New York, 1975.
20. G. Weiss, ed., "Hazardous Chemicals Data Book." Noyes Data Corporation, Park Ridge, New Jersey, 1986.
21. "Laboratory Waste Disposal Manual," 2nd ed. Manufacturing Chemists Association, Washington, D.C., 1973.
22. *NBS Handb. (U.S.)* **111** (1972).
23. R. Jenkins and D. J. Haas, *X-Ray Spectrom.* **2,** 135 (1973).
24. J. F. McPartland, ed., "National Electrical Code Handbook," 19th ed. McGraw-Hill, New York, 1987.

4
Methods of Sample Preparation

I. INTRODUCTION

The collection and preparation of samples for chemical and instrumental analyses are an important part of the analysis process. In sampling the objective is to obtain a relatively small quantity of solid or liquid that is representative of the material in the field sample. Samples submitted for analyses must first be identified and well labeled. Knowledge of the type, or the approximate composition, of a sample and the analytical data required is essential in selecting methods of sample preparation. If the sample is not a liquid, drying, crushing, and grinding operations must be considered as preparative steps before any of the analytical procedures can be applied. Losses of sample constituents in these operations should be minimal. If trace element concentrations are required, precautions to avoid contamination of the sample are essential. Powdered samples can be used directly in some x-ray fluorescence and neutron activation procedures. Liquids may be analyzed for some elements with minimal sample preparation. However, some form of concentration is commonly still necessary for trace element work. A number of procedures for the total or partial decomposition of solid samples will be described.

II. FIELD SAMPLING

The problem of obtaining representative quantities of samples of particular geological formations has been the subject of many studies. Detailed accounts are found in books by Govett (1), Johnson and Maxwell (2), and others. These sources should be consulted for detailed information on sampling. Only a very concise summary can be presented here.

A large enough sample of rock (or other solid matter) must be collected so that the constituents sought (including trace elements) are present in amounts which are representative of their concentrations in the main geological formation. Obviously, the volume of the rock sample must be large if the grain size of the rock is large. A large sample is required also if the concentration of the desired constituent is low and if the constituent is not distributed uniformly. Govett (1) has stated that the following formula yields the sample weight *(g)* in grams in relation to other important variables

$$g = \frac{2.75\pi d^3}{6}\left[\frac{(100 - P) \times 10^4}{PR_a^2}\right]$$

where d is the diameter (centimeters) of the grain, P the percentage of the mineral of interest, and R_a the relative standard deviation of the analytical method expressed as a percentage of P.

He concluded that a sample size of 1 kg is sufficient for grain sizes of 1–2 mm and for constituents of interest present over 0.2%. The normal range of analytical precision for exploration geochemistry was assumed. For grain sizes over 5 mm and for constituents of interest over 3%, a sample weight of 1 kg was also found to be adequate.

The reduction in the weight of a large field sample to a smaller weight which is more suitable for laboratory work has been studied by Gy (3) and others. Gy's guidelines for this process are expressed in the equations below.

$$W = C\mu^3/s^2$$

where w is the weight (grams) of the required sample, μ the size (centimeters) of the largest particles in the sample, s the standard deviation, and C a constant for the material being sampled. The units of C are grams per cubic centimeter.

$$\mathrm{C = clfg}$$

where l is the liberation factor of the critical component; $l_{max} = 1$ when the component is completely liberated, i.e., when it is present as fragments of pure mineral; l is never smaller than 0.03. The factor l is calculated from

$$l = \sqrt{d_l/d}$$

with d the diameter of the particle and d_l the maximum particle diameter ensuring complete liberation of the critical component. f is the shape factor. For all materials except for alluvial gold ores, $f = 0.5$. When the shape of the gold particles is spheroidal, $f = 0.5$. When the particles are flat or elongated, $f = 0.2$. g is the particle size distribution factor. For particles that are completely uniform in size $g = 1.0$. An average value for g is 0.25. c is the mineralogical composition factor; the units of c are grams per cubic centimeter.

$$c = \frac{1 - a_L}{a_L}[(1 - a_L)\lambda_c + a_L\lambda_g]$$

where the average mineralogical content a_L is expressed as a decimal fraction and is 0.1 for a 10% content of a particular mineral, λ_c is the density of the critical mineral component, and λ_g is the density of the gangue. The factor c can be calculated by using average values of $\lambda_c = 5.0$ and $\lambda_g = 2.6$.

More information about the applications of Gy's constant can be found in Ref. 4.

After the field sample has been reduced in quantity to an amount suitable for the analytical laboratory, one or more portions must be weighed out for the analysis. Again, grain size is of major importance in determining the weight of sample which should be taken for trace element determinations. To be representative, the subsample should contain 10^6–10^7 particles (1). If a 1-g sample with an average specific gravity is taken for analysis, the sample should be ground to 170 mesh or finer (1). A calculation of the amount of sample required can be made by deriving Ingamell's constant. The constant, K_S, is the weight of sample (grams) which will result in a sampling uncertainty not exceeding 1% at the 68% confidence level. For a constituent X in a sample

$$K_S = R^2 w = (100s)^2 w/\bar{X}^2$$

where R is the relative deviation (percent), w the sample weight (grams), $\bar{X}$ the mean of N analytical results, and s the standard deviation. The coefficient R is found by determining X in a series of 1-g subsamples. The assumption is that the analytical variance is not significant in comparison with R (2).

III. DRYING

In general, rock samples need not be dried prior to crushing and grinding. Samples of soil, silt, sediment, and organic matter will contain varying proportions of moisture. They are dried at a temperature of 100–105°C. If mercury is to be determined, the drying temperature should not exceed 80°C. Drying such samples at ambient temperature is preferred if the samples can be protected from contamination and time is not a factor.

IV. CRUSHING, GRINDING, AND SIEVING

Large pieces of rock are broken down into smaller fragments suitable for grinding by means of crushers. Sometimes it is necessary to break very large chunks in rock trimmers prior to crushing. Jaw crushers are widely used to produce particles small enough to pass through a −10 mesh (2 mm) sieve. Crushers whose jaws are steel-plated may contaminate samples with iron from wear on the jaws. Tuff (5) studied contamination of silicate rock samples in crushing and grinding operations. Iron, manganese, cobalt, chromium, and vanadium contaminated samples when a steel crusher and a steel pulverizer were used. When iron and steel alloying elements are of interest, the use of crushers fitted with ceramic plates is recommended.

The large amount of −10 mesh crushed material must be reduced to a sample of more convenient size in a manner which will yield material of the same composition as the whole. The operation called coning and quartering is often employed for this purpose. The crushed sample is placed near the center of a large sheet of glazed paper supported on a table. One corner of the paper is raised so that the sample rolls toward the opposite corner. It is important to have the sample tumble over itself in order to promote mixing. Then the other corners are raised in succession to repeat the operation. Finally, the sample is gathered near the center of the paper and is divided into quarters by means of a ruler or large spatula. Opposite quarters are removed from the paper. The remainder (i.e., one-half of the original) is mixed and quartered again. The operation is repeated often enough to yield a final sample of about 50 g. This amount can be ground up or part of it set aside as a reserve.

The reduction of a sample by means of a riffle is an alternative to coning and quartering. The crushed sample is poured in a steady stream along the length of the riffle, while portions of the sample are collected from the chutes on either side. The portion collected on one side of the riffle is selected and the remainder discarded. The operation is repeated with the retained portion. This time, the portion collected on the opposite side of the riffle is retained. As the quantity of retained sample decreases, smaller riffles become more convenient to use. Again, the objective is a reduction in sample size to a final weight of about 50 g.

The coarse particles of the crushed samples must be reduced in size so that samples can be subjected to physical and chemical methods of analysis. Particle size reductions are achieved by grinding in mortars or in ball mills. Any contamination of samples from abrasion and degradation of the grinding surfaces must be avoided as much as possible. Agate or mullite mortars are recommended for grinding by hand. Tungsten carbide mortars may contribute tungsten, tantalum, and cobalt (5) to samples. Because hand grinding is a laborious and time-consuming operation when many samples must be treated, some laboratories use ball mills. A number of small units can be mounted on a mechanical shaker to speed up the operation.

For most chemical analyses, material in the 100–200 mesh range is adequate. An 80 mesh sieve size is recommended for soil and sediment samples. However, before soils and sediments are sieved, any large lumps or clumps should be broken down. In this operation, sometimes called disaggregation, the lumps are subjected to gentle pressure in a mortar and pestle. Then all of the sample is sieved and the +80 mesh fraction is discarded.

In sieving samples, brass screens should not be used because of contamination problems. Stainless steel screens are satisfactory as long as they are not soldered. For work involving trace element analyses, noncontaminating sieves can be constructed from nylon mesh and cross sections of plastic cylinders. Any large particles of rock samples retained on the top or coarse sieve must be subjected to additional grinding in order to maintain a representative sample. Some minerals are more resistant than others to grinding. If some of the sample is discarded, the overall composition is altered.

The equipment used for crushing, grinding, and sieving should be cleaned on a regular basis to prevent contamination by sample carryover. Brushing of the working surfaces and receivers or air blowing is usually sufficient. However, a steel crusher and a steel pulverizer which had been used for the preparation of sulfide ores appeared to contribute lead and sulfur to a silicate rock sample even after the equipment had been thoroughly cleaned (5). Sieves should be examined frequently to make certain that the mesh area is intact and that there are no tears around the frame. Grinding and crushing surfaces should be checked for evidence of excessive wear.

V. POWDER SAMPLES

Loose and pressed powders, 200–325 mesh, are frequently used for x-ray fluorescence work, although fused disks are also popular forms for quantitative analysis. Powders are also used in the direct determination of some elements by means of neutron activation. The direct elemental analysis of solid samples (i.e., powders) by means of atomic absorption and ICP–AES has been the subject of considerable research and work is continuing to find suitable methods of sample introduction (6). With the exceptions of spark and arc AES, no approach to direct solid-sample analysis has received acceptance. The main drawbacks in solid-sample analysis by means of furnace–AAS include nonspecific absorption, the requirement for samples smaller than 2 mg, and the need for standards of almost the same chemical compositions and physical characteristics as the samples.

ICP–AES is particularly attractive for solid-sample analysis because the high temperature of the plasma results in efficient vaporization, atomization, and excitation. There should be less dependence on the sample matrix. The stability of the ICP allows good precision of measurements provided that the introduction of the solids does not upset its operation. Work to date has not resulted in procedures as reliable as those involving liquid samples. However, a procedure for the determination of metals in clay introduced as a suspension in water is included in this book.

A. Loose Powders

X-ray fluorescence analyses of loose powders offer the advantage of speed in sample preparation when large numbers of geological exploration samples are to be analyzed. An automatic sequential x-ray spectrometer requires only 30–100 s for the determination of a designated element. The time per element is longer for semiautomatic and manual spectrometers because the changes in wavelength settings require more time. Automatic multichannel analyzers can be set to measure concentrations of up to 30 elements in periods of 30 s to 3 min. Thus, sample preparation may be the time-limiting part of the total procedure. After the grinding and sieving operations have been completed, the only additional step is the transfer of the samples to metal or plastic cups. Some of the limitations of loose powder methods are outlined below. The relative accuracy is 10–15%.

X-ray spectral intensities depend on the physical state of the sample. For loose powders, particle shape and size, particle size distribution, and packing density are factors which contribute to accuracy and reproducibility of analyses. Segregation is a problem because many samples are not homogeneous and consist of particles of different sizes and densities. To minimize particle size effects, the sizes of the largest particles in a sample should be less than the critical thickness of the sample for the analyte line of longest wavelength being measured. This means that the samples should be ground to pass a 325 mesh sieve. The packing density of the samples in the holders should be as uniform as possible, and the surfaces exposed to the incident x-rays should be level. In practice, these conditions are difficult to meet. Powder standards of the same physical characteristics as well as comparable chemical composition should be used for quantitative work.

In spite of the disadvantages of working with loose powders, the advantages of sensitivity and speed of sample handling should not be overlooked. Internal standards and diluents can be added provided that mixing is satisfactory.

B. Pressed Powders

Some of the disadvantages of loose powders in x-ray work are overcome when the samples are pressed to form pellets or briquets. Provided that the same weight of sample and the same final pressure are used in sample preparation, uniform packing densities are achieved and a better surface is formed. However, the effects of particle size on x-ray intensities are still present. If a vacuum path is required for the x-ray beams in the spectrometer, the samples must be pressed under vacuum. The mechanical strength of the briquets is increased with the addition of a binder to the samples before pressing. The same proportion of binder together with any diluent and internal standard must be added to all of the samples as well as to calibration standards. After thorough mixing of the sample and additives, the briquets are formed in a cylindrical die by means of a hydraulic press. Final pressures are in the range of 14×10^4 to 70×10^4 kPa. If the binder contains a thermosetting resin, the briquets are warmed on a hot surface to complete the preparation. Calibration standards can be used for long periods without deterioration.

Organic as well as inorganic samples can be analyzed as powders or briquets. For example, six trace elements in organic-rich soils have been measured in briquets prepared from dried, finely ground samples (7). Calibration standards were made from starch to which known quantities of the elements had been added.

Relative errors in the range of 6–10% have been reported for the determination of trace elements in pressed powders (8, 9). A critical comparison of sample preparations as glass disks and pressed powders has been reported by Bower and Valentine (10). They concluded that pressed powders have almost no advantages over glass disks.

C. Thin Films

A sample ground to a very fine powder and distributed uniformly over a support such as filter paper or plastic film is called a thin film. In x-ray work, the effects of inhomogeneity and absorption enhancement are much reduced for this kind of sample preparation. The intensities of the emitted x rays are related directly to the thickness and density of the sample. If only a very small quantity of sample is available for analysis, the preparation of a thin film offers advantages. A small quantity of powder can be held between two pieces of Mylar film and mounted in a frame. Adhesive tape is also a suitable support. Precipitates and suspended solids can be collected on paper or membrane filters. Solution samples are evaporated on plastic films or filter papers. Filter papers loaded with ion-exchange resins are useful in concentrating ionic species from solution. In the preparation of any type of thin film, the most important factors are uniformity of sample thickness, absence of absorption of x rays (from the sample), and reproducibility of the procedure for sample preparation. Standards must be prepared in the same manner as the samples. Blanks must also be determined.

VI. DECOMPOSITION OF SOLID SAMPLES

Most of the instrumental methods of analysis described in this book require some kind of chemical treatment of samples to obtain a form suitable for qualitative and quantitative analyses. Materials which contain major amounts of organic matter must be treated so that the organic substances are destroyed without losses of the elements of interest. Wet and dry oxidations of organic matter are discussed in subsequent sections. Dolezal *et al.* (11) have written a book on the subject of decomposition of samples for inorganic analyses.

Many procedures have been described for the decomposition of geological materials. They include digestion and leaching with acids, fusion with fluxes, sintering, and pyrolysis. The procedures can be classified under the headings of total or strong, and partial or weak. In total or strong decompositions the objective is the liberation or release of all elements from their compounds. Partial decompositions result in the release of elements from specific fractions of the sample or of elements which are weakly bonded, adsorbed, and complexed in any of the fractions. The terms are relative, and the analyst must decide whether a particular procedure will be suitable for the work in hand. The choice of a procedure depends on a number of factors, including (a) the kind of instrument to be used for the analyses, (b) the number of elements to be measured, (c) the number of samples, and (d) the laboratory facilities. For example, lithium borate fusion is a popular decomposition method when major and minor elemental concentrations of samples are to be measured by means of x-ray fluorescence, flame atomic absorption, and ICP–AES. However, accurate measurements of trace concentrations of elements are often more difficult to achieve in the presence of the large concentrations of salts resulting from fusions. Not only do the salts themselves interfere in furnace atomic absorption and emission methods of analysis, but also the fusion may introduce larger levels of impurities than can be tolerated. In addition, dilution factors may be too large. Because acids of high purity are available and excess acid can be evaporated, treatment of samples with acids is generally chosen for trace element work.

Partial decompositions are popular when the elements sought can be selectively dissolved in a solution of an extractant and the remainder of the sample is not attacked. Often, partial decompositions are easier to perform and require less time and equipment than total decompositions. However, variables such as extraction time and temperature, reagent strength, and pH become very important. The subject will be discussed in more detail later.

A. Treatment of Organic Materials

1. Ashing

Botanical samples consist of many different types of organic compounds which often vary greatly in their behavior toward the reagents used for sample decompositions. Because of this variation it is important to realize that no one method will be suitable for the decomposition of all botanical samples.

Basically, there are three approaches to the decomposition of organic samples: wet ashing, dry ashing, and fusions. Wet ashing involves treatment of the botanical sample with a mixture of acids including an oxidizing agent. The latter may be an acid or a salt. Dry ashing is the treatment of the sample (a) at elevated temperature (usually above 450°C) in air to remove the organic matter as oxides of carbon or (b) at 50–100°C under reduced pressure in an oxygen plasma discharge. In fusions the sample is mixed with a reagent capable of fluxing the sample. This approach is generally reserved for samples low in organic constituents.

The potential advantages and disadvantages of dry and wet ashing can be summarized briefly. Wet ashing involves more operator time but generally takes less overall time. A small sample size must generally be used. Because a relatively large volume of reagents must be employed, contamination due to reagents' impurities can be a problem. Dry ashing requires a longer overall time but operator involvement is minimal. Relatively large samples can be treated if high-temperature ashing is employed. (Plasma ashing at low temperature is applicable to only very small samples.) In high-temperature ashing, because containers must be left open to the atmosphere, contamination from airborne substances and components of the ashing apparatus may be a problem. Some of the more volatile metals (and compounds of the metals), e.g., cadmium, lead, zinc, mercury, and arsenic, may be lost. Losses may also occur by high-temperature incorporation of analyte into the walls of the container and the oxidized sample residue.

2. Volatility Losses

There is considerable confusion in the literature concerning loss of elements because of the presence and formation of volatile compounds during sample ashing. In carrying out a volatility loss study, it is essential to be sure that any measured losses are actually due to volatilization and not to the incorporation of analyte into either the container wall or an insoluble residue. The authors believe that this problem has invalidated a number of published studies. Radioisotopes can be particularly useful in such study.

Mercury and many of its compounds are volatile at relatively low temperatures. Thus when mercury is to be determined it is important to take special precautions during sample decomposition. In this regard it is essential to ensure oxidizing conditions throughout the decomposition procedure.

The covalent hydride-forming elements arsenic, selenium, tin, and tellurium are readily lost as hydrides at relatively low temperatures. Thus, as in the case of mercury, oxidizing conditions should be maintained throughout the decomposition step.

The presence or formation of metal halides can be a problem in sample decomposition. Many metal halides are volatile in the range of 250–650°C. It is important to realize that many organic samples contain chloride in one form or another.

3. Loss to Residues in Container Walls

It is common to find that analyte can be incorporated in container walls during ashing at high temperature. The mechanism is not clear but the reaction between a metal oxide and silicate can form glasses which are not easily attacked by mineral acids. The presence of sodium chloride, a common constituent of organic samples, is believed to add to the severity of this problem (13).

Many organic materials contain appreciable quantities of silicon. During ashing procedures silicon often remains in a residue as silica or a silicate. Such residues can readily trap metals.

4. Dry Ashing

Although dry ashing generally requires a long period of time, it is attractive because operator time is low, large sample sizes can be employed, and, unless a reagent addition is necessary, contamination from reagents is low.

Oxidizing conditions should be maintained as much as possible throughout the ashing periods, and the temperature employed depends on the presence or absence of chlorides and the identity of the metals of interest.

Addition of aluminum or magnesium nitrate is sometimes recommended. This is to aid in maintaining oxidizing conditions, to speed the decomposition, and to minimize interaction of analytes with the container wall.

Contamination during dry ashing can be a serious problem. The sample must be open to the air over an extended period. This means that dust particles from the air can easily be trapped by the sample. Also, loose scale from furnace walls and heating elements can be a serious source of contamination. Dry ashing should be done in as clean an area as possible. Blanks must always be carried through the dry ashing procedure.

Organic matter frequently contains appreciable levels of substances which form an acid-insoluble residue during dry ashing. Most common among these is silicon. The residue thus obtained can readily trap trace elements so that they will not be recovered during the subsequent acid leach step. If total decomposition is required it is crucially important not to discard residues from dry ashing until it is certain that no trace elements have been trapped. To put difficultly soluble residues into solution, hydrofluoric acid treatment or fusion with potassium bisulfate or lithium metaborate is recommended.

Organic materials vary in the temperature required for complete ashing. A typical temperature range for ashing organic matter is 450–600°C. Ashing times also depend on the nature of the material and range from 2 to 24 h. If cadmium, lead, and zinc must be determined, temperatures much higher than 450°C must be avoided.

5. Wet Ashing

A variety of acids and acid mixtures have been proposed for the decomposition of organic samples. In most cases it is essential to employ an oxidizing agent such as an acid or another constituent mixed with an acid to obtain complete decomposition. When the possible loss of an element by volatility is a potential problem, the decomposition should be done under reflux or in a closed container.

6. Decompositions under Reflux

Problems with losses of mercury during decompositions of organic samples led to the use of a refluxing system. It is obvious that the mercury can be lost at room temperature as the metal. Thus, prevention of the mercury loss would appear to involve minimizing the presence of Hg° during decomposition. However, Hg^{2+} salts such as $HgCl_2$ can also be lost at relatively low temperatures. Many workers advise that for the most accurate work samples for determination of mercury should be decomposed in a refluxing apparatus. This equipment consists of a round-bottom flask joined to a refluxing condenser by a reservoir with a two-way stopcock. The reflux condenser is in turn connected to a splash head. All connections are made with ground-glass joints. The round-bottom flask is fitted with a thermometer. Decompositions when arsenic, selenium, and tellurium are to be determined can also be done in such an apparatus.

7. Decompositions in Closed Vessels

When problems with loss of volatile constituents are anticipated it is possible to use closed decomposition vessels. When liquids are placed in such vessels and heated, a pressure builds up. It is possible to purchase commercially a variety of pressure decomposition vessels, commonly called "bombs," which can be used for this purpose. These usually consist of a Teflon insert or inner vessel contained in a metal outer vessel. The apparatus is then capable of being sealed. This prevents loss of sample through rupture of the Teflon reaction vessel.

Because of the expense involved with having individual bombs for each determination, the authors, on the advice of Dr. A. Brzezinska, use in their laboratory a household pressure cooker to contain seven heavy-walled Teflon closed-vessel decomposition containers. This apparatus is shown in Chapter 9, Fig. 9.3. In this way the pressure built up inside the Teflon vessels can be compensated by pressure inside the pressure cooker. Samples in Teflon bombs can also be digested by placing the bombs in a microwave oven. The best power levels and digestion times must be found for each type of sample.

8. Types of Decomposition Vessels

It is often necessary to heat oxidizing acid mixtures to fumes of a high boiling point acid such as sulfuric acid. For this purpose platinum evaporation dishes were commonly employed up to the 1970s. The expense of these dishes coupled with the more recent availability of suitable plastics (particularly Teflon) has greatly decreased their use. At this time borosilicate and pure silica glass beakers and Teflon dishes are commonly used for this purpose. In the authors' laboratory Teflon has proved to be suitable even for use with perchloric acid. The size of the container is dictated by the amount of sample, but a 100–250-ml vessel is a generally useful size.

9. Comparisons of Decomposition Procedures

An interesting comparison of sample treatment procedures for the analysis of biological samples by ICP–AES was made by De Boer and Maessen (14). These researchers adopted the following criteria for the procedural assessment.

A. Preparation of sample solution
   (1) Resulting dilution factor of the original sample
   (2) Range of feasible sample sizes
   (3) Experimental labor intensity
   (4) Risk of contamination
   (5) Time required per sample
   (6) Safety of the procedure

B. Suitability of sample solution for ICP analysis
   (1) Accuracy and precision of determinations
   (2) Occurrence of matrix and acid effects

The sample decomposition procedures which were studied were as follows.

Direct Approach

- 0.65 g sample mixed with 6.5 ml 2% nitric acid
- 24 h contact time at room temperature
- Centrifugation and filtration

Solubilization in Tetramethylammonium Hydroxide (TMAH)

- 0.05 g sample mixed with 1.5 ml 10% aqueous TMAH solution
- 2 h on water bath at 65°C
- Dilution to 5 ml

Acid Digestion Methods

Digestion with nitric and perchloric acids in microwave oven

- 0.15 g sample, 5 ml 65% nitric acid, and 1 ml 70% perchloric acid in conical flask
- Flask in microwave-heated oven (2450 MHz, 600W)
- Heating until near dryness, ~20 min
- Dilution to 10 ml

Digestion with hydrogen peroxide and sulfuric acid

- 1 g sample mixed with 1.5 ml 96% sulfuric acid
- Dropwise addition of 9–16 ml 50% hydrogen peroxide during 1 h heating up to 300°C
- Dilution to 5 ml

Digestion in Teflon-lined steel bombs

- 0.25 g sample and 3 ml 65% nitric acid in 23 ml capacity bomb
- 2 h heating in oven at 140°C
- Dilution to 5 ml

Ashing Methods

Low-temperature ashing

- 1 g sample portions
- Oxygen pressure: 2 torr
- Net 27 MHz power: 70 W
- Exposure time: 20 h
- Ash dissolved in 2% hydrochloric acid
- Final volume: 5 ml

Muffle furnace ashing

- 1 g sample heated for 24 h in temperature-programmed muffle furnace
- Final temperature: 520°C
- Ash dissolved in 30% hydrochloric acid
- Dilution and filtration
- Final volume: 5 ml

Table 4.1
Assessment of Procedures, Numerical Values

Sample treatment procedures	Dilution factor of sample (ml/g)	Maximum amount of sample (g)	Duration of procedure (h)	Concomitant metals content of sample solution (mg/ml)	Acid content of sample solution (%, w/w)
Solubilization (TMAH)	100	0.05	2	0.1	—
Acid digestion (HNO_3, $HClO_4$)	70	0.08	0.3	0.2	2 ($HClO_4$)
Acid digestion (HNO_3, bomb)	20	0.25	2	0.6	30 (HNO_3)
Extraction (diluted HNO_3)	10	0.65	24	1.3	2 (HNO_3)
Acid digestion (H_2SO_4, H_2O_2)	5	1	1	2.6	31–36 (H_2SO_4)
Low temperature ashing	5	1	12	2.6	3 (HCl)
Muffle furnace ashing (520°C)	5	1	24	2.6	6 (HCl)

Results

The dilution factor, maximum amount of sample applicable, duration of the procedure, concomitant metals content of the solution, and acid content of solutions are given in Table 4.1.

Table 4.2 shows comparisons of the certified values with the results obtained on NBS Bovine Liver SRM when a variety of sample treatments were used (14). The solutions were analyzed by means of ICP–AES.

Table 4.2
Elements in Bovine Liver SRM 1577[a]

Sample treatment procedure	Content (μg/g) and 95% confidence limits					
	Mn	Zn	Cu	Fe	Cd	Pb
Extraction with diluted HNO_3 at room temperature	10.6±0.3	132±3	62±3	154±6	0.30±0.02	0.40±0.13
Solubilization in TMAH	9.7±0.3	131±2	103±5	256±6	ND[b]	ND
Digestion with HNO_3 and $HClO_4$ in microwave-heated oven	10.2±0.2	131±2	191±2	264±3	0.30±0.06	ND
Digestion with H_2SO_4 and H_2O_2	10.3±0.3	134±3	188±2	263±3	0.30±0.02	0.28±0.10
Digestion with HNO_3 in Telfon-lined steel bomb	10.3±0.2	133±3	201±3	268±3	0.31±0.02	ND
Low-temperature ashing	10.0±0.3	131±2	190±3	260±3	0.29±0.02	0.37±0.08
Muffle furnace ashing	10.0±0.2	132±3	183±5	258±6	0.31±0.02	0.33±0.03
NBS certified value	10.3±1.0	130±13	193±10	268±8	0.27±0.04	0.34±0.08

[a]Analyses by ICP–AES.
[b]ND; Not determinable.

Low-temperature ashing and the hydrogen peroxide/sulfuric acid procedure yielded the best accuracy, but low-temperature ashing is favored from a labor intensity point of view.

B. Acid Attack

Inorganic acids are divided into two groups. In the oxidizing group are nitric acid, perchloric acid, and hot concentrated sulfuric acid. Nonoxidizing acids include hydrofluoric, hydrochloric, hydrobromic, phosphoric, and dilute sulfuric acids. Carbonates, phosphates, chlorides, borates, and some sulfates and sulfides are readily soluble in acids. Oxide minerals are often soluble in nonoxidizing acids. Hydrofluoric acid decomposes most silicates, but hydrochloric acid is also useful for the dissolution of basic silicates.

In many procedures, samples are treated with acids in open test tubes, beakers, and dishes inside fume hoods. Very reactive materials such as carbonates must not be treated with concentrated acids since foaming and spattering will be severe. Such materials are moistened with water, and then acid is added slowly. When the quantitative determination of silicon in silicate minerals is not required, samples are treated with hydrofluoric acid either alone or in combination with other acids in open vessels.

As mentioned above, hydrofluoric acid is used extensively in procedures for silicate analysis. The acid, either alone or in combination with other acids, must not be used in glass equipment. Beakers, flasks, etc. made from polyethylene, polypropylene, or a fluorocarbon are satisfactory. Teflon equipment is used extensively, and it can be heated up to about 240°C. A number of elements in addition to silicon form volatile fluorides, which are lost in total or in part from hot solutions. These include boron, arsenic, antimony, and germanium. Some silicate minerals such as zircon, beryl, kyanite, cassiterite, and chromite resist decomposition in open vessels. The rate of acid attack is increased if the reactions take place at elevated temperatures and pressures. Closed Teflon vessels and Teflon-lined bombs are popular for this work. The safe limits of temperature and pressure must not be exceeded, and the vessels must always be cooled again before they are opened. Details are given in the appropriate analytical procedures. In the authors' laboratory, several small closed Teflon vessels are heated together inside a metal pressure cooker on a hot plate, as noted earlier.

Hydrofluoric acid is often used in combination with other acids (nitric, perchloric) to obtain oxidizing conditions during the decomposition. Heating perchloric or sulfuric acid to dryness is a means of expelling fluoride (or fluosilicic acid) when the presence of fluoride is detrimental to subsequent work. Perchloric acid is preferred since some sulfates have limited solubilities. Nitric acid can be substituted for perchloric, but the elimination of fluoride is not as effective because of the lower boiling point of the acid. Fluorides attack glass nebulizers and spray chambers of atomic absorption and ICP–AES instruments.

Solutions of nitric and perchloric acids attack clay minerals and liberate most of the trace elements. These solutions also attack some silicates and the base metals are dissolved. Aqua regia and other acids are used in some procedures to dissolve trace elements adsorbed on acid-resistant solids. Aqua regia and also nitric acid alone is used to dissolve some sulfides and arsenides.

A typical acid digestion procedure with mixed acids is recorded below.

1. Transfer 0.2 g of powdered rock into small Teflon beakers or test tubes.
2. If the sample contains a significant proportion of carbonate, moisten it with a few drops of water before adding acid. Add 4 ml of nitric acid and 2 ml of perchloric acid and swirl to mix. Then add 10 ml of hydrofluoric acid.
3. Warm the containers on a hot plate or in a heating block so that the temperature increases to approximately 150°C over a period of 3–4 h. Then increase the temperature to approximately 200°C. Evaporate all the acid.
4. Cool the samples. Dissolve the salts in 2 ml of 6 *M* nitric acid by warming. Cool again and dilute to 25 ml with water.

C. Fluxes

A number of fluxing agents have been studied for the decomposition of silicates and other geological materials. The two fluxes of most importance in the work described here are lithium tetraborate and lithium metaborate. They are widely used in x-ray fluorescence, flame atomic absorption, and ICP–AES procedures. These fluxes will dissolve almost all minerals. Mixtures of meta- and tetraborates have been recommended for fusions at different temperatures (15). The melts are cast in molds to form disks suitable for x-ray work. In addition, the disks are soluble in dilute acids for analysis of major and minor components of geological materials. Fusions are rarely used for analyses of trace elements because of the disadvantages described previously. However, a ratio of flux to sample of 2 : 1 (or a 3 : 1 dilution of the sample) has been used for the determination of trace elements in rocks (16). A critical comparison of glass disks and pressed powders as methods for sample preparation has been reported (10).

1. Glass Disks for X-Ray Fluorescence

Fusion of samples of geological materials offers several advantages. All difficulties associated with inhomogeneity, particle size, and density of solids are overcome. Matrix effects are greatly reduced if the flux-to-sample ratio is at least 10 : 1. Larger proportions of flux will reduce matrix effects further, but at the expense of reduced sensitivities for minor and low atomic number elements such as sodium, magnesium, and phosphorus. Another advantage of glass beads is that the smooth surfaces usually do not require any further treatment. Lithium salts are more frequently used than sodium salts. Lithium tetraborate has a melting point of about 920°C and is very good for the dissolution of basic oxides. Some workers recommend the addition of other components such as lithium fluoride, lithium carbonate, and sodium nitrate or ammonium nitrate. (Nitrates oxidize sulfides.) High-absorption diluents such as lanthanum oxide can be added, as well as an internal standard. Flux-to-sample ratios as low as 2 : 1 have been used successfully, but some workers have reported that the disks are more difficult to prepare.

The samples are fluxed in platinum–gold crucibles. Platinum–rhodium, vitreous carbon, and graphite crucibles can also be used. The mixture is fused over a Meker burner for a period of up to 15 min. Agitation of the contents from time to time during fusion is necessary in order to obtain complete solution of the sample. Cooling and annealing of the poured glass pellet is important. Rapid cooling results in a mechanically unstable glass. If the rate of cooling is too low, crystallization of the glass may occur. Directions for preparing a lithium borate disk are given below. Some equipment, for example, a Claisse Fluxer (17), is designed to fuse up to six samples simultaneously.

1. Transfer 1.00 g or less of powdered sample to a clean platinum–gold crucible. Add 10.00 g of lithium tetraborate and 0.30 g of lithium fluoride. If the sample contains sulfide, add 0.50 g of ammonium or sodium nitrate.
2. Heat the crucible over a Meker burner to fuse the mixture. Swirl or agitate the contents several times during a heating time of 10–15 min.
3. Pour the melt into an aluminum ring (diameter ~ 3 cm) on a smooth, polished aluminum surface maintained at 450°C (hot plate).
4. After 2 min push the pellet onto a Transite plate at 450°C. Remove the Transite plate from the hot plate, and allow it to cool to room temperature.

Standards are prepared in the same manner. The disks are durable and they can be used many times without deterioration.

2. Fusions for Subsequent Dissolution

As noted previously, lithium metaborate, which melts at 845°C, and lithium tetraborate, with a melting point of 920°C, are both used for sample dissolution. The former is better suited for attacking and dissolving acidic oxides. A procedure for a typical fusion with metaborate is given below.

1. Weigh a 0.5 g sample of powdered rock into a crucible of approximately 30 ml capacity. Suitable crucibles are made of platinum, platinum–gold alloy, and graphite. Graphite crucibles must be manufactured from high-purity graphite, and they have a limited lifetime. Ignite the graphite crucible and cover at 1000°C for 15 min, and cool prior to adding the sample. Leave any graphite dust in the crucible.
2. Add anhydrous lithium metaborate to the crucible and mix the contents well. The ratio of flux to sample can vary from 3 : 1 to 10 : 1. If resistant minerals such as zircon are present, a large ratio must be used for a successful attack. A low ratio (3 : 1) is preferred for analyses of minor and trace elements.
3. Fuse the mixture in a furnace at 1000°C or over a Meker burner for about 30 min. Swirl the contents once or twice during the period.

4. Allow a fusion in a metallic crucible to cool to room temperature. Immerse the crucible in a solution of 165 ml of water and 10 ml of nitric acid. (An internal standard can be added at this point.) Place a small Teflon-coated stirring bar inside the crucible, and stir the contents immediately. The solids will dissolve in 1–2 h.
5. Pour a fusion in a graphite crucible directly into a dilute acid solution as described in step 4, and stir the contents. Examine the cool crucible for any retained material. Transfer any residue to the sample beaker. Filter the solution to remove graphite particles.

D. Sintering

The decomposition of many minerals resistant to attack by acids and fluxes can be accomplished by means of sintering with sodium peroxide (18). At temperatures below 500°C, a sodium peroxide sinter for 2 h will decompose the most resistant minerals. A much shorter period of about 20 min is sufficient for many minerals. The procedure is suitable for determination of major and trace elements in minerals and rocks by means of flame atomic absorption. Details of a recommended procedure are given below.

1. Use fresh, dry, powdered sodium peroxide. The peroxide must be mixed quickly with the dry sample. If the sulfur content of the sample is less than 5%, platinum or platinum–gold crucibles are suitable. Otherwise, nickel, iron, or zirconium should be used, depending on the elements to be determined.
2. Weigh a 0.5 g sample of powdered rock and transfer it to a small bottle with a ground-glass stopper. Then transfer 4.0 g of sodium peroxide to the bottle. Mix the contents by shaking.
3. Pour the premixed sample into a crucible. Heat the crucible at 480°C for 2 h.
4. Cool the crucible. Extract the salts with cold water and add more water to bring the volume to slightly below the final volume required. Add 5 *M* hydrochloric acid until a clear solution results. Make the final acid concentration 0.3 *M*. If the samples contain large proportions of manganese, add 2 or 3 drops of hydrogen peroxide to the acidified solution in order to dissolve the manganese.
5. Boil the solution to release free oxygen in order to prevent bubbles forming during nebulization of the sample.

E. Partial Dissolution

In the past 10–15 years, there has been increasing interest in partial dissolution methods in geochemical exploration. When ore elements and pathfinder elements are released from their primary sources to more readily accessible sites by the action of water or by other means, they are held in relatively labile forms on the substrates. The labile forms of the elements can be dissolved by extractants which do not attack and decompose the entire sample. This empirical approach has frequently resulted in greater contrasts between background and anomalous values for elements of interest.

An extensive review of partial dissolution techniques has been made by Chao (19). He stated that partial dissolution in analyses of samples should be of most use with geological materials collected from the weathering zone.

The most important feature of a partial dissolution procedure is the ability of a chemical (extractant) to dissolve a specific component or components of the sample. However, claims that individual components of a sample can be extracted and that the released metals are genetically related to the individual components should be viewed cautiously.

Partial dissolution methods include single dissolutions and sequential multiple dissolutions. In the latter, several chemical extractants are applied in a sequence. The treatments are arranged so that the mildest ones are applied first and the drastic treatments later.

Examples of extractants are water, aqueous solutions of salts, buffers, strong and weak inorganic acids, organic acids, oxidants, and reductants. Whether an extraction can be classified as selective or nonselective depends not only on the reagent (extractant) but also on the nature of the sample and other experimental conditions. The concentration of the extractant, the duration of the contact time, the temperature, and the ratio of extractant to sample all play a part.

In geochemical work, some of the fractions of samples that provide useful information are water-soluble constituents, exchangeable metals, manganese oxides, iron oxides (amorphous and crystalline), sulfides, carbonates, and organic matter. Silicates and some minerals are not attacked to any significant degree by any of the types of extractants listed above. These are sometimes designated as "residue." Residues can be treated with combinations of strong acids and/or fluxes as described previously.

The dissolution of sulfides is an important part of analyses of rocks. A discussion of a number of procedures for dissolving sulfides in the presence of silicates appears in Chapter 6, together with a recommended procedure. Water-soluble constituents including chloride, fluoride, calcium, magnesium, sodium, and potassium have been determined in other work on rocks (20, 21).

Partial extractions have been used extensively in studies of soils and sediments. The usual procedure is to mix the sample with the extractant for a specified time, separate the phases, and then analyze the liquid phase. Metals of most interest have been copper, lead, nickel, and zinc, although other elements can also be determined in the extracts. The term "loosely bonded metals" is sometimes applied to these elements. However, an extractant such as cold dilute HCl may attack some oxides, sulfides, and other minerals as well as carbonates.

Hydrogen peroxide, alkaline sodium hypochlorite, and alkaline pyrophosphate are used to oxidize or to extract organic matter with the release of associated metals. Sodium hypochlorite does not attack the oxides of iron and manganese as strongly as hydrogen peroxide. The effectiveness of sodium and potassium pyrophosphate is due to the ability of pyrophosphate to remove calcium, iron, and alumina, which stabilize organic matter in soils and sediments.

Trace elements are often associated with precipitated manganese and iron oxides. The selective dissolution of the oxides and the determination of any trace metals have been studied by many workers. A procedure for the determination of a number of trace metals in oxalic acid extracts is contained in Chapter 6.

Manganese oxides dissolve rapidly under moderately reducing neutral or acid conditions. They can be separated from iron oxides by means of hydroxylamine hydrochloride solutions. A typical concentration is 0.1 *M* and a pH of 2. The attack on iron oxides is minimal.

Iron oxides present a more difficult problem because of their occurrence in amorphous and crystalline forms. Since the proportion of each form is of interest, researchers have tried to find conditions for their selective dissolution. Ammonium oxalate solution (pH 3) dissolves the amorphous phase. A solution of a strong reducing agent such as sodium dithionite attacks both phases.

There are three main objectives in using selective sequential extractions in geochemical exploration: (a) The enhancement of geochemical contrasts between background and mineralized areas is very important. (b) The study of the manner of occurrence of trace elements in sediments and soils can provide useful data. (c) Sequential extractions can help to differentiate between environmental and lithological effects related to mineralization.

A number of schemes for sequential extractions of samples of various types have appeared in the literature. As noted earlier, the milder extractants are generally applied first. If a significant proportion of organic matter is present, it is oxidized at or near the beginning of the sequence. Any precipitates formed in an extraction must be redissolved before the next step in the procedure. A good example of a procedure for sequential extraction is shown in Table 4.3. It was used for the fractionation of copper in soils and sediments (12, 22).

Table 4.3
Sequential Extraction Procedures[a]

No.	Extractant	Phase decomposed
1	0.1 *M* hydroxylamine hydrochloride in 0.01 *M* HNO_3	Manganese oxide
2	0.25 *M* hydroxylamine hydrochloride in 0.25 *M* HCl	Iron oxides, amorphous
3	Sodium dithionite[b]	Iron oxides, crystalline
4	Potassium chlorate/HCl	Sulfides
5	HF/HNO_3	Silicate residue

[a]Source: Ref. 12.
[b]Any precipitate formed must be redissolved.

REFERENCES

1. G. J. S. Govett, "Handbook of Exploration Geochemistry." Elsevier, Amsterdam, 1983.
2. W. M. Johnson and J. A. Maxwell, "Rock and Mineral Analysis," 2nd ed. Wiley, New York, 1981.
3. P. M. Gy, "Sampling of Particulate Materials: Theory and Practice." Elsevier, Amsterdam, 1979.
4. D. J. Ottley, *World Min.* **40** (1966).
5. M. A. Tuff, *Adv. X-Ray Anal.* **29,** 565 (1986).
6. J. C. Van Loon, *Spectrochim. Acta, Part B* **38B,** 1509 (1983).
7. R. C. Leake and D. Peachey, *Trans. Inst. Min. Metall., Sect. B* **82,** 25 (1973).
8. L. Leoni and M. Saitta, *X-Ray Spectrom.* **6,** 181 (1977).
9. B. Schroeder, G. Thompson, M. Sulanowska, and J. N. Ludden, *X-Ray Spectrom.* **9,** 108 (1980).
10. N. W. Bower and G. Valentine, *X-Ray Spectrom.* **15,** 73 (1986).
11. J. Dolezal, P. Povondra, and Z. Sulcek, "Decomposition Techniques in Inorganic Analysis." Am. Elsevier, New York, 1968.
12. T. T. Chao and P. K. Theobald, *Econ. Geol.* **71,** 1560 (1976).
13. T. T. Gorsuch, "The Destruction of Organic Matter." Pergamon, Oxford, 1970.
14. J. L. M. De Boer and F. J. M. J. Maessen, *Spectrochim. Acta, Part B* **38B,** 739 (1983).
15. H. Bennett and G. J. Oliver, *Analyst (London)* **101,** 803 (1976).
16. J. T. Hutton and S. M. Elliot, *Chem. Geol.* **29,** 1 (1980).
17. R. Le Houillier and S. Turmel, *Anal. Chem.* **46,** 734 (1974).
18. J. A. Corbett, W. C. Godbeer, and N. C. Watson, *Proc. Australas. Inst. Min. Metall.* **250,** 51 (1974).
19. T. T. Chao, *J. Geochem. Explor.* **20,** 101 (1984).
20. J. C. Van Loon, S. E. Kesler, and C. M. Moore, *in* "Geochemical Exploration" (M. J. Jones, ed.), p. 429. Institution of Mining & Metallurgy, London, 1973.
21. W. D. Goodfellow and J. L. Wahl, *J. Geochem. Explor.* **6,** 35 (1976).
22. R. E. Learned, T. T. Chao, and R. F. Sanzalone, *J. Geochem. Explor.* **15,** 563 (1981).

5
Field Methods

I. INTRODUCTION

From an analytical chemistry point of view it is highly desirable to do geochemical exploration analyses in a well-equipped modern laboratory. However, there are instances when it is necessary to have methodology for geochemical exploration which can be used in the field. Experimental conditions in the field are always somewhat less than ideal. Also, methodology suitable for field use involves compromises which would not exist in laboratory methodology. Contamination, an ever present hazard in trace metal laboratories, is a greater problem in the field.

Classically, colorimetric methods were used for field analysis. In this regard, almost every old-time exploration geochemist is familiar to some degree with dithizone methods for determination of copper, lead, and zinc in the field. Many have had the misfortune to have used these in a damp tent on rainy days when field geochemistry could not be done. Colorimetric methods for field geochemical exploration are still frequently employed today.

In recent years, with the miniaturization resulting from development of solid-state electronics, it has become possible to buy some types of instruments which operate from relatively low-power sources (batteries or low-power generators). Such equipment may often be suitable for field use.

One of the most extensive offerings of equipment specially designed for field work is the portable laboratories sold by Hach Company (Loveland, Colorado). This equipment was developed for water and soil analysis, mainly in response to environmental and agricultural requirements. Several portable laboratory packages exist. Each contains reagents and reaction vessels for specific chemical tests together with measurement equipment and an analytical methods manual. Examples of equipment are digital titrators, color comparators, spectrophotometers, and pH and conductivity meters. The instrument(s) included in a particular portable laboratory offering depends on applications and price. Reagents and equipment for performing a particular analysis are also available in a single test kit. These test kits contain prepackaged reagents (usually in single doses) and reaction vessels together with color comparators. The methodology has been designed to be as simple as possible, requiring a minimum amount of wet chemistry.

Specific ion meters and pH meters suitable for field use are now available from a number of companies and can be purchased from scientific suppliers such as Orion Research Inc. and Sargent-Welch Scientific. Selective ion electrodes are becoming better and better (more selective and having better detection limits). The fluoride electrode has over the years proved to be of greatest utility in geochemical exploration.

There are a few offerings of larger scale instrumentation being designed for field and laboratory use. Of particular interest for base and precious metal exploration is the Scintrex Ltd. (Concord, Ontario, Canada) atomic absorption spectrometer. This device contains a tungsten strip atomizer which requires relatively low power and is crucially important. Background correction is carried out using the Zeeman effect.

Several workers refer to the use of energy-dispersive x-ray analyzers in the field. Such equipment is mounted in a truck or van. The detector requires cooling in liquid nitrogen or air. Because of these requirements the present authors question whether such analyzers can be of any great practical utility in field geochemical operations.

In this chapter methods are given for the most commonly determined elements in geochemical exploration. Emphasis is placed on colorimetric methodology because this type of approach is practical for even small budgets.

In some laboratories, instrumentation for trace element analysis may not exist. Thus, the following colorimetric methods could also be employed in these laboratories for geochemical exploration.

II. COLORIMETRIC METHODS

Much of the following methodology was developed at the U.S. Geological Survey and was published in a special volume in 1963 (1). Most of the methods appearing in the volume were intended for field work. In this regard, the methods have been designed to minimize the complexity and numbers of steps. The authors state that usually 30 to 50% of the "correct" value is obtainable. Minimum determinable amounts (in solid samples) of mercury, molybdenum, nickel, lead, copper, zinc, and uranium are (in parts per million) 2, 4, 25, 25, 10, 25, and 4, respectively.

Some of the reagents required in the methodology are potentially hazardous, particularly under field conditions. Thus, care is required to prevent exposure to excessive levels of poisonous solvent vapors and contamination of skin and clothing by hazardous chemicals. Frequently, personnel with little training in analytical chemistry are utilized in the field for chemical analysis. Such workers must be carefully instructed and supervised to prevent relatively hazardous situations from occurring.

Reagents and water used for chemical tests must be free from significant levels of the elements of interest. In this regard, waters available from field sources should be purified by means of demineralizers. Demineralizers for field use are available from many scientific suppliers. Great care is essential to prevent contamination of reagent grade chemicals and purified waters by dirt, dust, smoke, etc.

Choice of suitable apparatus is an important requirement in planning a field chemical analysis program. Some of the more rudimentary equipment includes a balance (or calibrated scoop), test tube racks, water baths (or drilled test tube heating blocks), heaters (e.g., stove), drying oven, sieves, demineralizer, and storage bottles for reagents and purified water.

Two types of reaction test tubes are essential. Both should be constructed of borosilicate glass. Test tubes for acid digestion are thin-walled and relatively cheap. Test tubes suitable for fusions must have thicker walls.

Sieves can be constructed from 80 mesh screen. This should preferably be made from nonmetallic material, e.g., nylon cloth.

High-density polyethylene containers should be used to store standard solutions. Concentrated acids should be purchased in containers which have the minimum chance of shattering. Also, it is well to purchase acids in relatively small-volume containers to minimize the hazard if one is broken.

Balances are much superior to calibrated scoops for sample measurement. The latter give different weights depending on the density of the sample being measured. In a study by McCarthy (2), the error in using a scoop was ± 15%. This may not be a problem when all requirements of the analysis and error sources have been considered.

III. COLOR STANDARDS

Stable standard colors for comparison with the unknown can be made or purchased. Hach Co. supplies "color cubes," which consist of a series of five different color intensity, fixed cubes in a single cuvette attached to an empty cuvette. The latter is used to contain the reacted unknown. Hawkins, Canney, and Ward (3) recorded a method for the preparation of plastic standards which can be used in a similar manner. This procedure is as follows:

Prepare a set of the liquid standards that are to be matched in plastic. For standards that are of the same color but differ in intensity throughout the range of values covered by the method, pour about 3 tablespoonfuls of clear Castolite into a paper cup and add small amounts of dye until the color matches that of the liquid standard having the most intense shade. Prepare the remaining plastic standards by dilution of this colored resin with appropriate amounts of clear resin to produce a series of the proper intensities. For standards that differ in color with different concentrations of metal, each standard must be prepared separately.

Take a glass tube of the same dimensions as those containing the standard solutions to be used as color guides. Clamp it in a vertical position and pour in about 1 teaspoonful of clear resin mixed with the hardener and cold-setting promoter. Allow the mixture to gel for about 10 min.

Mix the hardener and cold-setting promoter with the colored resin corresponding to the lowest standard and pour this mixture into the tube until the volume of the layer is approximately equal to the volume of the organic layer in the liquid standard. Allow it to gel. After the colored layer has gelled, add a layer of clear resin and allow it to gel.

Continue the process of pouring alternating clear and colored layers until the whole series of standard colors has been matched; then pour a clear layer about 1.9 cm thick on top of the last colored layer.

Allow the gelled plastic to cool to room temperature, then cure the plastic rod by heating the tube for 30 min in a water bath at 70°C. (Some heat is liberated by the plastic as it gels. Unless the plastic is cooled to room temperature prior to curing, the casting will crack while it is being cured.)

Take the tube from the water bath and add about 5 ml acetone. Remove the casting from the glass tube by inverting the tube and tapping its mouth sharply against a solid surface. Wipe the rod with tissue to remove a thin adhering layer of syrupy resin.

A. Copper–Biquinoline Standards

No commercially available dye satisfactorily duplicates in plastic the color of the copper–biquinoline complex. The authors (3) found that a complex formed by the reaction of tin(IV) with 4,5-dihydroxyfluorescein (gallein) gives the resin a red color closely resembling that of the copper complex. This red complex is prepared by allowing stannic ion to react with 4,5-dihydroxyfluorescein in an acetate-buffered aqueous medium at a pH of 2–2.5. The complex is extracted by a binary mixture of *o*-dichlorobenzene and cyclohexanone.

A difficulty experienced with the use of this compound as a dye is that the color tends to fade slightly in the curing. To correct this, the resin should be made slightly deeper in color than the standard it is to match; the necessary correction must be determined experimentally. Once the plastic has been cured, however, the color is resistant to fading.

B. Zinc Dithizonate Standards

The dyes that are commercially available for coloring Castolite can be used in preparing the plastic standards for zinc. The best results are obtained by using dilute alcohol or acetone solutions of these dyes to color the resin. The dyes are diluted as described below:

Green: 50 mg green dye dissolved in 10 ml acetone

Blue: 4 drops concentrated blue dye in 19 ml acetone

Violet: 4 drops concentrated violet dye in 10 ml acetone

Red: 4 drops concentrated red dye in 10 ml acetone

Table 5.1 shows the dye mixtures required to obtain a fair approximation of the colors of the zinc standards. The standard rod is prepared according to the general procedure.

C. Mercury Dithizonate Standards

To prepare plastic standards for mercury, use a saturated ethyl alcohol solution of Orange II to color the resin. About 7 drops of Orange II dye in 1 tablespoonful of clear resin are required to match the standard with the highest metal content. Dilute portions of this preparation with sufficient amounts of clear resin to match the other members of the standard series. Cast each standard separately in a pointed 15 ml centrifuge tube to conform as nearly as possible to the dimensions of the liquid standards, which are prepared and used in 60 ml separatory funnels.

NOTE: Because plastic standards are not able to compensate for variations in color intensity due to slight differences in reagent strength and the color-developing procedure, it is important to include an aqueous standard to be used as a control within each sample set.

Table 5.1
Dye Mixtures Required to Obtain Approximate Colors of the Zinc Standards

Metal content of standard (μg)	Amount of dye to be added to 1 tablespoon clear resin	
	Number of drops	Color of dye
0	2	Green
1	1	Green
	4	Bluc
	2	Violet
2	2	Blue
	5	Green
	2	Red
3	7	Violet
	1	Blue
	4	Red
4	1	Blue
	7	Violet
	6	Red

Source: Ref. 3.

IV. SAMPLE PREPARATION (SAMPLES CONTAINING ORGANIC MATTER)

As in the laboratory, there are two basic approaches to sample ashing when organic matter must be removed. These are dry and wet ashing methods. When power is available in the field (e.g., generators), modified laboratory methods based on ovens and hot plates can be employed. If sufficient power is not available, then barbecues or braziers can be used for dry ashing.

V. DRY ASHING

If an oven is available, dry ashing can readily be performed. To improve sample throughput in the ashing process, plant material should be pelletized (1). This also results in a more uniform ash. Prior to the pelletizing, the plant material should be rinsed in purified water to remove as much surface dust as possible. This is followed by drying and grinding. The pelletized samples are then ashed at 500°C.

When dry ashing cannot be done in an oven, the following method by Ward *et al.* (1) can be employed.

A modified charcoal brazier or barbecue has been found satisfactory for field ashing of vegetation, producing an ash comparable in quality to that obtained by use of a muffle furnace in the laboratory.

The bottom of a charcoal brazier is covered with aluminum foil (see Fig. 5.1). A layer of coarse gravel 2.5 to 10.1 cm deep is placed in the bottom of the brazier. A covered aluminum can 7.6 cm in diameter and 10.1 cm in depth, with a 1.2 cm hole cut in the center of the cover, is used as a container for each sample. To keep the aluminum cans from melting from the intense heat of burning charcoal, each is held in the brazier by a 14 gauge wire sleeve of 4 by 4 mesh, fashioned 7.9 cm in diameter by 15.2 cm in depth, and with a 14 gauge cross-wire insert 6.6 cm from one end. Twelve or more such sleeves are placed, one for each sample can, in the charcoal brazier, utilizing the space to the fullest advantage. Commercial

Figure 5.1. Cross section of brazier.

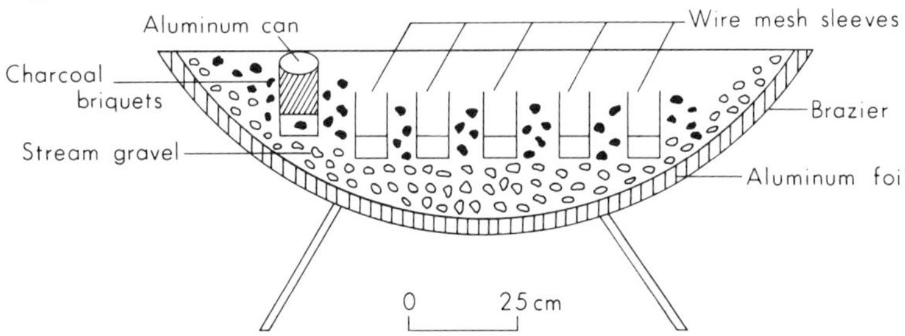

charcoal briquets supply the fuel. One briquet is placed in the bottom of each sleeve, and each sleeve is then surrounded by briquets.

With pruning shears, cut the samples of vegetation into small pieces so that 50–75 g of each sample will fit into its aluminum container. Place the covers on the containers, place the containers in the sleeves in the brazier, and ignite the charcoal. When the excess smoke has ceased pouring from the hole in the cover of each container, remove the cover and stir the sample with a glass rod. Let containers stand open until the organic material has oxidized; remove the containers from the sleeves.

NOTE: The entire procedure takes about 2 h. Since the coals around the sleeves are already burning, a new batch of samples may be placed in the unit and the operation repeated.

VI. WET ASHING

Many of the laboratory wet ashing procedures, given elsewhere in this book, could be modified for use in the field. The following procedure in Ward *et al.* (1) is recommended as a rapid method.

Reagents and Equipment

- Hydrogen peroxide, 30%
- Nitric acid, ACS grade, concentrated
- Sulfuric acid, ACS grade, concentrated
- Flask, 500 ml Erlenmeyer, having a ground-glass joint that fits a 30.4 cm air condenser

Procedure

Prepare a 3–5 g sample pellet and transfer the pellet to a flask containing 25 ml of sulfuric acid and 20 ml of nitric acid. Fit the flask with a 30.4 cm air condenser and boil the mixture gently. Continue gentle boil, adding small amounts of nitric acid at frequent intervals until liquid is yellow. Cautiously add hydrogen peroxide dropwise until all color is removed. Cool; dilute the colorless solution to 100 ml with 0.5 *M* sulfuric acid.

NOTE: The ashing time by this procedure is about 20–30 min.

VII. ANALYTICAL PROCEDURES

A. Determination of Copper, Lead, and Zinc in Soils and Rocks (1)

Comments on the Method

This method involves a pyrosulfate fusion. The fusion is carried out in a 16 by 150 mm test tube. The fusion mixture is dissolved in 6 *M* hydrochloric acid. Suitable aliquots of the sample solution are treated separately for zinc, lead, and copper. The color reagent for zinc and lead is dithizone. The compound 2,2′-biquinoline is used for copper. Because of the extreme sensitivity of this method, great care is required to prevent problems due to contamination.

Reagents

- Dithizone stock solution, 0.01%: Dissolve 0.005 g dithizone (diphenylthiocarbazone) in 500 ml carbon tetrachloride (CCl_4). Shake the mixture intermittently over a period of about 1 h to effect solution.
- Dithizone test solution, 0.001%: Dilute 50 ml dithizone stock solution (0.01%) to 500 ml with CCl_4.
- Hydrochloric acid, 6 *M:* Add 100 ml metal-free water to 100 ml concentrated reagent grade hydrochloric acid.
- Potassium pyrosulfate ($K_2S_2O_7$), fused powder, reagent grade.

NOTE: Dithizone solutions are readily oxidized in the presence of light. The problem increases for more dilute solutions. Thus, a strong stock solution is prepared.

Interferences are an ever present problem with colorimetric methods. In the determination of lead at pH 8.5, interferences due to copper, zinc, and other elements are prevented by the addition of potassium cyanide. Additional interferences can be caused by tin (as Sn^{2+}), thorium, and bismuth (all usually present in only very small amounts). Interference in the zinc determination at pH 4.5 is minimized by the addition of sodium thiosulfate. Lead and Sn^{2+} would still interfere if present. If 10 $\times$ greater amounts of cadmium, cobalt, nickel, or copper are present, an interference will be experienced (this is unlikely). Organic matter is the worst interference problem in the copper determination. Samples containing large amounts of organic matter must be oxidized by either the dry or the wet procedure before the copper is dissolved.

Lead reagents

- Lead buffer solution: Transfer 50 g ammonium citrate [$(NH_4)_2HC_6H_5O_7$], 10 g potassium cyanide (KCN), and 8 g hydroxylamine hydrochloride ($NH_2OH \cdot HCl$) into a large separatory funnel; add 800 ml metal-free water and shake the funnel until the reagents are dissolved. Add 2 ml thymol blue indicator and then concentrated ammonium hydroxide until the pH of the solution is 8.5 (color is blue). Remove any contaminating heavy metals by the following procedure.
  - Transfer 50 ml dithizone solution (0.01%) to the separatory funnel, shake, allow the CCl_4 phase to collect in the bottom of the funnel, and discard the CCl_4 phase. Continue this process with additional 50 ml increments until the discarded organic phase is green.
  - Some dithizone is dissolved in the buffer solution, imparting to it a yellow color. This is removed by successive extractions with 50 ml portions of chloroform ($CHCl_3$). Chloroform dissolved in the buffer solution is removed by

shaking the buffer solution with 100 ml CCl_4 and discarding the organic phase. Dilute the aqueous buffer solution to 1 liter with metal-free water.

- Potassium cyanide solution (0.1% aqueous). Caution: Dissolve 1 g potassium cyanide (KCN) in 1 liter metal-free water.
- Standard lead solution, 100 μg/ml: Dissolve 0.016 g lead nitrate [$Pb(NO_3)_2$], dried for 1 h at 110°C, in 100 ml metal-free water containing 1 ml concentrated nitric acid.
- Standard lead solution, 10 μg/ml: Add 10 ml standard lead solution (100 μg/ml) to a 100 ml volumetric flask containing 1 ml concentrated nitric acid and about 10 ml metal-free water. Dilute to 100 ml with metal-free water.
- Thymol blue, 0.2% aqueous: Dissolve 0.2 g sodium salt of thymol blue in 100 ml metal-free water.

Copper reagents

- 2,2′-Biquinoline solution, 0.02%: Add 0.2 g 2,2′-biquinoline to about 900 ml isoamyl alcohol in a beaker. Warm on a steam bath in a well-ventilated room away from open flame until the biquinoline is dissolved. Allow the solution to cool, and dilute to 1 liter with isoamyl alcohol. The solution should be colorless; if it is yellow, the reagent is impure and should not be used.
- Copper buffer solution: Dissolve 400 g sodium acetate ($NaC_2H_3O_2 \cdot 3H_2O$), 100 g sodium tartrate ($Na_2C_2H_4O_6 \cdot 2H_2O$), and 20 g hydroxylamine hydrochloride ($NH_2OH \cdot HCl$) in 1 liter metal-free water. If necessary, adjust the pH of this solution with hydrochloric acid or sodium hydroxide solution to between 6 and 7, using pH test paper. To check for copper contamination, place 10 ml of buffer in a culture tube, add 1 ml of the biquinoline reagent, cap, and shake for 2 min. Absence of a pink color in the organic layer indicates that the solution is satisfactory for use. If found to be contaminated, the buffer solution may be cleaned by the procedure given for the lead buffer solution above.
- Hydrochloric acid, 6 *M:* Add 500 ml concentrated reagent-grade hydrochloric acid to 500 ml metal-free water.
- Standard copper solution, 100 μg/ml: Dissolve 0.200 g copper sulfate ($CuSO_4 \cdot 5H_2O$) in 500 ml 0.1 *M* hydrochloric acid.
- Standard copper solution, 10 μg/ml: Transfer 10 ml standard copper solution (100 μg/ml) to a 100 ml volumetric flask, and dilute to 100 ml with 0.1 *M* hydrochloric acid.

Zinc reagents

- Zinc buffer solution: Dissolve 125 g sodium thiosulfate ($Na_2S_2O_3 \cdot 5H_2O$) in about 400 ml metal-free water in a large separatory funnel. Remove heavy metals by extracting them from the solution with dithizone solution (0.1%) in one or more 50 ml portions, as follows: After the addition, the funnel is shaken and the CCl_4 layer allowed to settle to the bottom. If this layer shows any color except the characteristic clear green of dithizone, heavy metals are present and are reacting with the dithizone to form colored dithizonates. In that case, the layer should be drained away and another 50 ml portion dithizone (0.01%) added. When the lower layer is green and remains green, the excess dithizone must be removed from the buffer solution by extracting it with successive 50 ml portions CCl_4. The process is the same as that described above, except that now the portions are added, shaken, then discarded until the lower layer is colorless.

  - Dissolve 300 g sodium acetate ($NaC_2H_3O_2$) in 400 ml of metal-free water, then add 60 ml of glacial acetic acid. Remove heavy metals by the procedure described above.
  - Combine above solutions and dilute to 2 liters with metal-free water.

- Standard zinc solution, 100 μg/ml: Dissolve 0.100 g of reagent grade 30 mesh zinc metal in 10 ml concentrated hydrochloric acid and dilute to 1 liter with metal-free water.
- Standard zinc solution, 10 μg/ml: Transfer 10 ml standard zinc solution (100 μg/ml) to a 100 ml volumetric flask and dilute to volume with metal-free water.

Equipment

- Culture tubes, 16 by 150 mm and 22 by 175 mm
- Caps, polyethylene, size S-3, to fit 22 by 175 mm tubes
- Scoop, 0.1 g and 0.5 g
- Spatula
- Pipets, volumetric, 1 ml graduated in tenths and 2 ml
- Separatory funnels, 125 ml
- Separatory funnel rack
- Graduated cylinders, 25 ml glass-stoppered
- Rubber bulb (used to provide suction for pipetting poisonous solutions such as potassium cyanide)

Standard Solutions

Lead standard solutions

Prepare lead standards of 0, 1, 2, and 3 μg by pipetting suitable aliquots from a standard lead solution (10 μg/ml) into 125 ml separatory funnels containing 10 ml lead buffer. Proceed as for samples below. As these liquid standards for lead usually begin to fade after approximately ½ day, they should be prepared fresh twice daily.

Zinc standard solutions

Prepare zinc standards of 0, 1, 2, 3, and 4 μg by pipetting into each of five 22 by 175 mm culture tubes, respectively, suitable aliquots from a standard zinc solution

(10 μg/ml). Add 8 ml zinc buffer solution to each of the tubes and proceed as for samples below. These standards are usually stable for at least 2 days.

Copper standard solution

Prepare copper standards of 0, 0.2, 0.4, 0.8, 1.5, 3, and 6 μg by pipetting into seven 25 by 200 mm culture tubes suitable aliquots from a standard copper solution (10 μg/ml). Add 10 ml copper buffer solution and proceed as for samples below. If kept tightly stoppered to prevent evaporation of the isoamyl alcohol, these standards are usually stable for several months.

Procedure

Weigh or scoop a 0.1 g sample and transfer it to a 16 by 150 mm culture tube. Scoop 0.5 g $K_2S_2O_7$ flux into the culture tube and shake the tube to mix the sample and flux. Fuse the mixture for about 2 min after the flux melts. Remove the tube from the flame and rotate it so that the melt cools in a thin film on the side of the tube. When the tube is cool, add 3 ml of 6 *M* hydrochloric acid to it and place it in a hot-water bath until the melt has disintegrated. To hasten disintegration of the melt, shake the tube intermittently or crush the melt with a glass rod. Remove the tube from the water bath and dilute the sample solution to 10 ml with metal-free water. Shake the tube vigorously to mix the sample solution. Take suitable aliquots from this single sample solution for copper, lead, and zinc estimations.

Estimations

Estimation of lead

Transfer a 2 ml aliquot of the sample solution to a 125 ml separatory funnel containing 10 ml lead buffer solution. Add 2 drops thymol blue indicator solution to the funnel; then add concentrated ammonium hydroxide dropwise until the pH of the solution is 8.5 to 9, as indicated by a change from yellow to blue. Add 5 ml dithizone solution (0.001%) and shake the separatory funnel gently for about 115 s. Drain the CCl_4 layer into a 25 ml glass-stoppered graduated cylinder containing 10 ml potassium cyanide solution (0.1%). Shake the cylinder gently for 5 s. Compare the color of the CCl_4 layer with the similarly prepared standard solutions. If the sample color is above the highest standard, repeat using a smaller aliquot.

Estimation of zinc

Transfer a 2 ml aliquot of the sample solution to a 22 by 175 mm culture tube containing about 8 ml zinc buffer solution. Add 5 ml dithizone solution (0.001%) and tightly cap the culture tube. Shake the tube vigorously for 30 s. Compare the color in the CCl_4 layer with similarly prepared standard zinc solutions. If the sample solution color is above the highest standard, repeat using a smaller aliquot.

Estimation of copper

Transfer a 2 ml aliquot of the sample solution to a 22 by 175 mm culture tube containing 10 ml copper buffer solution. Add 2 ml of 2,2′-biquinoline solution to the tube. Tightly cap the tube and shake vigorously for 45 s. Allow the layers to separate, and compare the color of the isoamyl alcohol layer with similarly prepared standards of known copper content. Repeat, if necessary, using a smaller aliquot.

B. Determination of Copper, Lead, and Zinc in Plants (1)

Procedure

The plant material must be ashed by one of the procedures given previously. If dry ashing is employed the ash must be ground to 200 mesh after mixing.

Fuse a 0.1 g portion of the ash with 0.5 g potassium pyrosulfate. Dissolve the fusion in 3 ml of 6 *M* hydrochloric acid. Dilute the solution to 10 ml. React aliquots of the solution in an identical manner to the reaction of rock and soil solutions using the above procedure.

C. Determination of Cobalt in Geological Materials (4)

Comment on the Method

The sample is decomposed by fusion with potassium bisulfate. Tri-*n*-butylammonium tetrathiocyanatocobaltiate is formed and the blue complex is extracted with amyl alcohol. A pH range of 4.5–5.2 is maintained to minimize interferences. Up to 10% each of iron, aluminum, and manganese can be tolerated. If calcium is above 5%, low results for cobalt may result from coprecipitation of cobalt with calcium sulfate. The concentration range for cobalt is 1–250 ppm, and this can be increased by the use of a 0.1 ml aliquot.

Reagents

- Potassium bisulfate, fused powder.
- Hydrochloric acid, concentrated.
- Hydrochloric acid, 6 *M:* Mix 200 ml of concentrated acid with 200 ml of water, or use the constant-boiling mixture.
- Hydrochloric acid, 0.5 *M:* Mix 20 ml of concentrated acid with 420 ml of water.
- Buffer solution. A: Dissolve 170 g of sodium acetate (trihydrate), 40 g of sodium pyrophosphate (decahydrate), and 5 g of hydroxylamine hydrochloride in water; add 140 ml of 6 *M* hydrochloric acid and dilute to 1 liter with water. B: Dissolve 360 g of potassium thiocyanate in water and dilute to 1 liter. Mix A and B in the ratio 9 : 1 as required each day.
- Tri-*n*-butylamine, 10%: Mix together 90 ml of *n*-amyl alcohol or isoamyl alcohol and 10 ml of tri-*n*-butylamine.
- Standard cobalt solutions:
  - 100 μg cobalt/ml: Dissolve 202.4 mg of cobaltous chloride (hexahydrate) in 0.5 *M* hydrochloric acid and dilute to 500 ml with the 0.5 *M* acid.
  - 20 μg cobalt/ml: Dilute 20 ml of the 100 μg/ml solution to 100 ml with 0.5 *M* hydrochloric acid.

- ○ 5 μg cobalt/ml: Dilute 25 ml of the 20 μg/ml solution to 100 ml with 0.5 *M* hydrochloric acid.

Procedure

Weigh 0.25 g of sample into a test tube (16 by 150 mm). Add 1 g of potassium bisulfate, mix, and fuse until a quiescent melt is obtained. Leach with 5 ml of 0.5 *M* hydrochloric acid on a sand tray or in a boiling-water bath.

When cold, pipet a 2 ml aliquot into a test tube (16 by 150 mm) calibrated at 10 and 12 ml and containing 10 ml of buffer solution. Add 0.5 ml of 10% tri-*n*-butylamine solution. Cork the tube and shake it vigorously for 1 min. Compare with a standard series. If the color in the solvent phase exceeds that of the highest standard, use an aliquot of 0.1 ml to 10 ml of buffer solution and dilute to 12 ml with 0.5 *M* hydrochloric acid before proceeding with the extraction.

$$\text{Cobalt (ppm)} = 20 \times \mu\text{g of matching standard/ml of aliquot}$$

Standards

To 23 test tubes (16 by 150 mm), each calibrated at 10 and 12 ml and containing 10 ml of buffer solution, add respectively 0, 0.25, 0.5, 1.0, 1.5, 2.0, 2.5, 3.0, 3.5, 4.0, 4.5, and 5.0 μg of cobalt using a standard solution containing 5 μg cobalt/ml, and 6.0, 7.0, 8.0, 9.0, 10.0, 12.0, 14.0, 16.0, 18.0, 20.0, and 25.0 μg of cobalt using a standard solution containing 20 μg cobalt/ml. Dilute to 12 ml with 0.5 *M* hydrochloric acid. Add 0.5 ml of 10% tri-*n*-butylamine. Cork the tubes and shake them vigorously for 1 min.

D. Determination of Gold in Geological Materials (4)

Comment on the Method

The sample is treated with hydrochloric/nitric acid to dissolve the gold as chloroauric acid. After evaporation of excess acid, gold is coprecipitated with tellurium. The metals are redissolved and the gold is reacted with brilliant green and extracted with toluene. The colors of the standard series range from colorless (zero) to intense blue; they are stable for only 3 h.

The concentration range is 0.05–4 ppm.

Reagents

- Hydrochloric acid, concentrated.
- Hydrochloric acid, 2 *M:* Mix 80 ml of the concentrated acid with 360 ml of water.
- Hydrochloric acid, *M:* Mix 40 ml of the concentrated acid with 400 ml of water.
- Hydrochloric acid, 0.5 *M:* Mix 20 ml of the concentrated acid with 420 ml of water.
- Nitric acid, concentrated.

- Aqua regia: Mix 300 ml of concentrated hydrochloric acid with 100 ml of concentrated nitric acid.
- Acid mixture: Mix 500 ml of aqua regia with 500 ml of water.
- Copper–tellurium solution: Dissolve 2 g of cupric sulfate (pentahydrate) and 210 mg of tellurium tetrachloride in 100 ml of 2 *M* hydrochloric acid.
- Ferric chloride solution: Dissolve 100 g of ferric chloride (hexahydrate) in 2 *M* hydrochloric acid, and dilute to 1 liter with this strength of acid.
- Stannous chloride solution: Dissolve 200 g of the dihydrate in 2 *M* hydrochloric acid, and dilute to 1 liter with this strength of acid.
- Brilliant green, Color Index Reference No. 42040.
- Ethyl alcohol, absolute.
- Brillian green solution, 1%: Dissolve 1 g of brilliant green in 100 ml of ethyl alcohol.
- Brillian green solution, 0.05%: Dilute 5 ml of the 1% solution to 100 ml with toluene.
- Potassium dichromate solution: Dissolve 5 g of solid reagent in 100 ml of water.
- Toluene, sulfur-free grade: Shake 1 liter for 2 min with 10 ml of potassium dichromate solution, discard the aqueous phase, and wash the toluene four times by shaking with successive 500 ml volumes of water. Filter the toluene through a Whatman No. 1 filter paper to remove water.
- Standard gold solutions:
  - 5 mg gold/ml: Dissolve 504 mg of sodium chloroaurate (dihydrate) in 25 ml of *M* hydrochloric acid and dilute to 50 ml with water.
  - 50 μg gold/ml: Dilute 1 ml of the 5 mg/ml solution to 100 ml with 0.5 *M* hydrochloric acid.
  - 1 μg gold/ml: Dilute 2 ml of the 50 μg/ml solution to 100 ml with 0.5 *M* hydrochloric acid.

The 50 μg/ml and 1 μg/ml solutions must be prepared freshly as required.

Procedure

Weigh 1 g of sample into a test tube (19 by 150 mm, with rim) calibrated at 5 and 10 ml. Add 5 ml of acid mixture and evaporate to dryness in an oven. When cold, add 5 ml of 2 *M* hydrochloric acid and boil gently for 15 min. Add 5 ml of 2 *M* hydrochloric acid, mix, and leave to settle.

Pipet a 5 ml aliquot of the clear solution into a beaker (100 ml) calibrated at 50 ml and dilute to 50 ml with 2 *M* hydrochloric acid. Add 2 ml of copper–tellurium solution. Add 5 ml of ferric chloride solution and mix well. Add 10 ml of stannous chloride solution. Bring to the boil, and maintain just below boiling for 30 min. Add 1 or 2 drops of stannous chloride solution to check that precipitation is complete.

Filter through a filter stick (porosity 3) while the solution is still hot, and wash the beaker and precipitate with 2 *M* hydrochloric acid. Dissolve the precipitate in 1 ml of hot aqua regia. Rinse the precipitation beaker with another 1 ml of hot aqua regia. Pour the solution through the filter stick and wash the stick with a few milliliters of water. Evaporate to dryness in a test tube (19 by 150 mm, with rim) calibrated at 2.5 and 10 ml, using an oven. Dissolve the residue in 2.5 ml of 2 *M* hydrochloric acid by warming gently. Dilute to 10 ml with water.

Add 1 ml of 0.05% brilliant green solution. Cork the tube and shake it vigorously for 30 s. Allow the phases to separate; then transfer the toluene phase with an automatic pipet to a test tube (19 by 150 mm) calibrated at 10 ml and containing 10 ml of 0.5 *M* hydrochloric acid and 0.05 ml of 1% brilliant green solution. Cork the tube and shake it vigorously for 2 min. Compare with a standard series.

If the intensity of color in the toluene phase exceeds that of the highest standard, use an aliquot of 0.5 ml and begin again at the second paragraph in the Procedure.

$$\text{Gold (ppm)} = 10 \times \mu\text{g of matching standard/ml of aliquot}$$

Standards

To 11 test tubes (19 by 150 mm) calibrated at 10 ml add respectively 0, 0.05, 0.1, 0.2, 0.3, 0.4, 0.5, 0.75, 1.0, 1.5, and 2.0 μg of gold, using a standard solution containing 1 μg gold/ml. Dilute to 10 ml with 0.5 *M* hydrochloric acid. Proceed as described in paragraph four of the procedure. Store in the dark when not in use. This standard series is stable for only 3 h.

E. Determination of Nickel in Soils, Rocks, and Plants (1)

Comments on the Methods

Potassium pyrosulfate is used to fuse a sample. The fusion thus obtained is dissolved in 6 *M* hydrochloric acid. The reagent α-furildioxime forms a yellow-colored complex with nickel. An interference is experienced when the cobalt : nickel ratio exceeds 12 : 1. Copper and iron (as Fe^{2+}) can cause trouble, but only if the ratio of metal to nickel exceeds about 100 : 1. A pH of 8.5–9.5 is used for best color development. Vegetation samples must be ashed (see Section VII,B) prior to their fusions.

Reagents

- Potassium pyrosulfate ($K_2S_2O_7$), fused powder.
- Hydrochloric acid, 2 *M*.
- Nickel buffer solution: Dissolve 60 g ammonium citrate [$(NH_4)_2HC_6H_5O_7$] in about 200 ml metal-free water. Add 130 ml concentrated ammonium hydroxide, then dilute to 1 liter with metal-free water.
- α-Furildioxime solution, 0.2%: Dissolve 0.2 g α-furildioxime in 15 ml absolute ethanol and dilute to 100 ml with benzene.

- Standard nickel solution, 100 μg/ml: Dissolve 0.2 g clear uneffloresced crystals of nickel chloride ($NiCl_2 \cdot 6H_2O$) in 500 ml of 0.1 *M* hydrochloric acid. From this solution prepare standard solutions (10 μg/ml and 1 μg/ml) prior to preparing the series of nickel standards.

Preparation of Calibrating Solutions

A series of nickel standards for 0, 0.5, 1, 2, 4, 8, 15, 30, 60, and 100 μg of nickel is recommended. To prepare the standards, place 1 ml of 2 *M* hydrochloric acid and 5 ml of nickel buffer in each of a series of 16 by 150 mm culture tubes, add to each the required amount of nickel solution to produce the series suggested above, and then follow the appropriate sections of the following procedure.

Equipment

- Culture tubes, 16 by 150 mm, each marked at 5 ml and 10 ml capacities
- Scoops, 0.1 and 0.5 g
- Pipets: 5 ml automatic, 2 ml automatic, 2 ml volumetric, 1 ml graduated in tenths

Procedure

Weigh or scoop a 0.1 g sample into a 16 by 150 mm culture tube. Add 0.5 g of potassium pyrosulfate to the tube, heat to fuse the mixture, and continue heating for about 2 min after the contents melt. Remove the tube from the flame and rotate the tube so that the melt cools in a thin film on the side of the tube. When the tube is cool, add 5 ml of 2 *M* hydrochloric acid and place the tube in a hot-water bath until the melt disintegrates. Allow the tube to cool, then dilute the sample solution to 10 ml with water. Transfer a suitable aliquot (not more than 2 ml) of the sample solution to a 16 by 150 mm culture tube that contains 5 ml nickel buffer solution. Add 1 ml α-furildioxime solution, stopper the tube, and shake for at least 1 min. Compare the color of the benzene layer with colors of benzene layers in similarly prepared standards.

F. Determination of Molybdenum in Soils and Rocks (1)

Comments on the Method

The sample is sintered with a mixture of sodium carbonate and potassium nitrate. The sinter cake is leached with warm water. (Water prevents the solubilization of potential interferents, particularly iron and copper.) An aliquot of the resulting solution is treated with 1 *M* hydrochloric acid dropwise until the pH is reached where the pink color of phenolphthalein just disappears. An amber-colored solution for analysis is obtained by addition of potassium thiocyanate. The method can detect down to 5 ng of molybdenum. Tartrate is added to minimize tungsten interference. Large amounts of rhenium, uranium, vanadium, and niobium may interfere. If excessive amounts of ferric ion are leached by the water, stannous chloride solution is added to reduce the ferric to ferrous, which does not appreciably interfere.

Reagents

- Flux: Mix equal parts by weight of sodium carbonate (Na_2CO_3) and potassium nitrate (KNO_3) and grind to pass an 80 mesh sieve.
- Hydrochloric acid, concentrated, reagent grade.
- Isopropyl ether: If ether is not peroxide-free, on the day it is to be used shake it with one-tenth its volume of equal amounts of stannous chloride ($SnCl_2$) and potassium thiocyanate (KSCN) reagents.
- Phenolphthalein indicator solution, 1%: Dissolve 1 g phenolphthalein in 100 ml alcohol.
- Potassium nitrate solution, 10%: Dissolve 10 g in 100 ml water.
- Potassium thiocyanate solution, 5%: Dissolve 5 g salt in 100 ml water. Prepare fresh daily.
- Sodium tartrate ($NaHC_4H_4O_6 \cdot H_2O$), powdered, reagent grade.
- Standard molybdenum solution, 100 μg/ml: Dissolve 0.075 g pure molybdic acid anhydride in 1 *M* sodium hydroxide, dilute with water, add 1 *M* hydrochloric acid until solution is acid, and then dilute to 500 ml with water.
- Standard molybdenum solution, 1 μg/ml: Prepare daily by diluting 1 ml standard solution (100 μg/ml) to 100 ml with water.
- Stannous chloride reagent, 10%: Dissolve 10 g $SnCl_2 \cdot 2H_2O$ in 17 ml concentrated hydrochloric acid, and dilute with water to a volume of 100 ml.

Preparation of Calibrating Solutions

Transfer, respectively, 0, 0.1, 0.2, 0.4, 0.8, 1.5, 3, and 6 μg molybdenum to a series of culture tubes, then follow the sample procedure below.

Equipment

- Culture tubes, 16 by 150 mm
- Flask, volumetric, 100 ml
- Scoop, 0.1 g
- Mullite mortar and pestle
- Pipets, 1 and 2 ml, graduated
- Stevens extractors (These are made by constricting near one end a glass tube 120 mm long by 7 mm i.d. and packing the resulting small bulb with fine glass wool to serve as a filtering medium.)

Procedure

Add 0.1 g sample to a 16 by 150 mm culture tube, then add 0.5 g flux. Shake the tube thoroughly to mix the sample and flux. Heat and rotate the tube over a burner to effect a sinter. Cool tube, and add 4 ml metal-free water. Put a stoppered Stevens extractor into the culture tube and place the tube in a boiling-water bath for 5 min. Remove the tube from the water bath and cool.

Pipet a 1 ml filtered aliquot into a 16 by 150 mm culture tube marked at 5 ml. Add 1 drop phenolpthalein solution to the aliquot, and then add 1 *M* hydrochloric acid, drop by drop, until the pink color disappears. Add 0.5 ml potassium nitrate solution and 0.2 g powdered sodium tartrate and shake to dissolve. Bring the volume of the solution to 5 ml with water. Add the following reagents and shake the tube after each addition:

- 0.5 ml concentrated hydrochloric acid
- 0.3 ml potassium thiocyanate
- 0.5 ml stannous chloride solution

Allow the tube to stand 1 min and then add 0.5 ml isopropyl ether; shake the tube vigorously for 10 s. Compare the amber color of the organic layer with those obtained from standard solutions.

Organic-Rich Soils

Stanton *et al.* (5) studied two methods of decomposing organic-rich soils for analysis of molybdenum. A fusion in potassium pyrosulfate yielded on average low results. An attack by nitric and perchloric acids followed by evaporation of the acid and then dissolution of the residue in 6 *M* hydrochloric acid yielded good results.

G. Determination of Molybdenum in Plants (1)

Comments on the Method

Lithium nitrate is used as a flux to complete the oxidation of organic matter after a pre-dry ashing over a stove. The fusion is then leached with 1 *M* hydrochloric acid. An amber-colored complex, formed by the addition of stannous chloride (as a reductant) and potassium thiocyanate, is extracted into isopropyl ether and used in the colorimetric determination of molybdenum. Potential interferences, mentioned in the preceding molybdenum procedure, are generally not present in sufficient quantities to cause problems. The method can be used to detect molybdenum down to the 250 ng level.

Reagents

- Ammonium hydroxide solution, 1 *M:* Dilute 6.6 ml concentrated ammonium hydroxide to 100 ml with metal-free water.
- Hydrochloric acid, concentrated, reagent grade.
- Hydrochloric acid, 1 *M:* Dilute 89.3 ml concentrated hydrochloric acid to 100 ml with metal-free water.

- Isopropyl ether, practical grade, free of peroxides: If the ether is not peroxide-free, shake it with one-tenth its volume of a solution containing equal volumes of stannous chloride and potassium thiocyanate.
- Lithium nitrate ($LiNO_3$), reagent grade.
- Phenolphthalein solution, 1.0%: Dissolve 1 g phenolphthalein in 100 ml alcohol.
- Potassium nitrate solution (KNO_3), 10%: Dissolve 10 g of the salt in 100 ml water.
- Potassium thiocyanate solution (KSCN), 5%: Dissolve 5 g of the salt in 100 ml water.
- Standard molybdenum solution, 100 μg/ml: Dissolve 0.075 g pure molybdic acid anhydride in 3 ml of 1 *M* sodium hydroxide solution; dilute with water, make slightly acid with 0.1 *M* hydrochloric acid, and make up to 500 ml with water.
- Standard molybdenum solution, 2 μg/ml: Dilute 2 ml standard solution (100 μg/ml) to 100 ml with water. Prepare fresh daily.
- Stannous chloride solution, 10%: Dissolve 10 g $SnCl_2 \cdot 2H_2O$ in 17 ml concentrated hydrochloric acid and dilute to 100 ml with water.

Equipment

- Microbalance, torsion type, capacity 75 mg. A scoop, previously calibrated by weighing a measured amount of ash, can be used under field conditions in place of the balance.
- Dishes, evaporating, 60 ml platinum.
- Scoop, 0.25 cm^3; for measuring lithium nitrate.
- Tongs, platinum-tipped.
- Pipets: 1 ml graduated in 0.1 ml, 1 ml graduated in 0.01 ml, 2 ml graduated in 0.5 ml, 5 ml graduated in 0.1 ml.
- Tubes, culture, 16 by 150 mm; calibrated at 5 ml mark.

Procedure

Ash the fresh plant material by heating in a platinum dish over the stove. Mix the ash. Fuse 25 mg ash with 1 scoop of lithium nitrate in a calibrated culture tube. Dissolve the whitish melt in 1 ml of 1 *M* hydrochloric acid; add 1 drop indicator and 1 *M* ammonium hydroxide dropwise to faint pink. Add water to 5 ml mark.

To a series of 11 culture tubes add, respectively, the following amounts of molybdenum standard (2 μg/ml): None to first tube, 0.5 μg to second tube, 1.0 μg to third tube, and continue, increasing the amount in each tube by 0.5 ml increments; the last tube should contain 5 μg of molybdenum. Add water to the 5 ml mark. To both samples and standards add 0.6 ml concentrated hydrochloric acid, 0.5 ml potassium thiocyanate, and 1 ml stannous chloride. Shake the tube after each addition. To all samples and standards, add 1 ml isopropyl ether, stopper the tubes, and shake each for 15 s. Within 30 min compare the amber-colored ether layers over the sample solutions with those obtained over standard solutions.

H. Determination of Mercury in Soils and Rocks (1)

Comments on the Method

A mercury iodide sublimate is separated from the sample by heating with ammonium iodide. This sublimate is dissolved in an acetate buffer solution. A distinctive colored solution is obtained for mercury by the addition of dithizone in the presence of EDTA and thiocyanate. The last two reagents complex potential interfering elements, namely bismuth, copper, gold, silver, and tin. The lowest detectable amount of mercury is 2.5 μg/g in the sample.

Reagents

- Ammonia (NH_4OH), 0.4 *M*, metal-free.
- Ammonium iodide (NH_4I), reagent grade.
- Ammonium thiocyanate (NH_4SCN), reagent grade.
- Buffer, sodium acetate–acetic acid solution: Generally, the buffer is prepared by mixing 10 ml of 2 *M* acetic acid with 5 ml of 2 *M* sodium acetate; however, as different batches of sodium acetate may have either a greater amount of alkali than the stoichiometric requirement or a variable amount of water, the volume ratio of 2 *M* acetic acid and aqueous sodium acetate must be adjusted to produce the desired pH. The correct volume ratio can be determined by titrating a given volume of 2 *M* acetic acid with the aqueous sodium acetate until the pH of the mixture is 4.
- Complexing agent: A mixture of 5 g hydrazine sulfate and 10 g EDTA.
- Dithizone, 0.01%: Dissolve 0.01 g purified reagent in 100 ml reagent grade chloroform ($CHCl_3$).
- Dithizone, 0.0015%: Mix 15 ml dithizone solution (0.01%) in chloroform with 85 ml Soltrol.
- Ethylenediaminetetraacetic acid, disodium salt (EDTA).
- Mercury standard A, 1000 μg/ml: Dissolve 0.1354 g reagent grade mercuric chloride ($HgCl_2$) in 100 ml 0.5 *M* sulfuric acid.
- Mercury standard B, 10 μg/ml: Dilute 1 ml standard A to 100 ml with 0.5 *M* sulfuric acid. This standard is stable for at least 3 months at ordinary temperatures.
- Soltrol: A high-boiling nonvolatile solvent composed mostly of isoparaffinic hydrocarbons and marketed by Phillips Petroleum Co. In most areas this solvent can be used as shipped; however, if desired, solvent can be freed of trace metals by shaking as much as 250 ml once with 10 ml of 0.5 *M* sulfuric acid and then

removing the acid by shaking successively three times with 50 ml portions of metal-free water.

Preparation of Calibrating Solutions

To each of four 60 ml separatory funnels add 10 ml acetate buffer, 0.1 g EDTA, and the appropriate amount of standard B—none to first funnel, 1 μg mercury to second, 2 μg to third, 4 μg to fourth. Proceed as below.

Equipment

- Culture tubes, 18 by 150 mm
- Scoops: 0.5 g and 0.1 g
- Pipets, automatic, 5 ml plunger type
- Separatory funnels, 60 ml, Squibb type

Procedure

Mix 2 small scoops (0.2 g total) sample and 1 small scoop NH_4I in a culture tube and heat the mixture at low heat for 1 min. Cool. Invert tube and dump the loose solid material. Holding the tube upright, add 10 ml buffer, 1 large scoop of complexing agent, and 1 small scoop of ammonium thiocyanate and dissolve. Transfer the contents of the tube to a separatory funnel and add 1 ml dithizone solution (0.0015%). Shake funnel 1 min. Drain and discard aqueous phase. Add 5 ml water and shake funnel 10 s. Drain and discard aqueous phase. Add 5 ml 0.4 *M* metal-free ammonia and shake funnel 5 s. Drain and discard aqueous phase. Compare the colors of the organic layers with the colors of standards.

I. Determination of Tin in Geological Materials (4)

Comment on the Method

The sample is decomposed by heating with ammonium iodide. Cassiterite is converted to stannic iodide. The reaction of tin with gallein takes place in a buffered solution with a pH of 2.0 to 2.5. Gallein also reacts with antimony, molybdenum, titanium, and tungsten, but only the antimony complex has a color similar to that of tin. The presence of iodide prevents gallein from reacting with copper, iron, lead, and manganese. The standard series ranges from green at zero through gray to purple and pink.

The concentration range is 0.5–100 ppm.

Reagents

- Ammonium iodide, fine crystals.
- Hydrochloric acid, concentrated.
- Hydrochloric acid, 1 *M:* Mix 40 ml of the concentrated acid with 400 ml of water.

- Sodium hydroxide, pellets.
- Buffer solution: Dissolve 26 g of sodium hydroxide in 400 ml of water, and when this solution is cold mix it with a cold solution of 106 g of chloroacetic acid and 20 g of hydroxylamine hydrochloride in 400 ml of water. Dilute to 1 liter with water and mix. This solution should have a pH value of 2.55 ± 0.1.
- Ethyl alcohol, absolute.
- Stock reagent solution: Dissolve 0.1 g of gallein in 100 ml of ethyl alcohol by warming gently, then filter through a Whatman No. 1 paper (11 cm diameter). Dissolve 0.03 g of methylene blue in 200 ml of water, warming gently. Combine these two solutions in equal proportions.
- Dilute reagent solution: Mix 10 ml of the stock reagent solution with 400 ml of buffer solution and 90 ml of *M* hydrochloric acid.
- Gelatin solution: Dissolve 0.5 g of gelatin in 100 ml of water by heating gently.
- Standard tin solutions.
  - 100 μg tin/ml: dissolve 50 mg of tin powder in 50 ml of concentrated hydrochloric acid and dilute to 500 ml with water.
  - 5 μg tin/ml: dilute 5 ml of the 100 μg/ml solution to 100 ml with *M* hydrochloric acid.

Procedure

Weigh 1 g of sample into a test tube (18 by 180 mm). Mix thoroughly with 1 g of ammonium iodide. Heat, with frequent rotation and agitation of the tube, until all ammonium iodide has sublimed. Allow to cool and add 5 ml of 1 *M* hydrochloric acid. Leach on a sand tray or in a boiling-water bath, and allow to settle.

Pipet a 1 ml aliquot of the supernatant solution into a test tube (18 by 180 mm) calibrated at 5, 10, 15, and 20 ml. Dilute to 5 ml with buffer solution. Warm gently to reduce iodine and leave to cool. Add 0.1 ml of stock reagent solution.

Mix, and allow to stand for 10 min. Compare with a standard series.

If the test solution is more pink than the highest standard, add successive 5 ml volumes of dilute reagent solution until the color is gray, mix well, and leave for 10 min before comparing with the standard series. If the test solution is still more pink than the highest standard, use an aliquot of 0.1 ml and also add 0.5 ml of 1 *M* hydrochloric acid.

$$\text{Tin (ppm)} = \frac{\mu\text{g of matching standard} \times \text{ml of final solution}}{\text{ml of aliquot}}$$

Standards

To 11 test tubes (18 by 180 mm) calibrated at 5 ml, add respectively 0, 0.25, 0.5, 1.0, 1.5, 2.0, 2.5, 3.0, 3.5, 4.0, and 5.0 μg of tin using the standard solution containing 5 μg tin/ml. To each of the first six tubes add 0.5 ml of hydrochloric acid.

Add 0.2 ml of gelatin solution to all standards. Dilute to 5 ml with buffer solution. Add 0.1 ml of stock reagent solution. Mix, and allow to stand for 10 min before using.

NOTE: Samples must not be left in contact with residual solid matter since tin can be lost from solution by adsorption or precipitation. If the samples are retained overnight after the tin has been dissolved, then they must be filtered.

J. Determination of Titanium in Geological Materials (4)

Comment on the Method

The sample is decomposed with hydrofluoric acid, and the titanium is fused with potassium bisulfate. The color with Tiron is formed in the pH range 4.3–9.6. The interference of iron is overcome by reduction of ferric ion with dithionite. However, chromium, molybdenum, tungsten, and vanadium interfere.

The concentration range is 100 ppm to 20%.

Reagents

- Hydrofluoric acid.
- Potassium bisulfate.
- Sulfuric acid.
- Sulfuric acid, 1.7 *M:* Cautiously add 100 ml of the concentrated acid to 1 liter of water and mix well.
- Tiron solution: Dissolve 4 g of the solid reagent in 100 ml of water.
- Buffer solution: Dissolve 40 g of ammonium acetate in water, add 15 ml of acetic acid (glacial), and dilute to 1 liter with water. The pH value of this solution should be 4.5.
- Sodium dithionite.
- Standard titanium solutions:
  - 500 μg titanium/ml: Weigh 3.68 g of potassium titanium oxalate (dihydrate) and 8 g of ammonium sulfate into a Kjeldahl flask, add 100 ml of concentrated sulfuric acid, and gradually heat to boiling. Boil for 10 min, then cool and carefully pour the solution into about 700 ml of water. Cool, and dilute to 1 liter with water.
  - 10 μg titanium/ml: Dissolve 0.8 g of ammonium sulfate in about 20 ml of water, add 10 ml of concentrated sulfuric acid, and cool. Add 2 ml of the 500 μg/ml solution and dilute to 100 ml with water.

Procedure

Weigh 0.1 g of sample into a platinum dish (50 ml). Moisten the sample with water and add 10 ml of hydrofluoric acid. Stir the contents of the dish with a platinum stirring rod. Evaporate to dryness on a steam bath. Add 1 g of potassium bisulfate and fuse. When cool, add 20 ml of 1.7 *M* sulfuric acid and warm to dissolve the residue. Transfer the cold sample solution to a volumetric flask (100 ml). Rinse the dish with two 10 ml volumes of 1.7 *M* sulfuric acid and add them to the flask. Dilute to 100 ml with 1.7 *M* sulfuric acid and mix well.

Transfer a 2 ml aliquot to a graduated cylinder (25 ml). Add 5 ml of Tiron solution. Add concentrated ammonia solution dropwise until the color changes from blue–green to violet.

Add 15 ml of buffer solution. Dilute to 25 ml with water and mix. Add 50 mg of sodium dithionite, stopper the cylinder, and shake it to dissolve this salt.

After 5 min, but within 15 min, compare with a standard series.

If the color of the test solution is more intense than the color of the highest standard, repeat from paragraph two with a smaller sample aliquot.

Standards

Transfer suitable aliquots of standard titanium solution (10 μg titanium/ml) containing 1–60 μg of titanium to graduated cylinders (25 ml) using as a reagent blank one additional cylinder to which no titanium has been added. Add 5 ml of Tiron solution. Add concentrated ammonia solution dropwise until the pH value is within the range 5–7. Proceed as described in paragraph three of the Procedure.

K. Determination of Copper, Cobalt, and Nickel in Soils by Paper Chromatography (1)

Comments on the Method

The sample is fused in potassium pyrosulfate. Then the sample is leached with a solution of nitric acid and hydrochloric acid. A volume of 0.01 ml of sample is transferred to a filter paper. A solution of methyl ethyl ketone, hydrochloric acid, and water is used to develop a paper chromatogram of the above mixed metal solution. The chlorides of copper, nickel, and cobalt are thus separated. The procedure for drying the 0.01 ml sample spot on the paper prior to its insertion into the solvent is crucial in obtaining the correct hydrochloric acid and water content. No interferences are mentioned.

Reagents

- Potassium pyrosulfate ($K_2S_2O_7$), reagent grade, powdered.
- Hydrochloric acid, concentrated, reagent grade.
- Nitric acid, concentrated, reagent grade.
- Nitric acid–hydrochloric acid solution: Dilute 25 ml concentrated nitric acid with about 100 ml metal-free water, add 250 ml concentrated hydrochloric acid, and dilute with solution to 500 ml with water.
- Ammonium hydroxide, reagent grade.
- Methyl ethyl ketone (2-butanone), dry, reagent grade.

- Rubeanic acid solution, 0.1%: Dissolve 0.1 g rubeanic acid (dithiooxamide) in 60 ml ethyl alcohol and dilute to 100 ml with water.
- Standard copper, cobalt, nickel (500 μg each element per milliliter): Dissolve 0.4050 g nickel chloride hexahydrate ($NiCl_2 \cdot 6H_2O$), 0.4037 g cobaltous chloride hexahydrate ($CoCl_2 \cdot 6H_2O$), and 0.3930 g cupric sulfate pentahydrate ($CuSO_4 \cdot 5H_2O$) in 200 ml nitric acid–hydrochloric acid solution.
- Solvent mixture: Transfer to a 25 ml graduated cylinder 15 ml methyl ethyl ketone, 3 ml concentrated hydrochloric acid, and 2 ml water. Mix and transfer to a 600 ml beaker and cover with a watch glass. Prepare 15 min before required because the vapor in the beaker must be in equilibrium with the solvent mixture to obtain good separation of the metals in the chromatogram.

Preparation of Calibrating Chromatogram

Transfer aliquots of 0.1, 0.2, 0.4, 0.8, 1.6, 3.0, and 6.0 ml, respectively, of the standard copper–cobalt–nickel solution to seven 10 ml volumetric flasks, one aliquot to each flask; add 2.5 g of potassium pyrosulfate to each flask, dilute to volume with the acid mixture, warm to dissolve the solids, and shake each to mix the contents. Place 0.01 ml aliquots of these standards on the successive strips of the Whatman CRL/1 paper one aliquot to a strip, and follow procedure below as for samples. The numbers of micrograms of each metal on the seven strips are 0.05, 0.1, 0.2, 0.4, 0.8, 1.5, and 3.0, respectively, and correspond to 50, 100, 200, 400, 800, 1500, and 3000 ppm of the metal in the sample.

Equipment

- Filter paper, Whatman CRL/1: This paper comes in specially designed rectangular sheets (21.3 by 11 cm) with 11 slots cut in the paper parallel to the short side, leaving 12 strips 1.5 cm wide joined at the top and bottom.
- Test tubes, 18 by 150 mm.
- Beakers, glass, 600 ml and 50 ml.
- Watch glasses or Petri dishes, 125 mm diameter.
- Pipet, measuring 10 ml and calibrated in 0.1 ml.
- Pipet, micro, 0.01 ml.
- Atomizer, all-glass, 50 ml.
- Cylinder, graduated, 25 ml.

Procedure

Transfer to an 18 by 150 mm test tube 0.2 g sample, add 0.5 g potassium pyrosulfate, mix, and heat until a quiescent melt is obtained. Allow the melt to cool, add 2.0 ml of the nitric acid–hydrochloric acid solution, and heat for 10 min in a boiling-water bath that is shallow enough to leave the upper end of the test tube cool. Remove the tube from the bath and allow to cool. Prolonged standing or centrifuging may be required to ensure a clear supernatant solution above the potassium sulfate that crystallizes.

Prepare the solvent mixture as described above under Reagents, transfer to a 600 ml beaker, and cover with a watch glass.

With a micropipet, transfer 0.01 ml of the clear sample solution to a location near the end of one of the strips of a sheet of Whatman CRL/1 paper. Aliquots may be similarly placed on the remaining strips of each sheet of paper except on the outer strips, which cannot be used. Form a cylinder from the sheet of filter paper, and fasten the upper corners with a paper clip. Place this cylinder, with sample spots lowermost, into a dry 600 ml beaker floating in a bath of boiling water. After 3 min has elapsed, transfer the paper cylinder to the beaker containing the solvent and cover with a watch glass. The solvent will diffuse slowly upward through the paper; 30 min or more is required for it to reach the top of the strips. When the solvent has almost reached the top of the strips, remove the paper cylinder and allow it to dry in air for about 5 min. Place the cylinder in a 600 ml beaker containing a 50 ml beaker of ammonium hydroxide. Cover and leave 2 or 3 min. Remove the paper clip, spread the chromatographic sheet on a clean paper blotter, and spray both sides with the rubeanic acid solution. Again form the sheet into a cylinder, and place it in a 600 ml beaker containing a 50 ml beaker of ammonium hydroxide. Cover and leave for 2 or 3 min. Compare the colored bands with appropriate standards, and multiply by 1000 the micrograms of metal found in the sample to obtain parts per million.

L. Determination of Uranium in Soils and Rocks by Paper Chromatography (1)

Comments on the Method

Uranium is leached from soils and rocks using a nitric acid–aluminum nitrate solution. Excessive amounts of organic matter must be removed by ashing. Uranium is separated from potential interferences by paper chromatography using an ethyl acetate–nitric acid–water solution as the solvent. A characteristic brown color for uranium is obtained by reacting the uranium in the spot on the chromatogram with potassium ferrocyanide. The intensity and width of the color band are used to estimate the uranium content. This procedure can be used for uranium in amounts of 4 to 1200 μg/g in the sample.

Reagents

- Nitric acid, concentrated.
- Nitric acid–aluminum nitrate solution: Dissolve 60 g aluminum nitrate $[Al(NO_3)_3 \cdot 9H_2O]$ in 100 ml 4 *M* nitric acid.
- Standard uranium solution, 1000 μg/ml: Dissolve 0.211 g uranyl nitrate hexahydrate $[UO_2(NO_3)_2 \cdot 6H_2O]$ in 100 ml nitric acid–aluminum nitrate solution.
- Ethyl acetate.

- Solvent mixture: Place 30 ml ethyl acetate, 3.5 ml concentrated nitric acid, and 10 ml water in a 600 ml beaker, cover the beaker with a watch glass, and let stand for 20 min before using.
- Aluminum nitrate, crystalline.
- Hydrofluoric acid, 48%.
- Potassium ferrocyanide solution [$K_4Fe(CN)_6 \cdot 3H_2O$], 5%: Dissolve 5 g $K_4Fe(CN)_6 \cdot 3H_2O$ in 100 ml water.
- Magnesium nitrate [$Mg(NO_3)_2 \cdot 6H_2O$], crystalline.

Preparation of Calibrating Solutions

Prepare dilute uranium standard solutions by adding 0.1, 0.2, 0.4, 0.8, 1.5, 3, and 6 ml, respectively, of standard uranium solution (1000 μg/ml) to a series of 10 ml volumetric flasks—0.1 ml to the first flask, 0.2 ml to the second flask, and relative amounts to the remaining flasks—and making the volume up to a mark with HNO_3–$Al(NO_3)_3$ reagent. Place 0.01 ml aliquots from the dilute standard uranium solutions on chromatographic paper, one aliquot from one standard solution to each strip, and allow to dry ½ h in a desiccator. Place the paper in solvent and continue as with samples. These standard bands contain, respectively, 0.1, 0.2, 0.4, 0.8, 1.5, 3.0, and 6.0 μg of uranium.

Equipment

- Chromatographic paper, Whatman CRL/1
- Ignition tubes, 16 by 125 mm
- Cylinders, graduated, 50 and 10 ml
- Beakers, 600 ml
- Dishes: platinum and porcelain, both 50 ml
- Stirring rod, plastic
- Watch glasses, 125 mm diameter
- Bottle or atomizer for indicator spray
- Scoop, 1 g
- Flasks, volumetric, 250 and 10 ml
- Pipets: serological, 0.1 and 1 ml graduated; micro, 0.01 and 0.05 ml; volumetric, 1 and 2 ml
- Desiccator, inside diameter 150 mm

Procedure

Weigh or scoop 1 g sample and transfer it to a 16 by 125 mm ignition tube. For samples containing more than 1200 ppm uranium, 0.1 g can be used. Add 2 ml HNO_3–$Al(NO_3)_3$ solution and digest the sample for 1 h in a boiling-water bath.

Either centrifuge the sample or allow sufficient time for the insoluble residue to settle.

With a micropipet withdraw an appropriate aliquot (0.01 or 0.05 ml) of the supernatant sample solution; starting about 1.5 cm from the bottom of one of the strips on a sheet of Whatman CRL/1 chromatographic paper, spread the aliquot evenly across the strip. Follow the same procedure to place 10 sample aliquots on each sheet of paper, using the same size aliquot on all the strips of any one sheet. Fasten the top corners of the paper together with a paper clip to form a cylinder, place upright in a desiccator containing saturated $Mg(NO_3)_2$ solution, and leave for at least 30 min. After the aliquots have dried, remove the paper cylinder from the desiccator and place it upright (with the aliquots at the bottom) in the 600 ml beaker containing the solvent mixture. Leave the paper cylinder in the beaker until the solvent front has risen at least 2 cm above the area wetted by the sample aliquot, but not above the top of the slit. Remove the paper from the beaker, unfold, allow the ethyl acetate (solvent mixture) to evaporate, and spray the chromatogram with $K_4Fe(CN)_6$ reagent. Compare the brown ferrocyanide complex of uranium with a standard uranium chromatogram.

VIII. ATOMIC ABSORPTION SPECTROMETER FOR FIELD WORK

The atomic absorption instruments used for many of the methods in this book are large instruments designed for permanent installation in a laboratory where electrical services, cooling water, and cylinders of compressed gases are available. Because of their weight and complexity, they are not easily transported to and set up in remote locations. However, atomic absorption instruments can provide accurate data for many elements, and the availability of a portable unit for field work is desirable.

Castledine and Robbins have described an instrument for use in the field (6). It is manufactured by Scintrex Ltd., Concord, Ontario, Canada, as model AAZ-2. The electrothermal atomizer is a miniature tungsten filament. Tungsten was chosen because it requires less power than graphite tubes. Correction for background is a major concern in electrothermal atomization. In this instrument, the correction is accomplished by means of an electrically modulated magnetic field (Zeeman effect). The instrument weighs only 26 kg and has dimensions of 9 × 36 × 49 cm. It requires electrical power at 110/220 V ± 20%, 50/60 Hz, and an inert gas at low pressure. Argon is recommended. The consumptions are 500 W power and 0.5 liter/min gas.

The results of analyses of three standards, AN-G, BE-N, and MA-N, for five elements are shown in Table 5.2 (7). Good agreement with accepted values except for cobalt was obtained. The reason for low results for cobalt may be incomplete digestions of the rock.

Table 5.2
Comparative Results for the Geostandards ($\mu g/g$)

	AN-G			BE-N			MA-N		
		Geostandards[a]			Geostandards			Geostandards	
Element	This work	Range of values	Recommended value	This work	Range of values	Recommended value	This work	Range of values	Recommended value
Cadmium	0.02	0.031–0.18	—	0.09	0.089–0.21	—	2.0	1.8–2.2	2
Cobalt	8.2	10–56	25	34	43–75	61	0.1	0.35–38	1
Copper	25.0	8–27	19	79	43–109	72	97	76–187	140
Lead	2.0	0.62–13	2	4.2	3–17	4	15	10–86	29
Nickel	18.0	9–70	35	250	160–327	267	2.5	2–92	3

Source: Ref. 6.
[a]Geostandards values from Ref. 7.

The detection limits for a number of elements are shown in Table 5.3.

Directions for the determinations of a number of elements in geological samples are available in the applications manual for this instrument. A procedure for the determination of copper, lead, silver, nickel, cobalt, and zinc is included here. Samples are decomposed by means of a flux which melts at a low temperature (about 550°C). The flux is said to be more convenient to use in the field because large quantities of strong acids are not required. The flux does not attack glass, and a fusion can be completed in about 4 min.

Table 5.3
Some Detection Limits with the AAZ-2

Element	Wavelength (nm)	Detection limit[a] (g)	Equivalent concentration[b] (ng/ml)
Ag	338.3	6×10^{-13}	0.06
Au	242.8	5×10^{-12}	0.5
Cd	228.8	5×10^{-13}	0.05
Co	240.7	1×10^{-11}	1.0
Cu	324.8	4×10^{-12}	0.4
Ni	232.0	4×10^{-11}	4
Pb	283.3	8×10^{-12}	0.8
Zn	213.9	2×10^{-12}	0.2
Zn	307.6	4×10^{-9}	400

Source: Ref. 6.
[a]Measured with Zeeman modulation.
[b]Concentration in a 10 μl sample.

A. Determination of Copper, Lead, Silver, Nickel, Cobalt, and Zinc in Geological Materials (8)

Comment on the Method

A sample is decomposed by fusion and then is dissolved in dilute acid. The metals are complexed by ammonium pyrrolidinecarbodithionate (APDC) and extracted into methyl isobutyl ketone (MIBK). The pH of the aqueous phase must be in the range 2–4 to ensure that nickel and cobalt are extracted. A mercury bromide matrix modifier prevents volatilization losses of metals during the ashing cycle. Results of analyses of eight Chinese geological reference samples (GSD-1 to GSD-8) were satisfactory.

Equipment

- Model AAZ-2 atomic absorption spectrometer (Scintrex).
- Glass test tubes (25 by 200 mm): The tubes are closed with Teflon-lined screw caps so that fusions and extractions can be carried out in the same containers.
- Propane burner, or a heated aluminum (or steel) block with holes bored to fit the test tubes.

Reagents

- Scintrex Geoflux.
- Chelate solution: Dissolve 0.5 g of APDC in 10 ml of water (distilled, deionized). Purify the solution by extraction with 10 ml of MIBK.
- Extraction buffer: Dissolve 5.0 ml of acetic acid in 200 ml of water (distilled, deionized). Use sulfuric acid and/or sodium hydroxide to adjust the pH to the range 2–4. (Measure the pH with a meter or pH paper.) Transfer the solution to a separatory funnel. Add 1 ml of APDC and 30 ml of MIBK. Extract for 1 min. Collect the aqueous phase and discard the organic phase. Then extract the aqueous phase with 30 ml of MIBK for 1 min. Dilute the aqueous phase to 1 liter and adjust the pH again if necessary.
- Matrix modifier: This is a solution containing mercury at 1 mg/ml. Dissolve 0.42 g of $Hg(NO_3)_2 \cdot \frac{1}{2}H_2O$ in 250 ml of 5% (v/v) nitric acid. Transfer 20 ml of the solution to a separatory funnel. Add 0.4 ml of 0.1 *M* hydrobromic acid and 10 ml of MIBK. Extract for 2 min. Collect the organic phase, and use it as the matrix modifier.
- Ascorbic acid solution (in water), 20% (w/v).
- Ammonium thiocyanate solution (in water), 6 *M*.
- Element standard solutions: Prepare these by dilution with 0.1 *M* nitric acid from concentrated standards (see Chapter 3). The working ranges of the standards are shown in Table 5.4. Prepare each working standard by adding the appropriate volumes of the intermediate standards to 10 ml of the buffer solution.

Table 5.4
Standards for Calibration

	Element					
	Cu	Pb	Ag	Ni	Co	Zn
Intermediate standard, μg/ml	4	10	1	10	10	40
Working standards, ng/ml	0	0	0	0	0	0
	4	10	2	10	10	2[a]
	8	20	4	20	20	4[a]
	12	30	6	30	30	6[a]

Source: Ref. 8.
[a]Units for zinc in micrograms milliliter.

Instrument Parameters

AAZ-2 conditions are listed in Table 5.5. Hollow-cathode lamps are used for the elements, and the lamp currents are two-thirds of lamp manufacturers' rated maxima. The gas is 8% hydrogen in argon for all elements except zinc. Argon is recommended for zinc. The drying control is adjusted for MIBK solutions.

Procedure

Dry and crush the sample to at least 100 mesh. If the sample contains any sulfides, ignite it over a Meker burner. Mix 3 g of flux with 0.5 g of sample in a glass test tube. Heat the tube until the flux melts. Provide adequate ventilation for the evolved gas. When the evolution of gas has ceased, continue to heat the tube for about 2 min. Cool the sample, and add 25 ml of 1 *M* sulfuric acid. Centrifuge the sample. Transfer an aliquot of the clear liquid to a clean test tube. (For the analysis of zinc, follow the procedure below.) Dilute to 10 ml with extraction buffer. Verify

Table 5.5
Instrument Parameters (AAZ-2)

	Element					
Parameter	Cu	Pb	Ag	Ni	Co	Zn
Wavelength, nm	327.4	283.3	338.3	232.0	240.7	307.6
Ashing conditions						
Ramp rate	50	40	40	45	45	50
Temperature	6	2.5	5	5.5	6	3
Time	30	20	15	15	15	15
Atomizer condition: temperature	17	14	15.5	22	21	11.5
Gas flow	1.5	1.5	1.5	1.5	1.5	1.5
Zeeman modulation	High	High	High	High	High	High
Auto gain	On	On	On	On	On	On
Sample volume, μl	10	10	10	10	10	10

Source: Ref. 8.

that the pH is in the range 2–4. Add 10 ml of MIBK and 200 μl of APDC (chelate solution). Extract for 2 min.

Analyze the organic layer. Place 10 μl of matrix modifier and 10 μl of the solution onto the tungsten filament. Use the instrument conditions shown in Table 5.5. To each standard, add 10 ml of MIBK and 200 μl of APDC solution. Extract and analyze the standards in the same manner as the samples.

If analyses for nickel and cobalt are not required, 1 *M* sulfuric acid can be used in place of the buffer solution.

Determination of zinc

Use the sulfuric acid solution prepared above. Transfer an 8 ml aliquot of the clear solution after centrifuging to a clean test tube. Add 4 ml of 0.1 *M* nitric acid, 0.5 ml of 20% (w/v) ascorbic acid, and 100 μl of 6 *M* ammonium thiocyanate. Mix the contents. Add 2 ml of MIBK and extract for 2 min. Analyze the organic layer for zinc, as above, but do not use matrix modifier. Calibrate for zinc by carrying the working standards through the extraction procedure.

IX. PORTABLE MERCURY SPECTROMETER

Measurement of mercury vapor in the field is important because mercury is said to be a pathfinder element for deposits of gold, silver, antimony, and massive sulfides. Instruments for the detection and quantitative determination of mercury in air samples from test holes, or in vapor evolved from samples which have been heated, depend on the attenuation of the mercury emission line at 254 mm. However, provision must be made for correction for interferences from other substances in the air which absorb strongly in the ultraviolet region of the spectrum, as well as for the stability of the mercury lamp. Scintrex Ltd. manufactures a portable instrument designated as model HGG-3, which will be described briefly.

The source of the mercury radiation is a discharge tube which is mounted between the poles of an electromagnet. When the magnetic field is switched on and off rapidly, the interferences, i.e., background correction, can be accomplished by means of the Zeeman effect as described in Chapter 2. Radiation from the lamp passes through the sample cell onto the detector. When the magnetic field is off, the detector signal is a measure of the absorption of radiation by mercury and interfering substances. When the magnetic field is on, the signal corresponds to the absorption of radiation by interfering substances. In addition, a fraction of the radiation is split off prior to passing through the cell and is directed to a reference detector. In this manner, the long-term stability of lamp and the magnetic field can be monitored and the drifts eliminated. Power for the instrument is supplied by a battery pack of four 6 V batteries. The battery life under continuous operation is about 4 h, but intermittent operation extends the time. Provision must be made for recharging the batteries.

In operation, air is sampled from test holes by inserting a probe connected to the sample cell and drawing air through the cell by means of a hand pump. The peak meter reading is recorded. Mercury evolved from solid samples can be measured in the field if the samples are heated in closed test tubes. At the end of the heating period, the probe is connected to the test tube and the air is sampled as before. On heating, different naturally occurring mercury compounds release mercury vapor at different temperatures.

X. PORTABLE X-RAY EQUIPMENT FOR FIELD ANALYSES

Laboratory equipment is much too complex and bulky for work in the field. A number of instruments have been designed so that some of the advantages of x-ray methods can be retained in portable units. Some of the earlier models of commercial analyzers are lightweight (from 7 to 35 kg). The energy source is a radioactive isotope. The units contain filters and scintillation or proportional counters in place of goniometers. Only one element at a time is detected and measured, and the resolution of neighboring lines is poor (9).

In recently designed equipment, the objectives have been to improve both sensitivity and resolution and to provide for multielement analyses. In this respect, the equipment is better described as being of a mobile laboratory type. The units are mounted in trucks or cars. They require electrical generators and also a supply of liquid air or nitrogen.

Hoffmann *et al.* (10) described a mobile installation for multielement work. It included a cadmium-109 source, tin shielding, a two-position sample changer (for

Table 5.6
Detection Limits of Trace Elements in Rock, Soil, and Stream Sediment Samples for Mobile Tube-Excited ED–XRF and Radionuclide-Excited ED–XRF (ppm)

	Ni	Cu	Zn	Ga	As	Br	Rb	Sr
Tube ED–XRF								
35 kV	5	5	5	5	5	5	5	5
50 kV								
Radionuclide ED–XRF								
^{109}Cd	100	50	30	30	10	10	10	10
^{241}Am								

	Y	Zr	Nb	Pb	Th	Ba	La	Ce
Tube ED–XRF								
35 kV	5	5	5	5	10			
50 kV						70	70	70
Radionuclide ED–XRF								
^{109}Cd	10	10	10	20	50	100	100	100
^{241}Am						5	10	10

Source: Ref. 12

powdered samples), an Si(Li) detector, and electronic components. An alternating-current generator (gasoline powered) and a mobile liquid air machine were required.

A radionuclide-excited energy dispersive x-ray unit was used by Kramar and Puchelt (11) for multielement prospecting. Because of restrictions placed on the use of radioactive sources, a small x-ray tube has been substituted in a recent design by Kramar (12). A 1 mA/50 kV air-cooled pulsed rhodium tube was mounted together with an Si(Li) detector in the cabinet of a 20-sample changer. A 10 liter Dewar vessel contained sufficient liquid nitrogen to operate the detector for 6 days. Other components were a multichannel analyzer and a minicomputer. The total power requirement of 3.3 kW can be supplied by a 15 kVA generator which consumes 20 liters of gasoline per day. Several examples were given in which levels of niobium, nickel, copper, yttrium, and cerium in samples from three areas were measured. Limits of detection of a number of trace elements are listed in Table 5.6.

The papers cited should be consulted for details of the equipment.

REFERENCES

1. F. N. Ward, H. W. Larkin, F. C. Canney *et al.*, *Geol. Surv. Bull. (U.S.)* **1152** (1963).
2. J. H. McCarthy, *Congr. Geol. Int. Secc., 20th, 1956*, 363 (1959).
3. D. B. Hawkins, F. C. Canney, and F. N. Ward, *Econ. Geol.* **54,** 739 (1959).
4. R. E. Stanton, "Rapid Methods of Trace Analysis." Arnold, London, 1966.
5. R. E. Stanton, M. Mackler, and S. Newton, *J. Geochem. Explor.* **2,** 37 (1973).
6. C. G. Castledine and J. C. Robbins, *J. Geochem. Explor.* **19,** 689 (1983).
7. K. Govindaraju, *Geostand. Newsl.* **4,** 49 (1980).
8. J. D. Kinrade, H. Kostiok, and G. L. McCarthy, "Applications Manual for the Scintrex AAZ-2 Zeeman Modulated Atomic Absorption Spectrophotometer." Scintrex Ltd., Concord, Ontario, Canada, 1985.
9. T. K. Ball, S. J. Booth, E. F. P. Nickless, and R. T. Smith, *J. Geochem. Explor.* **11,** 277 (1979).
10. P. Hoffmann, K. H. Lieser, T. Hofmann, and R. Sommer, *X-Ray Spectrom.* **12,** 175 (1983).
11. U. Kramar and H. Puchelt, *J. Geochem. Explor.* **19,** 255 (1983).
12. U. Kramar, *J. Geochem. Explor.* **21,** 373 (1984).

6

Geochemical Methods for Soils and Related Samples

I. INTRODUCTION

Soils and sediment samples are used extensively in geochemical exploration. Both are relatively complex physical and chemical samples.

A. Soils

Soils consist of a mixture of mineral and organic matter contained in horizontal zones called horizons. The main components of soils are sand, silt, and clay. The physical and chemical properties of soils often vary considerably within these zones. A typical soil profile is shown in Fig. 6.1. As can be seen, the growth and subsequent decomposition of botanical species results in an organic-rich zone labeled A1. The organic content of soils then decreases quickly with depth.

The exchange capacity of soils is highest in the A1 layer, as is the free-silica content. Free Fe_2O_3 and alumina contents increase from A to B zones. Thus sand tends to be abundant in the A1 layer and to decrease with depth. The decrease in sand is accompanied by an increase in silt and clay in the lower layers.

Trace metal content of soils also varies with depth. Of great interest in exploration geochemistry is the fact that there often is a higher concentration of trace metals at the top of the B horizon. Thus this section of the soil profile is frequently sampled in geochemical surveys.

A variety of factors affect the nature of the soil in a particular area. They include source material, climate, biological activity, topography, and age. Soils can then be classified on the basis of these factors. Classifications are complex and beyond the scope of this book.

Soils are also divided into soil zones on a worldwide basis. These zones are

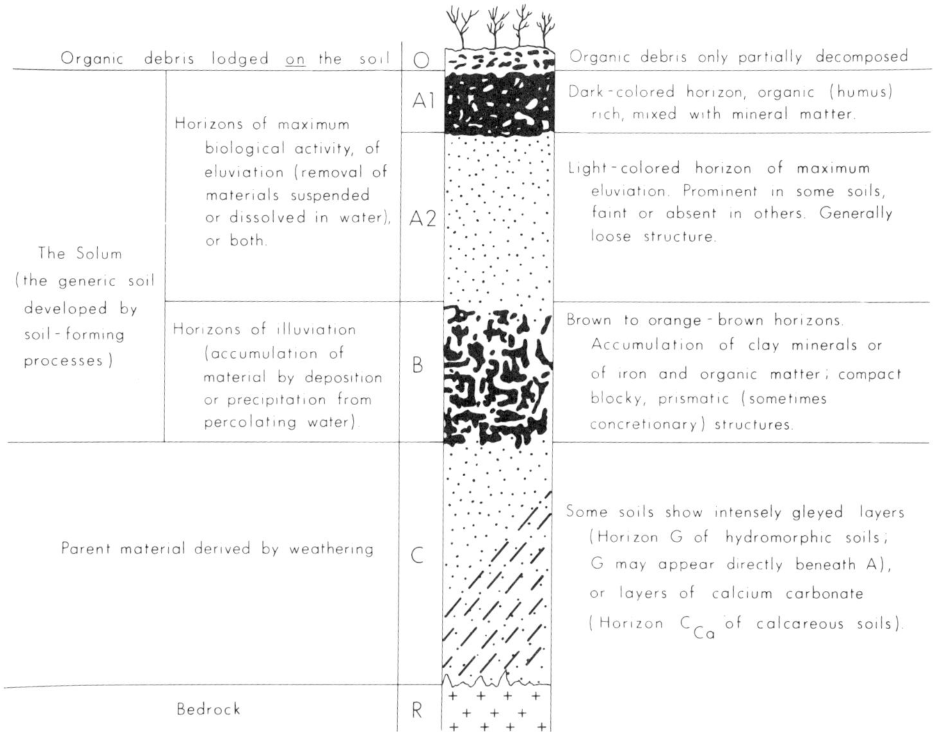

Figure 6.1. Hypothetical soil profile showing the principal horizons. (From A.W. Rose, H.E. Hawkes, and J.S. Webb, "Geochemistry in Mineral Exploration," 2nd ed. Academic Press, London, 1979.)

1. Arctic soils—dwarf shrub- and moss-covered soils of cold climates.
2. Podzolic soils—forest soils in humid temperate climates.
3. Grassland soils—soils in temperate climates with lower humidity.
4. Desert soils—sparsely vegetated soils found in temperate and tropical climates.
5. Tropical soils—forested and savanna-covered soils including dry to wet climates.
6. Mountain soils—usually stony soils containing characteristics of one or more of the above.

B. Sediments

A variety of sediments are sampled for exploration geochemistry. Commonly used are lake and stream sediments. The composition of the sediments depends on lake and stream morphology, biota types and abundances, drainage, glacial history of the area, and geographic and geological location. As with soils, the sediment is made up of mineral and organic matter.

C. Mineral Components

1. Hydrous Oxides

Manganese and iron, under oxidizing conditions, occur primarily as oxides, oxyhydroxides, and oxide hydrates. These exist in varying degrees of crystallinity. They accumulate in the sediment fraction after precipitation. During precipitation they scavenge trace metals, which then accumulate in the hydrous oxide fraction. The affinity of these hydrous oxides for trace elements depends on the identity of the metal. For example, copper is only weakly scavenged by manganese oxides, whereas zinc is strongly scavenged.

2. Clay Minerals

Clay mineral distribution is strongly dependent on the identity of the source materials. In the shield areas of Canada, the most common clays in sediment are illite, chlorite, and kaolinite. Clay minerals accumulate trace metals by adsorption and ion exchange.

D. Organic Components

The main and most important organic components of sediments are humic substances. These are of widely varying composition and are called humic and fulvic acids. Various mechanisms have been proposed for the accumulation of trace metals by humic substances. The most important are undoubtedly chelation, ion exchange, and surface adsorption.

E. Methods of Analysis

A wide variety of approaches have been proposed for the geochemical analysis of soils and sediments. These include total digestions (fusions and acids), strong acid attack, weak acid digestion, and the use of selective extractants. Which method is chosen depends on the samples, the elements being determined, and the information desired.

1. Total Digestions

Total digestions are now frequently employed prior to atomic absorption or ICP emission spectrometry. Frequently a wide range of elements (e.g., lead, copper, zinc, manganese, cadmium, cobalt, chromium) is determined. For this purpose acid mixtures including hydrofluoric acid are commonly employed. Recently it has been shown that if these decompositions are done in a closed Teflon vessel, they occur more effectively and more quickly. If organic matter content is high it is important that perchloric and nitric acids be added to the acid mixture.

Fusions, particularly using borate fluxes (e.g., lithium metaborate), have proved to be very effective for soil and sediment decomposition. It is important that the fusing agent be relatively low in the elements of interest. Fusing agents have, through the years, been serious sources of contamination. Blanks must therefore be run with each sample set.

Although it is not a total decomposition, the use of a slurried sample for quantitative analysis of clays has been reported (1). The slurry was aspirated directly into an ICP. The extremely high temperature of the ICP must result in a very high percent atomization of the sample.

2. Strong Acid Attack

Very commonly a mixture of nitric and hydrochloric acids is used to attack soil and sediment samples. A mixture of 1 part nitric and 3 parts hydrochloric acid is recommended. Such a method will remove metals which have been absorbed, ion-exchanged, coprecipitated with hydrous oxides, and chelated by some organic species. It will not attack the metals in most silicates, nor will it decompose completely many organic species. Large batches of samples can be treated in this manner.

3. Mild Acid Digestion

Many workers believe that a mild acid attack, often done near room temperature, is sufficient to remove a significant fraction of the more loosely bound metals. Results thus obtained have been shown to be adequate for geochemical exploration purposes.

Bradshaw *et al.* (2) discussed two very useful techniques in geochemical exploration: soil profile sampling and the use of different strengths of acids (and/or organochelating agents). In their study, they outlined dispersion mechanisms around ore bodies in residual soil—tropical, alpine glacial, and continental glaciated areas. They demonstrated that anomalies resulting from mineralized and nonmineralized rock can be distinguished in all of the three areas by using soil profile sampling and acid extractions based on different strengths of acids. Their conclusions apply only to metals which have a high percentage of hydromorphic dispersion, e.g., copper, lead, and zinc.

"The stronger chemical attacks, when used on sediments, tend to accentuate the effects of rock composition and consequently relatively reduce the effect of mineralization." Bradshaw *et al.* go on to point out that in some areas mineralization will not be easily seen when total attack is used on stream sediments because large changes in metal content of the country rock may mask mineral anomalies, which are relatively small by comparison. They recommend particularly the use of a weak extractant such as hot 0.5 *M* hydrochloric acid or cold EDTA. The latter was particularly effective in extracting only loosely bound metal from clays and other large surface area minerals. However, it is important to point out that in some areas of the world, metal is relatively tightly held in sediments even when hydromorphic movement is involved. In such cases stronger (but perhaps not "total") acid attack may be more useful.

Soil profile metal data can be particularly useful in distinguishing anomalies due to downslope seepage from those associated with mineralization. When underlying mineralization is present, an increase in metal content with soil depth is common. In the case of hydromorphically derived anomalies metal values decrease with soil depth.

4. Selective Extractants

There is some evidence that if selected fractions of soils and sediments are attacked, the data thus obtained will better highlight the anomaly. The present authors strongly advise caution in the interpretation of such data. The selective extractants used to date often attack more than the desired fraction. A more extensive discussion of this approach follows in the Procedure section.

5. Special Techniques

Several elements, e.g., arsenic, selenium, tellurium, and mercury, require that special precautions be observed during decomposition. In particular, these elements must be treated under oxidizing conditions at elevated temperature or losses will result. Oxidizing conditions are obtained using oxidizing acids and salts. An excess of the oxidizing agent must be ensured for the duration of the heating period. Fusions may also be applicable. In this regard, either sodium hydroxide or potassium hydroxide has commonly been used for arsenic, selenium, and tellurium determination.

II. DETERMINATION OF TOTAL METALS

A. Determination of Major and Trace Elements in Soils (3) by ICP–AES

Elements Determined

- Major—Si, Al, Fe, Ca, Mg, Na, K, Mn, Ti, P
- Trace—Be, Cd, Cr, Cu, Ni, Pb, Sr, Th, V, Y, Zn

Comment on the Method

Sample preparations for major and trace elements differ in that organic matter in the soils is destroyed prior to acid dissolution. The methods have been used for the analyses of 12 reference soil samples.

Equipment

ICP–AES direct reading instrument, Jarrell-Ash model 975 Plasma AtomComp.

Reagents and Apparatus

- Boric acid solution: Dissolve 44 g of boric acid in 1 liter of water.
- Teflon bombs.
- Teflon beakers, 100 ml.
- Muffle furnace.

Procedure

Major elements

Ash finely ground samples at 1050°C for 2 h in a muffle furnace. Determine the loss in weight. Weigh 0.1000 g samples into Teflon bombs. Add 2 ml of aqua regia and 3 ml of hydrofluoric acid to each. Heat the sealed bombs in a 140°C oven for 1 h. Cool the bombs. Add 50 ml of boric acid solution to each, and dilute to 100 ml with water. No particulate matter should remain. Aspirate these solutions for determinations of the major elements listed.

Trace elements

Weigh 1.000 g portions of the samples as received into 100 ml Teflon beakers. Moisten the samples with water, and then add 10 ml of nitric acid. Evaporate the samples to small volumes on a hot plate. Add 5 ml of nitric acid, 5 ml of perchloric acid, and 10 ml of hydrofluoric acid. Heat to fumes of perchloric acid, and continue heating for 30 min. Cool, and then add 10 ml of 1 : 1 hydrochloric acid. Boil the solutions for 10 min. Cool and dilute to 100 ml.

Apply background corrections for Cd, Mo, Ni, Pb, and Th. Interelement corrections are required because of spectral interferences with Cd and Pb.

Instrument operating parameters

ICP Operating Parameters and Analytical Lines, Jarrell-Ash Model 975 Plasma AtomComp[a]

Incident power	1.10 kW
Reflected power	<50 W
Entrance slit width	25 μm
Nebulizer type	Meinhard concentric
Nebulizer pressure	186 kPa
Argon flow rates	
Coolant	20 liter/min
Sample	0.5 liter/min
Plasma	0.5 liter/min
Sample aspiration rate	1.5 ml/min (pumped)
Observation height	18 mm

Element	Analytical line (nm)
Al	308.2
Ba	455.4
Ca	317.9
Cd	226.5
Cr	267.7
Cu	324.7
Fe	259.9
K	766.5
Mg	383.2
Mn	257.6

(continues)

Element	Analytical line (nm)
Mo	203.8
Na	589.0
Ni	231.6
P	214.9
Pb	283.3
Si	288.1
Sr	421.5
Th	401.9
Ti	334.9
V	292.4
Y	371.0
Zn	213.8

[a]Reproduced from Ref. 3, © 1980 by Williams & Wilkins.

B. Determination of Metals in Lake Sediments (4) by Flame Atomic Absorption

Elements Determined

Al, Ba, Be, Ca, Cd, Bi, Co, Cu, Fe, K, Li, Mg, Mn, Mo, Na, Ni, Pb, Sr, V, Zn

Comment on the Methods

The sample is decomposed by means of acids in a Teflon bomb or a Teflon dish. The bomb is necessary to ensure the quantitative retention of chromium in the acid digestion since chromyl chloride is volatile. When Teflon dishes are used, twice the volumes of acids are required and samples are fumed to dryness. Recoveries range from 91 to 101%, depending on the element.

Equipment

Flame atomic absorption instrument equipped with a Boling (three-slot) burner, a 3.1 cm burner (for air/acetylene flames), and a nitrous oxide/acetylene burner.

Reagents and Apparatus

- Teflon bombs or dishes
- Polypropylene bottles, 125 ml

Procedure

Dry the sample at room temperature. Crush the dried sample and then grind a representative 2 g sample to a powder which passes a 270 mesh sieve. Weigh a 0.100 g sample into a bomb. Add 4 ml of nitric acid, 1 ml of perchloric acid, and 6 ml of hydrochloric acid. Heat the sealed bomb to 140°C for 3.5 h. Cool and transfer the contents to a 125 ml polypropylene bottle which contains 4.8 g of boric acid in approximately 30 ml of water to dissolve the salts. Make up the volume to 100 ml, and use the solution for atomic absorption analyses. Prepare a blank solution in the same manner as the samples. All element standard solutions must contain 4% nitric acid, 1% perchloric acid, 6% hydrofluoric acid, and 4.8% boric acid.

Use the nitrous oxide/acetylene burner for Al, Ba, Be, Ca, Mg, Mo, Sr, and V. Use the Boling burner (air/acetylene flame) for the rest of the elements except for sodium and potassium. Use the 3.1 cm burner (air/acetylene flame) for sodium and potassium.

- Be, Cd, Cr, Co, Ti, Mg, Pb, Zn: Measure these elements by aspirating the sample solution, blank, and standards prepared above.
- Cu: Measure copper after making all solutions 2% (w/v) in ammonium chloride to overcome interference from iron.
- Cr, Mo: Make up all sample solutions and standards so that they contain 500 mg Al/liter (as $AlCl_3$).
- Al, Fe, Mn: Dilute the main sample solution by a factor of 10.
- Ba, Ca, Si: Make up all sample and standard solutions so that they contain 3000 mg Na/liter (as NaCl) to prevent ionization interference.
- K, Na: Make up all sample solutions and standards so that they contain 100 mg Cs/liter (as CsCl) to prevent ionization interference.

C. Determination of Metals in Marine Sediments (5) by ICP–AES

Elements Determined

- Major and minor—Al, Fe, Ca, Mg, Na, P
- Trace—Be, Co, Cu, Mn, Ni, Pb, V, Zn

Comment on the Method

Samples are dissolved by means of acids in Teflon bombs. The accuracy of the method has been demonstrated by analysis of two marine sediment reference materials. The method is not suitable for arsenic, cadmium, and molybdenum in these sediments because of inadequate sensitivity. Chromium and thallium are not completely dissolved.

Equipment

- ICP–echelle grating instrument for sequential and 20-channel multielement work
- DECLAB 11/03 data acquisition system
- Cross-flow-type nebulizer

Reagents and Apparatus

- Teflon bombs
- Teflon beakers, 100 ml
- Water bath

- Hot plate
- Stainless steel sieve, 120 mesh

Procedure

If the sediments are wet, remove water by freeze-drying. Break up any lumps, and sieve the material through a 120 mesh screen. Reject large particles, and blend the fine material. Weigh a 0.5 g sample into a Teflon bomb. Add 3 ml of nitric acid, 1 ml of perchloric acid, and 3 ml of hydrofluoric acid. Heat the sealed vessel in a boiling-water bath for 1–2 h. Cool the contents and transfer to a Teflon beaker. Evaporate the sample to dryness on a hot plate (low heat setting). Add 50 ml of 1 *M* hydrochloric acid to dissolve the salts. A small amount of material remains insoluble; allow it to settle overnight.

Prepare the element reference standard solutions in three groups. Make all solutions in 1 *M* HCl. Combine all major elements (Al, Fe, Ca, Mg, Na, Ti) in Group 1 standard. In Group 2 standard, combine B, P, As, Mn, Zn, Pb. In Group 3 standard, combine Cd, Co, Cr, Cu, Be, Mo, Ni, V. Use five to seven reference solutions for each element. Ranges of elements in marine sediments are shown below.

Major elements (%)		Minor and trace elements (μg/g)	
Si	20–50	Mn	200–2000
Al	3–10	P	500–2000
Fe	1–5	Cu	3–70
Ca	1–5	Zn	90–170
Mg	0.4–2	Cr	10–100
K	1–5	Ni	10–100
Na	1–6	As	13–24
Ti	0.3–0.6	Co	3–35
		Pb	10–50
		V	40–150

Reprinted with permission from Ref. 5. Copyright 1981 American Chemical Society.

Apply background corrections and corrections for spectral interferences as indicated in Tables 6.1 and 6.2. Operating conditions are summarized in Table 6.3. Detection limits for trace elements are given in Table 6.4.

D. Determination of 20 Elements in Stream Sediments by ICP–AES (6)

Elements Determined

- Major—Al, Fe, Ti, Mg, Ca, Na, K, Mn, Si
- Trace—Cu, Zn, Pb, Ni, Co, V, Mo, Cd, Cr, Ba, Sr

Table 6.1
Wavelengths and Background Measurement for Marine Sediment Analysis

Wavelength	Background measurement positions (nm)	
	1	2
B I 249.773	—[a]	—
Mo II 202.030	202.003	202.057
As I 193.696	193.678	193.714
Cd I 228.802	228.778	228.826
Zn I 213.856	—	—
P I 214.914	214.895	214.931
PB II 220.353	220.329	220.376
Co II 237.862	—	—
Ni II 231.604	231.585	231.625
Mn II 259.373	—	—
Al I 237.335	—	—
Fe II 259.940	—	—
V II 292.402	—	—
Cr II 267.716	267.688	267.745
Mg II 279.079	—	—
Cu I 324.754	324.719	324.789
Na I 330.237	—	—
Ti II 336.121	—	—
Be II 313.042	312.988	312.988
Ca II 317.933	—	—

[a]Background correction not performed.

Table 6.2
Correction Factors for Spectroscopic Interferences

Wavelength (nm)	Interferent	Correction factor[a] $(\text{mg/liter})_a/(\text{mg/liter})_i$	Source of interference
Al I 237.335	Fe	2.0×10^{-3}	Fe?
Zn I 213.856	Fe	1.35×10^{-4}	Fe I 213.859
Co II 237.862	Al	2.57×10^{-4}	Al I 237.841
	Fe	1.11×10^{-3}	Fe II 237.853
Mn II 259.373	Fe	1.11×10^{-3}	Fe II 259.373
V II 292.402	Al	4.8×10^{-5}	Al 292.452
	Fe	2.3×10^{-5}	Fe 292.435?
	Mg	5.4×10^{-5}	Mg?
Cu I 324.754	Fe	1.2×10^{-5}	Fe II 324.739

[a]Subscripts a and i denote analyte and interferent, respectively.

Table 6.3
Standard Operating Conditions

ICP	
Power	1.4 kW indicated incident power (<5 W reflected)
Plasma Ar	14 liters/min
Auxiliary plasma Ar	1 liter/min
Aerosol carrier Ar	0.7 liter/min
Spectrometer	
Observation height	12 mm above load coil
Entrance slit	100 × 0.5 μm
Exit slits	50 μm × 0.4 mm
Photomultiplier tube voltage	900 V

Comments on the Method

The stream sediments (− 180 μm fraction) are leached using 7 *M* nitric acid. Lithium and yttrium are added as reference elements (internal standards); lithium is used for the alkali elements and yttrium for the other elements. Background correction is employed only for the trace elements by means of a quartz refractor plate spectrum shifter. Detection limits for the trace elements are shown in Table 6.7 below and range from 0.04 to 2 ppm in the stream sediment. The proposed procedure was tested on synthetic solutions and on stream sediment samples analyzed by atomic absorption spectrometry. Generally acceptable agreements were obtained.

Reagents

In plasma work it is generally sufficient to use only two standards for each element, one high and one low (normally zero). Multielement standards can be used provided that spectral interference and chemical stability of the solutions are kept under control. In this work most standards are prepared from standard solutions for atomic absorption spectroscopy (BDH Chemicals Ltd.). Some are made from Specpure reagents (Johnson Matthey Ltd.). Double quartz-distilled water is used

Table 6.4
Estimated Detection Limits for Trace Elements in Sediments by This Procedure

Element	Detection limit (μg/g)	Element	Detection limit (μg/g)
As	5	Mo	1.5
Be	0.05	Ni	0.6
Cd	0.2	Pb	3
Co	0.5	V	0.3
Cu	0.1	Zn	0.1
Mn	0.05		

Table 6.5
Standard Solutions Used for Calibration[a]

1. Blank: 16 μg/ml of Li and Y in 0.35 *M* HNO_3
2. 2 μg/ml of Cu, Pb, Zn, Ni
3. 2 μg/ml of Sr, Ba, Cr, Cd
4. 2 μg/ml of V, Mo, Co
5. 100 μg/ml of Fe, Al
6. 100 μg/ml of Ca, Mg
7. 100 μg/ml of Si
8. 100 μg/ml of Mn, Na
9. 100 μg/ml of K, Ti

Source: Ref. 6.
[a]All standard solutions contain 16 μg/ml of Li and Y in 0.35 *M* HNO_3.

throughout. All standard solutions contain Li and Y at 16 μg/ml. Table 6.5 lists the standard solutions used for calibration.

Equipment

A Jarrell-Ash model 975 ICAP AtomComp with pneumatic cross-flow nebulizer was used in this work. The rf generator was a PlasmaTherm generator type HFP 2000 D operating at 27.13 MHz.

Operating system parameters

- Plasma parameters: forward power, 1.0 kW; reflected power, <10 W
- Argon flow rates: coolant, 15 liters/min; sample, 0.5 liter/min; plasma, 0 liter/min
- Sample uptake: approximately 1.6 ml/min
- Observation zone: 16 mm above the induction coil
- Exposure time: 12 s on line and 8 s on background (background correction only for trace elements)

The elements analyzed by the method include the major rock elements Si, Al, Fe, Ti, Mg, Ca, Na, K, and Mn and the trace elements Cu, Zn, Pb, Ni, Co, V, Mo, Cd, Cr, Ba, and Sr. The analytical lines used are shown in Table 6.6.

Procedure

Leach 1 g of the minus 180 μm fraction using 5 ml of 7 *M* nitric acid at 100°C for 3.5 h. After digestion dilute the solution to 20 ml and filter through a 0.02 mm nylon cloth. (Alternatively, centrifuge and decant.) Dilute 1 ml of this solution with 4 ml of a reference element solution containing 20 μg/ml Li and Y in water. The final solution thus contains 16 μg/ml Li and Y. Run the sample solution using two calibrating solutions. Use the ICP–AES conditions listed above.

Table 6.6
List of Analytical Lines

Analytical line		Wavelength (nm)
Li	I	670.7
Y	II	371.0
Cu	I	324.7
Zn	I	213.8
Pb	II	220.3
Ni	II	231.6[a]
Co	II	228.6
V	II	292.4
Mo	II	202.0
Cd	I	228.8[a]
Cr	II	267.7
Ba	II	493.4
Sr	II	421.5
Al	I	309.2
Fe	II	259.9
Ti	II	334.9
Mg	I	383.2
Ca	II	317.9
Na	I	588.9
K	I	766.4
Mn	II	257.6
Si	I	288.1

Source: Ref. 6.
[a]Second order.

The functioning of the system is controlled by watching the intensity readings of the reference lines of Li and Y. In addition, a control standard containing trace elements at 2 μg/ml, corresponding to 200 ppm in the sediment sample, is analyzed at certain intervals. A deviation of 5% is accepted.

Detection limits (concentrations that give a net line signal ratio equal to twice the standard deviation of the blank) are given in Table 6.7 for the trace elements.

E. Determination of Trace Metals in River Sediments (7) by Flameless Atomic Absorption

Elements Determined

As, Cd, Co, Cr, Cu, Mn, Ni, Pb, V, Zn

Comment on the Method

The following method was developed for the determination of As, Cd, Co, Cr, Cu, Mn, Ni, Pb, V, and Zn in coal fly ash, urban particulate furnace dust, and

Table 6.7
Detection Limits

Element	Detection limit (ppm)[a]
Cu	0.17
Zn	0.10
Pb	1.77
Ni	0.33
Co	0.23
V	0.25
Mo	0.22
Cd	0.07
Cr	0.29
Ba	0.18
Sr	0.04

Source: Ref. 6.
[a]In original sample.

iron ore in addition to river sediment. The proposed procedure was tested on U.S. National Bureau of Standards (NBS) and Eurostandards standard reference samples.

A lithium tetraborate fusion is employed. The fusion cake is dissolved in 5% hydrochloric acid followed by furnace atomic absorption determination. Although the precision and accuracy of the method compare favorably with those of an acid digestion procedure, the author claims superiority in speed of analysis. The procedure is given here because it is desirable to have an alternative to the acid digestion commonly used.

Reagents

All solutions were prepared with deionized water obtained from a Millipore Milli-Q system. All the glassware had previously been washed with 20% (v/v) nitric acid and rinsed with deionized water. The hydrochloric, nitric, and phosphoric acids used were Merck Suprapur, and the lithium tetraborate was analytical reagent grade. Standard solutions for atomic absorption (BDH) were used to prepare the working standards by diluting the concentrated chloride or nitrate solutions with lithium tetraborate solution (6 g/liter in 1% hydrochloric acid) in order to have the same salt concentration in the samples and standards. For arsenic, working standards and samples were prepared in 0.15 *M* nitric acid in the presence of 100 μg/ml nickel (as nitrate).

Equipment

The Perkin-Elmer model 503 atomic absorption spectrometer was fitted with a deuterium background corrector, an HGA-500 furnace, and a model 56 chart re-

Table 6.8
Recommended Conditions for the Spectrometry of Trace Elements in the Lithium Tetraborate Solution[a]

Element	Wavelength (nm)	Slit width (nm)	Ash (°C)	Atomize (°C)	Clean (°C)
As	193.7	0.7	1000	2700	2800
Cd	228.8	0.7	500	2000	2500
Cr	357.9	0.7	1200	2700	2800
Cu	324.7	0.7	1000	2500	2800
Co[b]	240.7	0.2	1000	2700	2800
Mn	279.5	0.2	1100	2700	2800
Ni[b]	232.0	0.2	1000	2700	2800
Pb	283.3	0.7	1000	2500	2800
V[b]	318.4	0.2	1500	2700	2800
Zn	213.9	0.7	700	2300	2500

Source: Ref. 7.

[a]Optimal instrument operating parameters held constant: drying temperature, 100°C; drying time, 45 s; ashing time, 60 s; atomization time, 5 s; cleaning time, 5 s; argon purge gas; continuous gas flow mode; volume injected, 20 μl.

[b]Pyrolytic graphite-coated tubes.

corder. Background correction was used for all measurements at wavelengths less than 320 nm. Hollow-cathode lamps were used except for arsenic, for which an electrodeless discharge lamp was employed.

The optimum conditions used are shown in Table 6.8. Particularly for arsenic, in order to reduce the volatility during ashing and to reduce the influence of matrix elements, matrix modification was used. The optimum program for this element (20 μl of sample injected with 100 μg/ml nickel as stabilizer in 0.15 *M* nitric acid) was: drying for 45 s at 100°C, ashing for 30 s at 1000°C, atomization for 5 s at 2700°C, cleaning for 3 s at 2800°C.

Procedure

Place about 0.3 g of sample (accurately weighed), 1.5 g of lithium tetraborate, and 78 mg of nickel nitrate (only for arsenic determination) in a 50 ml platinum crucible and thoroughly mix. Place the crucible in a muffle furnace at 1000°C for 1 h. When the fusion is complete, place the cooled crucible in a 100 ml beaker, insert a small Teflon-coated stirrer, and add 50 ml of 5% hydrochloric acid. Heat the solution for 15 min at 50–60°C on a magnetic-stirrer hot plate, and transfer it to a 250 ml polyethylene flask. Repeat the procedure with a second aliquot of acid and, when dissolution is complete, adjust the contents of the flask to volume with deionized water. Obtain the working blank by taking the same quantity of lithium tetraborate through the entire procedure.

F. Determination of Major and Trace Elements in Clay by ICP–AES (1)

Elements Determined

Al, As, B, Co, Cr, Cu, Cd, Ca, Fe, K, Mg, Mn, Mo, Na, P, Pb, Sr, V, Zn

Comment on the Method

About 20–25 samples can be analyzed per hour. Clay samples are dispersed in water and then aspirated into the plasma. Solution standards are used for calibrations. Dispersions of standard clays in water provide an alternative calibration procedure. Five American Petroleum Institute (API) clay standards were analyzed successfully.

Equipment

- Applied Research Laboratories model QA-137 direct reading spectrometer with an ICP source and a Meinhart concentric nebulizer
- PDP 1104 minicomputer

Reagents and Apparatus

- Freeze-drying unit
- Ultrasonic probe-type vibrator
- Muffle furnace
- Sand bath

Table 6.9
Operating Procedures for Routine Analysis Using the Applied Research Laboratories QA-137 ICP–AES Instrument

Incident power	1650 W
Reflected power	10 W
Entrance slit width	12 μm
Argon flow rates	
Coolant	10.5 liters/min
Sample carrier gas	1 liter/min
Plasma	1.5 liters/min
Sample aspiration rate	2.4 ml/min
Observation height	0.63 coil height
Nebulizer	Meinhart concentric
Operating temperature	24°C

Source: Ref. 1.

Sample Preparation

Disperse 100 g samples of clay in water using an ulstrasonic probe-type vibrator (3 min at 400 W). Separate a clay fraction of <2 μm by means of gravity sedimentation and centrifugation (8). Flocculate the separated aqueous dispersion with 1 *M* calcium chloride solution. Wash the precipitate with some water to remove excess electrolyte. Freeze-dry the solids.

Analysis

Weigh a 0.1–0.2 g sample of the dried solids into a 150 ml beaker. Add 100 ml of water. Disperse the clay by means of a 15 s treatment with the ultrasonic probe. Immediately aspirate the sample for ICP–AES analysis. Since the aspiration of a sample is 7–9% slower than that of a solution standard, a correction factor of 1.08 must be applied.

The spectrometer operating conditions are shown in Table 6.9. Spectral lines and detection limits for the elements are shown in Table 6.10.

Table 6.10
Spectral Lines, Detection Limits, and Standard Groups Used for Instrumental Calibration

Element	Spectral line (nm)	Detection limit[a] (ppm)	Group number	Standard concentration range (ppm)
Cu	324.75	1.0	1	0–10
Co	238.89	1.0	1	0–10
B	249.77	0.5	1	0–10
Cr	267.72	0.5	1	0–10
As	193.77	5.0	1	0–10
Pb	220.35	4.0	1	0–10
Cd	226.50	0.5	1	0–10
Zn	202.55	0.1	2	0–100
Mn	257.61	0.1	2	0–100
Mo	202.03	0.1	2	0–100
V	310.23	0.5	2	0–100
Sr	213.62	0.1	2	0–100
P	407.77	4.5	2	0–100
Ca	315.89	2.5	3	0–1000
Na	589.59	5.0	3	0–1000
Mg	279.08	3.0	4	0–1000
K	766.49	8.0	4	0–1000
Fe	259.94	1.5	5	0–2000
Al	308.22	5.0	6	0–2000

Source: Ref. 1.

[a]Extrapolated to indicate minimum concentrations in clay materials using dissolution technique described in this method.

Calibration

Prepare calibration curves for the elements using the same instrument conditions as for the samples. Use combined solution standards for groups of elements as shown in Table 6.10. The concentration ranges are contained in the same table.

Reference clay standards can also be used for calibration. No correction factor for viscosity differences is required.

G. Determination of Arsenic, Selenium, Chromium, Cobalt, and Nickel in Geochemical Samples by Furnace Atomic Absorption (9)

Comment on the Methods

Zeeman background correction was essential for the best results using furnace AAS for the analysis of five U.S. Geological Survey (USGS) standards GXR. If a deuterium arc is used, a severe overcompensation for selenium is obtained because of an iron spectral interference. Also, with arsenic at the 193.7 nm resonance a severe interference occurs in the presence of aluminum if the deuterium arc is employed for background correction. This interference, however, can be overcome if the 197.2 nm (twice less sensitive) line is used. Thus the deuterium arc background correction could be used except for selenium.

The samples are dissolved with hydrofluoric and perchloric acids in a 5 : 1 ratio. The present authors recommend very strongly that nitric acid be used whenever perchloric acid is employed. Thus 5 ml of nitric acid has been added as a modification to the procedure.

Matrix modifiers are employed. These are 20 μg Ni, 100 μg Ni, 50 μg $Mg(NO_3)_2 \cdot 6H_2O$, and 50 μg $Mg(NO_3)_2 \cdot 6H_2O$ for As, Se, Cr, and Co, respectively. None is needed for Ni. A L'vov platform is also employed. Results obtained on five USGS standard samples GXR were in satisfactory agreement with the accepted values.

Reagents

Standard solutions are prepared from 1000 mg/liter standards (Alfa Products, Ventron) by dilution with deionized water. The nickel matrix modifier solution is prepared by dissolving pure nickel cups (Perkin-Elmer part No. 303-0813) in 5.6% heated nitric acid (Ultrex). Magnesium nitrate solution is prepared from reagent grade $Mg(NO_3)_2 \cdot 6H_2O$ (Baker).

Water from a Continental water system was purified in a Millipore deionizer consisting of Milli-RO_4 and Milli-Q systems. The geological samples (GXR) were obtained from the U.S. Geological Survey.

Equipment

Experiments were performed on a Perkin-Elmer Zeeman/5000 equipped with an HGA-500 and with an AS-40 autosampler. To compare Zeeman and continuum

Table 6.11
Zeeman Graphite Furnace Conditions

Element	Modifier	Char temperature (°C)	Atomization temperature (°C)
As	20 μg Ni	1400	2500
Se	100 μg Ni	1400	2100
Cr	50 μg $Mg(NO_3)_2 \cdot 6H_2O$	1300	2500
Co	50 μg $Mg(NO_3)_2 \cdot 6H_2O$	1300	2400
Ni	none	1200	2500

source background corrections, some measurements were done on a model 4000 spectrophotometer equipped with an HGA-500 graphite furnace and an AS-40 autosampler. Peak absorbances and integrated absorbances were calculated on a Data System 10 and graphics were plotted on a Hewlett-Packard 7225A plotter. Pyrolytically coated graphite tubes (Perkin-Elmer part No. B0109-322) with solid pyrolytic graphite platforms (Perkin-Elmer part No. B0121-091) were used.

Procedure

Dissolve 0.5 g of the powdered GXR sample in a 6 ml solution (5 : 5 : 1) of nitric, hydrofluoric, and perchloric acids. Evaporate each sample to dryness and repeat the process two more times. Treat each sample with 3 ml of perchloric acid only and again evaporate to dryness. Dissolve the residue in 100 ml of 1% nitric acid. (Part of the calcium and aluminum fluorides remained undissolved in the final solution, but that does not affect the determination.) Coprecipitation does not occur and solutions with the sediments stored in plastic bottles are stable for at least 7 months.

All analyses reported here use dilutions in water of the original 0.5 g in 100 ml rock solutions that vary up to 100-fold to keep the resulting absorbances less than about 0.3*A*. Automatically pipet 10 μl of the diluted sample or standard solutions and 5 μl of the matrix modifer (As, Se, Cr, Co) onto the platform using the AS-40 autosampler.

Use the temperature conditions and modifiers given in Table 6.11. After 100 to 150 firings, replace the tube and platform. Clean the contact rings and windows to remove accumulated salt approximately once a week.

III. EXTRACTABLE METALS

A. Determination of Acid-Extractable Metals in Rocks and Soils by Flame Atomic Absorption (10)

Elements Determined

Cu, Ni, Zn, Pb, Cr, Mn, Co, Fe, Cd, Ag

Comment on the Method

"Acid-extractable metals" are defined as those loosely bound, absorbed, or precipitated on grain surfaces. This is a reliable method for the determination of relative metal contents when a number of samples are compared. The results do not represent total metal. The extraction efficiency varies from 60 to 95% depending on the metal and type of sample. If the concentration of a metal is close to its detection limit, background correction is essential. When silver is measured its concentration must be over 1 ppm and no hydrochloric acid should be used.

Equipment

A flame atomic absorption spectrometer equipped with a background correction unit was used.

Reagents and Apparatus

- 15 ml test tubes with a graduation mark at 10 ml
- Plastic-coated wire test tube holder for 48 tubes
- Water bath large enough to contain the test tube holder

Procedure

Weigh 0.25 g samples into 15 ml test tubes. Place the test tubes in the holder. Add 2 ml of nitric acid. After the reaction subsides, heat the test tubes in the water bath maintained at 85°C for 30 min. Then add 1 ml of hydrochloric acid (except if silver is required). Heat for 30 min. Cool the samples, and dilute them to 10 ml with water. Mix the contents by inverting the rack several times (with the tubes covered). Allow the solids to settle for several hours. Use the supernatant liquids for atomic absorption analysis.

Background corrections must be used when the concentrations of Co, Zn, Ni, Cd, Ag, and Pb are near the detection limits. Otherwise, only Co and Ag require background correction. If metal concentrations of samples are higher than the working ranges, dilute the samples or use less sensitive absorption lines.

Detection Limits

Element	Detection limits (μg/g)	Element	Detection limits (μg/g)
Cu	5	Pb	10
Ni	10	Co	10
Mn	2	Mn	10
Cd	2	Ag	10
Fe	5	Cr	10

B. Partial Extraction of Soils and Sediments Using Hot 0.5 *M* Hydrochloric Acid (Flame Atomic Absorption)

Elements Determined

Cu, Zn

Comment on the Method

As noted earlier in this chapter, strong acid attacks with concentrated acid mixtures sometimes fail to outline anomalies. This is particularly true when the metal content of country rocks varies greatly in an area and when the magnitude of the true anomaly is relatively small. A method which has received some acceptance in such cases is a leach using hot dilute (0.5 *M*) hydrochloric acid. The acid is added to a weighed sample in a test tube and heated near boiling for 20 min. Thus the method is also relatively rapid. The method has been tested for copper and zinc.

Equipment

Any flame atomic absorption unit, e.g., a Perkin-Elmer model 603, equipped with a high-solids burner and an acid-resistant nebulizer, is suitable. Test tubes are marked at the 15 ml mark and an aluminum block is drilled with holes large enough to hold the test tubes. A capacity of about 50 test tubes per run is desirable.

Reagents

Prepare sufficient 0.5 *M* hydrochloric acid from concentrated reagent grade (12 *M*) hydrochloric acid for the number of samples to be run. Four milliliters of 0.5 *M* hydrochloric acid is needed per sample.

Procedure

Weigh 0.5 g soil or sediment sample into each test tube. (This weight can be adjusted if necessary to suit the metal contents of the samples.) Add 4 ml of 0.5 *M* hydrochloric acid to each test tube. Place the test tubes in the aluminum block. Heat up gently to near boiling. Keep the solutions near boiling for 20 min. Cool and dilute to 15 ml with water. Cover each test tube with parafilm and invert it several times to mix the contents well. Clean and dry another set of test tubes. Filter the samples into the dry test tubes using dry Whatman No. 40 filter paper or the equivalent. (Alternatively, the samples can be set aside to settle and the supernatant liquids isolated by decantation.)

Run the samples for copper and zinc by atomic absorption or ICP emission spectrometry. Use the manufacturer's recommended conditions.

C. Determination of EDTA-Extractable Copper and Zinc in Soils by Flame Atomic Absorption

Elements Determined

Cu, Zn

Comment on the Method

Extractions of soils with chelating agents such as EDTA are used to estimate the more loosely bound metals. For some types of soils, particularly when the magnitude of the anomaly is small compared to the variation in strong acid-extractable metals, use of such extractants is advisable. Ten gram samples of soil are extracted with a neutral solution of 0.05 *M* EDTA. The metals in the extracts are determined by means of atomic absorption spectrometry. Calibration standards must be made up to contain 0.05 *M* EDTA. Detection limits for copper and zinc are about 2 ppm in the soil in each case.

Apparatus

Any modern atomic absorption spectrometer can be employed. It should be equipped with an acid-resistant nebulizer and high-solids burner. A blank solution of 0.05 *M* EDTA should be used initially to clean out the burner nebulizer system (by aspiration) and then between samples and standards. The instrumental parameters recommended by the manufacturer should be used. All filter papers, samples, and sample bottles must be prewashed with 0.05 *M* EDTA.

Reagents

Prepare a 0.05 *M* solution of reagent grade EDTA in water. Neutralize to pH 7.0 with purified ammonia.

Prepare calibration standards from 1000 μg/ml stock pure metal solutions by serial dilution in 0.05 *M* EDTA. The dilute standards should be prepared fresh daily.

Procedure

Weigh 10 g of soil samples ground to 100 mesh or finer into clean 100 ml high-density linear polyethylene bottles. Add 50 ml of 0.05 *M* EDTA solution to each bottle. Shake mechanically for 1 h. Filter sample solutions through dry 0.05 *M* EDTA-prewashed Whatman No. 40 filter paper into clean (0.05 *M* EDTA-washed) dry bottles. Run the samples on the atomic absorption instrument using the conditions recommended by the manufacturer. Aspirate blank 0.05 *M* EDTA for a short time after each sample or standard.

IV. SELECTIVE EXTRACTANTS

A. Selective Extractants for Soils and Sediments

Trace elements in soils and sediments are associated with various components of these samples. It may be possible to choose a selective extractant which will attack a specific component of the soil and release the trace elements therein. In this way a better contrast between background and anomalous samples might be obtained. At this time there is no unanimity of support for this view. Perhaps the main problem is that the selective extractants proposed to date are not as specific as would be desirable. That is, although a selective extractant attacks mainly one specific soil component, it may in addition attack other components to some degree.

Trace elements are fixed in soils and sediments as follows:

1. Sorption and coprecipitation by hydrous iron and manganese oxides
2. Sorption by clay minerals
3. Precipitation with organic matter
4. Processes involving hydrolytic reactions and both complexed and dissolved ions
5. Processes related to variations in physical and chemical nature of the sediment

The distribution of trace elements among these components varies with the identity of the element, the soil components present in a given sample, and the chemical environment of the samples.

1. Hydrous Oxides

Under oxidizing conditions both iron and manganese oxides are excellent scavengers of trace elements. The mechanisms of scavenging are coprecipitation, surface complex formation, ion exchange, and penetration of the crystal lattice. Under reducing conditions the hydrous oxides will redissolve. Considering zinc and copper, copper is only weakly scavenged by hydrous oxides (also favors iron over manganese oxide), whereas zinc is strongly scavenged.

2. Clay Minerals

The relative importance of clay minerals in trace element accumulation is not well understood. Factors which affect the selectivity of clays for different trace element cations are valence, hydrated ion radius, electronegativity, and free energy of formation. The cation-exchange capability of clays appears to be low compared to that of humic substances, which often coat clay minerals.

3. Organic Matter

The most important organic species as far as trace elements are concerned are the humic substances (humic and fulvic acids). These constitute a spectrum of compounds with a range of physical and chemical properties. They can exist as anionic species in water. Neutralization of the negative charge by reaction with a trace element cation or metal oxide colloids or by adsorption onto clay particles occurs. Interactions between humic substances and metal ions have been described as ion exchange, surface adsorption, and chelation. Both zinc and copper can strongly react with humic substances.

4. Choice of a Selective Extractant

A large variety of selective extractants have been proposed for geochemical exploration purposes. These include hydroxylamine hydrochloride, ammonium oxalate, sodium dithionite–sodium citrate, sodium pyrophosphate, and acetylacetone.

Lye (11) critically evaluated these extractants on lake sediment samples for the recovery of zinc and copper from various components. He found that hydroxylamine hydrochloride was better than pyrophosphate for the selective dissolution of hydrous manganese oxide; the latter seemed to dissolve some organically bound metal. Also, hydroxylamine dissolved relatively little hydrous iron oxide.

Oxalate extracts contain copper and zinc trapped in most of the clay components. Thus the oxalate extraction is not very selective. Likewise, dithionite was found to be poorly selective.

Acetylacetone and pyrophosphate were found to extract trace elements from organic matter. Acetylacetone proved to be more specific than pyrophosphate for organic matter because it decomposed less of the hydrous manganese oxide fraction. Both of these extractants were found to dissolve hydrous iron oxide.

The authors emphasize that the results obtained by following the methods should be interpreted with caution.

B. Determination of Copper and Zinc in Hydrous Manganese Oxide by Flame Atomic Absorption

Reagents

All chemicals were of reagent grade quality.

- Hydroxylamine hydrochloride, 0.1 *M:* Dissolve 6.95 g in distilled water. Add 10 ml of 1 *M* nitric acid. Dilute to approximately 900 ml. Adjust the pH to 2 and dilute to 1 liter.
- Acetylacetone, 5%: Add 50 ml of the concentrate to distilled water. Mix and dilute to 1 liter. Mix again.
- Standard metal solutions, 1000 μg/ml: Prepare each by dissolving 1 g metal in a minimum of nitric acid and then dilute to 1 liter. Final acid content should be 1%.

Equipment

A Perkin-Elmer model 603 atomic absorption spectrometer was used. Deuterium arc background correction was available. A single-slot air/acetylene high-solids burner was employed. Instrument settings were as follows:

Element	Wavelength (nm)	Slit	Lamp current (mA)
Cu	324.7	0.7	5
Zn	213.9	0.7	7

Procedure

Weigh 0.5 g of sample into a 50 ml plastic bottle. Add 25 ml of 0.1 *M* hydroxlyamine hydrochloride solution. Agitate on a mechanical shaker for 20 min. Filter the samples. Run on the atomic absorption unit using the conditions given in the Equipment section.

C. Determination of Copper and Zinc in Organic Matter by Flame Atomic Absorption

Weigh 1.0 g of sample into a 50 ml plastic bottle. Add 25 ml of 5% acetylacetone. Agitate on a mechanical shaker occasionally for 6 days. Transfer to a 50 ml centrifuge tube and centrifuge at 20,000 × *g* for 30 min. Transfer the supernatant liquid to a 50 ml plastic bottle. Wash the original bottle with 2 ml of extractant and transfer the liquid to the centrifuge tube with the original sample. Set aside these tubes containing residues and washings for 15 days. Shake occasionally during this period. Centrifuge again as above and filter the supernatant liquids thus obtained through a 0.45 μm membrane filter into the appropriate bottle (the one containing the original centrifugate). If a brick red precipitate is obtained in any sample, repeat the procedure using a smaller sample-to-extractant ratio. Run the solutions on the atomic absorption unit using the conditions given in the Equipment section.

D. Determination of Copper, Lead, and Zinc in Stream Sediments by Means of Five Selective Extractants and Flame Atomic Absorption (12)

Comments on the Method

Stream sediments are fractionated into eight size categories in the range 850 μm to <1 mm. Each fraction is extracted with five selective extractants in a sequential procedure. These are (in order of extraction) 0.5 *M* $MgCl_2$ at pH 7—exchangeable metals, 1 *M* NaOAc adjusted to pH 5.0—metals bound to carbonates, 0.04 *M* $NH_2OH \cdot HCl$ in 25% (v/v) HOAc—metals bound to iron and manganese oxides, 30% H_2O_2 adjusted to pH 2 with HNO_3 at 85°C followed by room temperature leach with 3.2 *M* NH_4OAc in 20% (v/v) HNO_3—metals bound to organic matter, and 5 : 1 HF : $HClO_4$—residual metals. (The present authors recommend that excess nitric acid be present during any perchloric acid treatment.) Iron and manganese are included because of their importance in oxide fractions. This work suggests that for the samples tested, two stream beds in the Gaspé Quebec region, an extraction with $NH_2OH \cdot HCl$ (hydroxylamine hydrochloride) gives high anomaly-to-background ratios. Grain size was not an important parameter in using this extractant.

Reagents

- Magnesium chloride: Prepare 0.5 *M* $MgCl_2$ solution and adjust the pH to 7.
- Sodium acetate: Prepare 1 *M* NaOAc and adjust the pH to 5.0 with acetic acid.
- Hydroxylamine hydrochloride: Prepare an 0.04 *M* $NH_2OH \cdot HCl$ solution in 25% (v/v) acetic acid.
- Hydrogen peroxide: Prepare a 30% solution and adjust the pH to 2 with nitric acid.
- Ammonium acetate: Prepare a 3.2 *M* solution in 20% (v/v) nitric acid.

Equipment

A Varian Techtron model 575-ABQ atomic absorption spectrometer was used with the manufacturer's recommended conditions.

Procedure

Sampling and preservation

Collect bottom sediment samples (1 kg) with hand gloves at each of the sites. Keep the samples at 4°C during transport to the laboratory, and then freeze them until needed for processing.

Trace metal analysis

Each subsample is subjected to a sequential extraction procedure designed to partition the particulate trace metals into the following five fractions:

Fraction 1: exchangeable metals. Extract the sediment sample for 10 min with 0.5 *M* $MgCl_2$.

Fraction 2: metals bound to carbonates. Leach the residue for 5 h with 1 *M* sodium acetate (NaOAc).

Fraction 3: metals bound to iron and manganese oxides. Extract the residue for 6 h at 96°C with 0.04 *M* $NH_2OH \cdot HCl$ in 25% (v/v) HOAc.

Fraction 4: metals bound to organic matter. Extract the residue at 85°C for 5 h with 30% H_2O_2 and then at room temperature with 3.2 *M* NH_4OAc in 20% (v/v) HNO_3.

Fraction 5: residual metal. Digest the residue with a 5 : 1 mixture of hydrofluoric and perchloric acids. (The present authors recommend addition of an excess of nitric acid for safety reasons.)

E. Determination of Copper, Zinc, Iron, Manganese, and Molybdenum Scavenged by Soil and Sedimental Organic Matter (Flame Atomic Absorption) (13)

Comments on the Method

Alkaline sodium hypochlorite is used to oxidize and dissolve organic matter (and sulfide minerals) in soils and lake and stream sediments. The copper, zinc, iron, manganese, and molybdenum thus released are determined by flame atomic absorption spectroscopy. A nitrous oxide/acetylene flame is required for molybdenum. Metals which are precipitated by the alkaline conditions are redissolved by a pH 3.0 ± 0.3 leach.

The only treatment prior to the leaching step is drying at 100°C followed by sieving through an 80 mesh sieve. The precision obtained on replicate determinations was ±25% at the 95% confidence level.

Reagents

- Sodium hypochlorite, 5% reagent grade: Adjust the pH to 9.5 by dropwise addition of 6 *M* hydrochloric acid.

- Acidic leach solution, pH 3.0: Adjust the pH of distilled water to 3.0 ± 0.3 by dropwise addition of 1.5 *M* hydrochloric acid.
- Molybdenum calibration standards: These must be made to contain aluminum in an amount similar to that added to the samples.

Equipment

Atomic absorption spectrometer equipped with air/acetylene and nitrous oxide/acetylene burners. Use the manufacturer's recommended conditions.

Procedure

Dry samples in an oven at 110°C. Then sieve (or pulverize for lake sediments) at 80 mesh. Place a 1.0 g split of −80 mesh material in a test tube (20 by 155 mm) and add 10 ml of reagent grade sodium hypochlorite, freshly adjusted to pH 9.5 with 6 *M* hydrochloric acid to oxidize organic matter. Homogenize the test tube contents at 2 h intervals for 10 h and then place in a water bath for 3 h to decompose excess hypochlorite. [Care must be exercised to avoid excessive temperatures (>80°C) because some elements, hydrolyzed under the alkaline conditions of the extraction, are susceptible to dehydration reactions. In later experiments the heating step was eliminated without altering the effectiveness of the extraction.]

Cool the sample to room temperature, centrifuge, and decant the supernatant liquid into a clean test tube. Repeat the procedure with an additional 5 ml of hypochlorite solution, and combine the supernatant with that from the first extraction. (Additional sodium hypochlorite may be required if the extraction continues to give brown, strongly colored solutions. In general, although many samples require up to three extractions to decompose the organic matter fully, more than 95% of the hypochlorite-leachable metal is dissolved during the first treatment.) Wash the residue from the extraction with sufficient distilled water so that, when decanted and combined with the solution from previous extractions, the final volume is 30 ml.

Metal hydroxide precipitates and basic salts formed in the alkaline solution are redissolved by adding 10 ml of distilled water to the sample residue and adjusting the pH to 3.0 ± 0.3 by dropwise addition of 1.5 *M* hydrochloric acid delivered from a buret.

Analyze the samples by standard atomic absorption procedures with an air/acetylene flame for copper, zinc, iron, and manganese. Determine the molybdenum content of the hypochlorite extracts also by atomic absorption spectrophotometry, using a nitrous oxide/acetylene flame. [Addition of aluminum (0.5 ml of 2.4% aluminum trichloride hexahydrate, w/v, to a 5 ml aliquot) as a releasing agent is necessary to suppress interferences in this determination. Precipitation of aluminum hydroxide is avoided by acidification of samples with 0.5 ml of 1.5 *M* hydrochloric acid.]

F. Determination of Secondary Iron and Manganese Oxides and Their Associated Metals in Soils and Stream Sediments by Means of Oxalic Acid (ICP–AES) (14)

Elements Determined

Al, Fe, Mg, Ca, Ti, Mn, Ag, As, Ba, Be, Bi, Cd, Ce, Co, Cr, Cu, La, Mo, Nb, Ni, Pb, P, Sb, Sn, Sr, V, W, Y, Zn

Comment on the Method

Oxalic acid leaching of soils and stream sediments is used to dissolve secondary iron and manganese oxides and their associated metals. The oxides occur as coatings on mineral surfaces, as fine particles, and as aggregates of colloidal dimensions. When they are associated with an ore body, the oxides are often enriched with trace elements as a result of surface and ground water movement. Amorphous aluminum and silicon oxides may also be important scavengers.

Equipment

- Applied Research Laboratories fixed-slit, 1 m polychromator quantometer with a spectral array of 32 elements
- Minicomputer DEC 1104 (dual floppy disk)
- Babington-type nebulizer

Reagents and Apparatus

- Oxalic acid, 5%
- Hydrochloric acid, 20%
- Hydrogen peroxide, 50%
- Whatman No. 44 filter paper

Procedure

Weigh a 1.00 g sample into a 150 ml beaker. Add 25 ml of 5% oxalic acid. Bring the contents to a boil on a hot plate, and boil for 30 min. Filter the solution through a Whatman No. 44 filter paper into a 150 ml beaker. Wash the residue on the paper with 25 ml distilled water. Add 5 ml of 50% hydrogen peroxide to the filtrate. Evaporate the filtrate to dryness. Add 20 ml of 20% hydrochloric acid to the residue and heat to dissolve the salts. Filter again through a Whatman No. 44 filter paper.

The spectral lines and detection limits and the operating parameters for the ICP are listed in Tables 6.12 and 6.13, respectively.

G. Determination of Copper and Nickel Sulfides by Flame Atomic Absorption

Comment on the Method

A number of methods have been proposed for the dissolution of sulfide minerals as a specific class of compounds. They generally rely on the oxidation of sulfide

Table 6.12
Spectral Array, Detection Limits, and Lower Limits of Determination for Soil Profile and Stream Sediments

Element	Wavelength (nm)	LQD[a] (μg/ml)	Range of lower limit of determination[b] (μg/g)	Lower limit of determination of routine method[c] (μg/g)
Al	396.1	0.058	1.16–1.74	ND[d]
Fe	259.9	0.028	0.56–0.84	1000
Mg	279.0	0.080	1.6–2.4	500
Ca	422.7	0.040	0.8–1.2	1000
Ti	334.9	0.007	0.14–0.21	20
Mn	257.6	0.003	0.06–0.09	20
Ag	328.0	0.015	0.30–0.45	1
As	193.7	0.460	9.2–13.8	500
Au	242.8	0.050	1.0–1.5	20
B	249.7	0.010	0.2–0.3	20
Ba	455.4	0.001	0.02–0.03	50
Be	313.0	0.0004	0.008–0.012	2
Bi	306.7	0.5	10–20	20
Cd	226.5	0.005	0.1–0.15	50
Co	345.3	0.200	4–6	10
Cr	283.5	0.015	0.3–0.45	20
Cu	324.7	0.020	0.4–0.6	10
La	398.8	0.016	0.32–0.48	50
Mo	287.1	0.038	0.76–1.14	10
Nb	309.4	0.030	0.6–0.9	50
Ni	231.6	0.040	0.8–1.2	10
Pb	220.3	0.150	3–4.5	20
Sb	217.5	0.270	5.4–8.1	200
Sn	189.9	0.200	4–6	20
Sr	407.7	0.0004	0.008–0.012	200
V	311.0	0.018	0.36– 0.45	20
W	239.7	0.260	5.2–7.8	100
Y	371.0	0.004	0.08–0.12	20
Zn	202.5	0.010	0.2–0.3	500
Ce	418.6	0.090	1.8–2.7	ND
P	213.6	0.140	2.8–4.2	ND
Ge	303.9	0.370	7.4–11.1	ND

Source: Ref. 14.

[a]Lowest quantitative determinable concentration is defined as the concentration of the element that will give a net signal equal to approximately 10 times the standard deviation of the background.

[b]Lower limit of determination (LQD × dilution factor) varies according to dilution factor.

[c]dc arc method.

[d] Not determined.

Table 6.13
Operating Parameters

Incident power	1.6 kW
Plasma gas	Argon
Plasma gas flow rate	1.5 liters/min
Coolant gas flow rate	10.5 liters/min
Carrier gas flow rate	0.6–0.7 liter/min
Solution uptake rate	4.7 ml/min
Observation height	15 mm above coil
Integration time	3–10 s period

Source: Ref. 14.

to release and dissolve the associated metals. Comparisons of 10 different chemical treatments of nine sulfide minerals have been reported (15). The hydrogen peroxide–ammonium citrate treatment described here is less destructive to the silicate minerals than treatments involving inorganic acids. Ammonium citrate acts as a buffer to control the acidity of the solution so that attack on other water-insoluble forms of the metals is minimized. A closely related method based on hydrogen peroxide–ascorbic acid proved to be moderately effective in dissolving sulfide minerals with the exception of stibnite and orpiment. A summary of the results of other sulfide dissolution methods is given at the end of the Procedure.

Reagents and Apparatus

- Hydrogen peroxide, 30%.
- Ammonium citrate, 10%.
- Reactant solution: Mix 65 ml of 10% ammonium citrate with 35 ml of 30% hydrogen peroxide.
- Plastic bottles, 50 ml.
- Test tube, 25 ml, with a graduation mark at 20 ml.
- Mechanical shaker which can accommodate a number of 50 ml bottles.

Procedure

Weigh a 0.1 g sample into a 50 ml plastic bottle. Add 10 ml of reactant solution. Shake the solution for 2 h. Transfer the sample to a 25 ml test tube. Add water to the 20 ml mark. Mix the contents, and then set the test tube aside for several hours to allow the sediment to settle.

Prepare dilute standards from the stock solutions. Use reactant solution for the final dilutions so that standards and samples have the same reactant concentrations.

Detection Limits

Copper, 25 μg/g; nickel, 25 μg/g.

Comparisons of Treatments of Sulfide Minerals

Descriptions of the procedures of treatments of sulfide minerals are shown in Table 6.14. Results of these treatments are contained in Table 6.15.

H. Determination of Organically Bound Gold by Flameless Atomic Absorption (16)

Comments on the Method

Organically bound gold is selectively extracted from soil and stream and lake sediment samples by means of two extractions with a 5% solution of sodium hypochlorite. Only a minimal amount of gold in other forms is dissolved by the proposed procedure for three samples tested (precipitated gold, MacInnis Lake Rock, and PTC-1). The present authors must stress the caution of Grégoire (36) that, depending on the state of subdivision, the level of attack of other gold may be different from that recorded in this paper. We find that appreciable levels of gold can be extracted with relatively mild oxidizing agents, depending on grain size and chemical form of gold.

Grégoire also demonstrated that adsorption of extracted gold onto inorganic components of the sample can be significant. In this work up to 30% of the organically bound gold could be lost. The precision of the method is better than 15%.

Table 6.14
Treatment of Sulfide Minerals

Treatment[a]	Procedure
H_2O_2–ascorbic acid	React with a solution of 10 ml 30% H_2O_2 and 25 ml 2% ascorbic acid overnight
$KClO_3$ + HCl followed by 4 *M* HNO_3 (10 min)	500 mg $KC1O_3$ + 10 ml conc. HCl for 30 min. Dilute with 10 ml H_2O. Separate residue, and boil 10 min with 4 *M* HNO_3
$KClO_3$ + HCl followed by 4 *M* HNO_3 (20 min)	Same as above with a 20 min boil in 4 *M* HNO_3
$KClO_3$ + HCl followed by 4 *M* HNO_3 (boiled dry)	Same as above with 4 *M* HNO_3 boiled dry. Wash residue with 10 ml H_2O
4 *M* HNO_3 (10 min)	Boil for 10 min in 4 *M* HNO_3
4 *M* HNO_3 (boil dry)	Same as above with 4 *M* HNO_3 boiled dry. Wash residue with 10 ml H_2O
Aqua regia	Mix with 12.5 ml aqua regia and heat to dryness. Wash residue with 5 ml conc. HCl and 5 ml H_2O
$KClO_3$ + HCl	500 mg $KClO_3$ + 10 ml conc. HCl. React 30 min. Dilute with 10 ml H_2O
$KClO_3$ + HCl (twice)	Same as above with a second treatment
Oxalic acid	Mix with 10 ml 3% oxalic acid 10 min. Dilute with 10 ml H_2O

Source: Ref. 15.
[a]Ground sulfide minerals were mixed in a 1 : 9 ratio with clay loam soil. A 500 mg sample of mixture was used.

Table 6.15
Dissolution of Sulfide Minerals: Result of Treatment[a]

Mineral	H_2O_2–ascorbic acid	$KClO_3$ + HCl followed by HNO_3: 10 min boil	20 min boil	Boiled until dry	4 *M* HNO_3: 10 min boil	Boiled until dry	Aqua regia	$KClO_3$ + HCl: Once	Twice	Oxalic acid
Galena	M	V	V	V	N	N	M	V	V	N
Chalcopyrite	M	E	V	V	M	M	E	M	V	N
Cinnabar	M	V	V	V	M	M	M	E	V	N
Molybdenite	M	M	V	V	M	V	E	N	N	N
Orpiment	N	V	V	V	E	V	V	E	V	N
Pyrite	M	V	V	V	V	E	E	M	E	N
Stibnite	N	V	V	V	M	V	E	V	V	N
Sphalerite	M	V	V	V	E	M	E	V	V	N
Tetrahedrite	M	V	V	V	M	M	M	V	V	N

Source: Ref. 15.
[a]N, Not effective (0–40% dissolution); M, moderately effective (41–70% dissolution); E, effective (71–86% dissolution); V, very effective (86–100% dissolution).

Reagents

All reagents used were of reagent grade. The 5% sodium hypochlorite solution (J. T. Baker, pH 11.5) was used without prior adjustment of pH. Gold standard solutions were prepared immediately prior to use by dilution of stock 1000 ppm gold(III) chloride solution (Aldrich Chemical Co.) with distilled water.

Equipment

All gold determinations were made on a Perkin-Elmer model 5000 atomic absorption spectrophotometer equipped with a model 500 graphite furnace.

Procedure

Treat 10 g of sample twice with separate 50 ml aliquots of 5% sodium hypochlorite solution. Allow the samples to remain in contact with the solution for 24 h. Warm the resulting mixture on a hot plate for 3 h and centrifuge at 2000 rpm for 10 min. Wash the residue twice with distilled water. Combine all leach solutions. Add 20 ml of 12 *M* hydrochloric acid to destroy any remaining hypochlorite. Evaporate to dryness. (Be careful not to bake the residue.) Dissolve the residue in 3 *M* hydrochloric acid. Then dilute with 3 ml hydrochloric acid so that the final volume of solution is 50 ml.

Extract this sample solution with 5 ml of methyl isobutyl ketone (MIBK) in a 125 ml separatory funnel. This is done by shaking the mixture for 10 min using a mechanical shaker. Discard the aqueous layer. Add 10 ml of 0.1 *M* hydrochloric acid (saturated with MIBK) to backwash any coextracted excess iron. Calibrating solutions are subjected to the same procedure.

Place 10 μl of the MIBK into the graphite furnace. Operate the furnace using the following parameters: dry, 100°C for 20 s; ash, 550°C for 20 s; atomize, 2400°C for 5 s. Do triplicate analyses of each extract. The 242.8 nm gold line is employed together with the manufacturer's recommended conditions for the spectrometer.

V. METHODS FOR MERCURY AND HYDRIDE ELEMENTS

A. Determination of Mercury in Soils and Rocks by Cold Vapor Atomic Absorption (17)

Comment on the Method

The sample is prepared by means of acid digestion and wet oxidation. The mercury is reduced and is then partitioned between the liquid phase and a fixed volume of air which is blown through an absorption cell. In this manner, frothing is not a problem, and sharper peak absorbances for mercury are obtained. The method has been validated by comparative analyses of a range of soils and rocks. For 1 g samples, the analytical range is 0.01–0.5 ppm.

Equipment

Atomic absorption spectrophotometer with a mercury vapor cell mounted in place of the burner. The special apparatus required is shown in Fig. 6.2. The sample vessel is the long-necked, 100 ml ground-glass-stoppered Kjeldahl flask (total volume including the long neck, about 200 ml) in which the acid digestion and wet oxidation are carried out. The flask is connected to the air train by flexible silicone rubber tubing fitted with Mohr spring clips to isolate the sample and the fixed volume of air with which mercury is partitioned by manual shaking of the flask and contents. An empty U-tube is used as a safety trap but no drier is necessary. Sample solutions prepared by the oxygen flask combustion method are decanted into the alternative vessel, a conical flask whose flat bottom makes it convenient for carrying out the partitioning operation by magnetic stirring. Isolation of the contents is achieved in this case by nylon stopcocks.

Standard Solutions

From a 1000 mg/ml mercury stock solution of mercury(II) chloride in 0.5 *M* sulfuric acid, a 10 mg/ml mercury solution in 0.5 *M* sulfuric acid stable for at least 1 month is prepared. Solutions containing 0.05, 0.1, 0.2, 0.3, and 0.5 μg Hg/ml are prepared daily by repeated dilution of the 10 mg/ml solution with 0.5 *M* sulfuric acid. Standard samples, analyzed with each set of analysis samples, are prepared by adding 1 ml of these solutions to Kjeldahl flasks with 10 ml aliquots of the nitric–sulfuric acid mixture and 20 ml of water and digesting in the same manner as the samples, using the same volume of potassium permanganate and hydroxylammonium chloride. Errors arising from volume differences and traces of mercury in the reagents

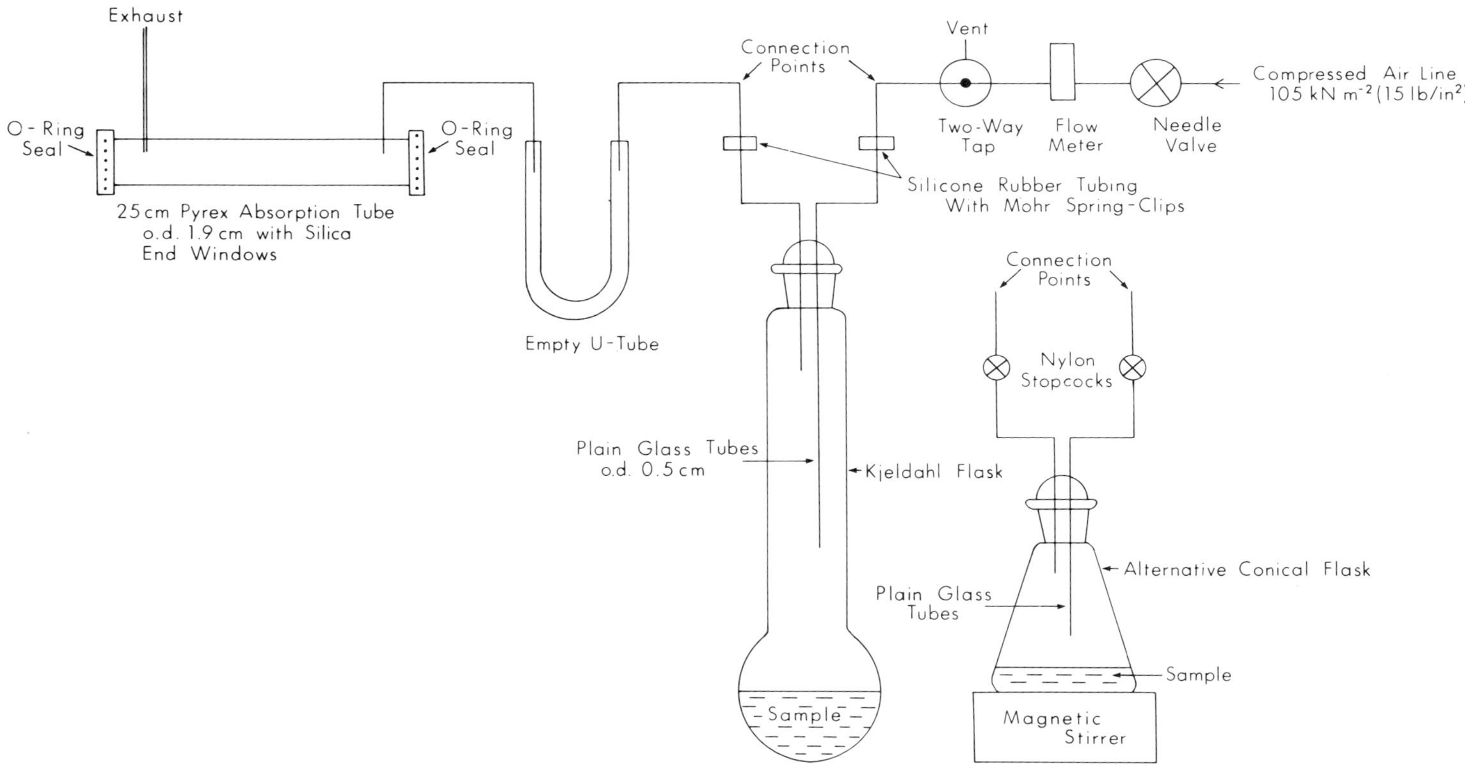

Figure 6.2. Apparatus for the determination of mercury (17).

are thus compensated. No loss of mercury occurred in the digestion of the standard solutions and the determined mercury contents did not differ from those in undigested standard solutions.

Method

Equipment cleaning

Pyrex glassware is washed in tap water, followed by 6 *M* hydrochloric acid, 8 *M* nitric acid, and several rinses in glass-distilled water. Ground-glass-stoppered Kjeldahl flasks are rinsed in tap water, brushed clean using Teepol, then rinsed thoroughly in tap water before being washed with acid and distilled water as above. Kjeldahl flasks in regular use are soaked once a week, or whenever stains develop, in a solution of Decon 90 (Decon Laboratories Ltd., Portslade, Brighton, UK), rinsed in tap water, and washed in acid and distilled water as above.

Sample preparation

Place the required weight of finely ground (<150 μm) air-dried (<30°C) soil or peat in a 100 ml Kjeldahl flask. Add 10 ml of a mixture of redistilled nitric acid and concentrated sulfuric acid (1 + 1) and swirl the contents to wet the sample thoroughly. The weight taken depends on the expected mercury content and the amount of oxidizable matter. It is generally 0.5–2.0 g, the sample size being limited by the oxidizing ability of the 40 ml of permanganate solution that the flask can conveniently hold; with peat, for example, 0.5 g is the maximum. Digest the sample unstoppered for 2 h in a water bath at 60°C, with swirling at intervals, only the bulb of the flask being immersed so that its long neck acts as an air condenser. Times of digestion from 1 to 3 h do not alter the recovery of mercury. Cool the flask, add 20 ml of glass-distilled water, cool the flask again, and add 5 ml aliquots of 6% potassium permanganate solution until the purple color is retained for at least 1 h. By grouping samples of similar organic content for analysis, the same volume of potassium permanganate solution can be used for each sample of a set. Leave the sample glass-stoppered at room temperature overnight. Add 20% (w/v) hydroxylammonium chloride slowly until the brown hydrated manganese oxides and the excess of potassium permanganate are reduced, giving a clear solution with a residue of insoluble material. Leave the unstoppered flask for 1 h to allow the evolved gases to be liberated; no loss of mercury is observed at this stage. Use this solution for the determination of mercury.

Procedure

Clean the apparatus before use and between samples by attaching a Kjeldahl flask, containing 10–20 ml of distilled water, to the air train. Attach the clips and shake the flask manually. After repeating with a fresh aliquot of distilled water, remove the clips, open the two-way tap, and pass air at 10 liters/min through the flask and measuring system until no absorption is obtained. Reduce the air flow to 3 liters/min, set the two-way tap to vent, and replace the clips.

Measure the mercury content of a sample by adding 5 ml of 20% $SnCl_2 \cdot 2H_2O$ in 6 *M* hydrochloric acid to the sample solution prepared as described above in

Table 6.16
Interferences in the Determination of Mercury

Added	Quantity	Effect[a]
Aluminum chloride	0.5 g	None
Calcium carbonate	0.5 g	None
Iron(II) sulfate	0.5 g	None
Magnesium oxide	0.5 g	None
Calcium phosphate	0.5 g	<10% depression
Copper (as sulfate)	250 μg	None
Silver (as nitrate)	100 μg	None
Sulfur	1000 μg	None
Selenium (as acid)	400 μg	90% depression[b]
Selenium (as acid)	40 μg	8% depression[b]
Selenium (as acid)	4 μg	None
Iodine (as KI)	100 μg	10% depression

Source: Ref. 17.
[a]0.1 μg of mercury.
[b]0.2 μg of mercury.

a Kjeldahl or conical flask, which is immediately fitted to the apparatus to prevent loss of mercury. After agitation for 2 min (manual shaking for Kjeldahl, magnetic stirring for conical flask), remove the clips (or open the stopcocks) and blow air through at 3 liters/min. Measure the absorption at 253.7 nm with an atomic absorption spectrophotometer of conventional design fitted with a pen recorder with a chart speed of 2.5 mm/min. Linear calibration curves of absorbance versus mercury content are obtained up to about 0.3 μg, with slight curvature from 0.3 to

Figure 6.3. Interference by gold (•), platinum (▲) and palladium (■) on the determination of 0.25 μg (full line) and 0.1 μg (dashed line) of mercury (17).

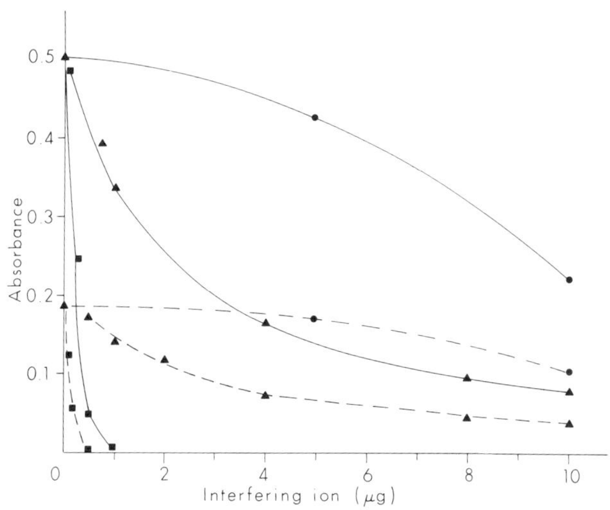

Table 6.17
Reproducibility of Complete Analytical Procedure Based on Acid Digestion and Wet Oxidation[a]

Sample	Sample weight (g)	Average Hg found[b] (μg)	s (μg)	s_r (%)
Soil 1	1.000	0.073	±0.005	±7
Soil 2	1.000	0.134	±0.008	±6
Soil 3	1.000	0.184	±0.012	±7
Peat 1	0.500	0.082	±0.004	±5

Source: Ref. 17.
[a] s, Reproductibilty; s_r, coefficient of variation.
[b] Average of 10 determinations.

0.5 μg. For a 1 g sample, the analytical range of 0.01–0.5 μg of mercury corresponds to 0.01–0.5 ppm mercury in the soil.

Interferences

The effects of several elements on the determination of mercury are listed in Table 6.16. Interferences of gold, platinum, and palladium on mercury are shown in Fig. 6.3. Molecular absorption and light-scattering interference effects were not significant in the studies of three soils and three rocks.

Reproducibility

The reproducibility and coefficients of variation are shown in Table 6.17.

B. Determination of Mercury in Rocks, Soils, and Sediments by Flame and Flameless Atomic Absorption (18)

Comments on the Method

This is an alternative to the cold vapor atomic absorption method. It is rapid, sensitive, and free of interference from major and trace elements found in geological procedures. Mercury is extracted from an acid digestate of the sample into methyl isobutyl ketone as the tetraiodomercurate. The mercury is determined by either the flame or carbon rod atomization method. Silver is used as a matrix modifier in the carbon rod atomizer. Eight standard reference materials have been analyzed satisfactorily.

Equipment

- Varian AA-6 atomic absorption spectrometer
- Varian model 63 carbon rod atomizer

- Varian model 53 automatic sample dispenser
- Deuterium lamp background corrector

Preparation of Sample

Weigh a 1.00 g sample (less than 100 mesh) into a 25 × 200 mm culture tube. Add 2 ml of concentrated nitric acid and 5 ml of concentrated hydrochloric acid. Heat the tube in a boiling-water bath for 20 min, and agitate the contents occasionally. Remove the tube, and adjust the volume to 20 ml with water. Cool the sample and then add 20 ml concentrated ammonium hydroxide. Mix the sample, and cool it again. Add 1.0 ml of 20% potassium iodide solution and mix. Next add 0.2 ml of 0.01 *M* silver nitrate solution (freshly prepared from a 1.0 *M* stock solution) and mix. Add 2 ml of methyl isobutyl ketone, stopper the tube, and shake the contents for 1 min. Centrifuge to separate the organic layer.

Determine the mercury in the organic layer by either the flame or carbon rod method. Measure peak heights. Inject volumes of 5 or 10 μl of solution from the automatic sample dispenser for the carbon rod atomizer. The operating conditions for the instrument are given in Table 6.18.

Calibration

Prepare some mercury-free solids from medium-textured soil or sediment ground to the same fineness as the sample. Heat the solids at 800°C for 4 h in a muffle furnace to drive off any mercury. Weigh 1.00 g samples into culture tubes. Add measured volumes of 10 or 100 μg/ml mercury solutions for carbon rod or flame atomization, respectively. Carry the standards through the procedure described above for the samples. The calibration curves are linear.

Use carbon rod atomization for mercury contents of 0.05–10 ppm in the sample. The carbon rod can be fired 200 times. For mercury concentrations of 1–200 ppm, use flame atomization.

Table 6.18
Instrument Operating Conditions

Source	Mercury hollow-cathode lamp
Wavelength	253.7 nm
Lamp current	2.5 mA
Slit width	0.5 nm
Flame atomizer	Fuel lean air/acetylene
Carbon rod atomizer	
Dry	100°C, 25 s
Ash	150°C, 5 s
Atomize	1200°C
Nitrogen flow meter	8.0

Source: Ref. 18.

Sensitivity

- Carbon rod atomizer, 2.2×10^{-10} g mercury (for 1% absorption)
- Flame atomizer, 0.73 μg/ml mercury (for 1% absorption)

C. Determination of Arsenic in Geological and Environmental Samples by Hydride Generation (Atomic Absorption) (19)

Comment on the Method

This is a rapid, sensitive, and accurate method which requires only a small amount of hydride generation apparatus. The fusion procedure decomposes most siliceous materials. It is also suitable for samples containing only small proportions of silicate and those containing large amounts of material forming sulfate precipitates. Results of analyses of standard reference materials have been in good agreement with accepted values. The precision is ±5%.

Equipment

- Commercial atomic absorption spectrometer, e.g., Perkin-Elmer model 603 with automatic background correction.
- Strip chart recorder.
- Absorption cell made of silica with dimensions 10 cm × 10 mm inner diameter (i.d.). It is heated by means of Chromel C resistance wire (0.7 ohm/30 cm). The windings must be close to but not touching each other. A thickness of 0.8 cm asbestos insulation is applied over the windings. The furnace is heated to 700–800°C by variac control.
- Silica tube 10 cm × 5 mm (i.d.) joined at right angles to the center of the atomizer. Nitrogen and hydrogen gas supplies are connected to the silica tube by polyethylene tubing. The absorption cell is fastened to the burner mount of the spectrometer and centered in the optical path.
- Arsine generator. This is a 50 ml plastic syringe fitted with a No. 20 stainless steel needle. Each syringe lasts for 100–200 injections.
- Sample digestion block made of aluminum (7.6 cm thick) with holes (2.54 cm in diameter, 6.2 cm deep) to hold 50 ml Pyrex Folin digestion tubes (calibrated).
- Nickel crucibles. These must be cleaned with steel wool rather than with acids. If acid cleaning is necessary, a blank fusion should be made before another sample is analyzed.

Instrument Conditions

The parameters are those recommended by the manufacturer. An arsenic electrodeless discharge lamp is the preferred source of the 193.7 nm line, but a hollow-cathode lamp can also be used.

The flows of nitrogen and hydrogen are 225 and 75 ml/min, respectively.

Solutions

- Sodium borohydride, 2% (w/v): Dissolve 2 g of the borohydride salt in 100 ml of water containing one pellet of potassium hydroxide. Filter the solution through a 0.45 μm porosity membrane. The sodium borohydride must be purchased in glass containers because metal containers contaminate the product.
- Hydrochloric acid: Must contain as small an amount of arsenic as possible. During this study, Fisher reagent grade acid proved to be the best.
- Arsenic standard solutions: Dissolve 1.320 g of primary standard arsenious oxide in 25 ml of 20% (w/v) potassium hydroxide solution. Neutralize the solution to a phenolphthalein end point with 20% (v/v) sulfuric acid. Transfer the solution to a 1 liter volumetric flask. Add more sulfuric acid to produce a final acid concentration of 1%. Then dilute to the mark with water. This solution contains 1000 μg/ml arsenic.

Prepare a stable stock solution of As(V). Transfer 100 ml of 1000 μg/ml arsenic(III) solution prepared as above to a 1 liter volumetric flask. Add 10 ml nitric acid, 2 ml sulfuric acid, and 2 g potassium persulfate. Heat the sample until white fumes are evolved. Cool and dilute to the mark with water. This stock solution (100 μg/ml) is stable for at least 3 months.

Working solutions of 1, 2, 5, 10, 20, 50, 75, and 100 μg/liter are prepared to contain 10% hydrochloric acid. They are stable for at least 1 month.

Preparation of Samples

Fusion method

Use this method for total arsenic in samples which contain appreciable siliceous matter. Place a 0.1–0.2 g sample in a 30 ml nickel crucible. For samples containing appreciable organic matter, add an equal weight of magnesium oxide. Add 1 or 2 drops of ethanol, mix to a slurry with several drops of water, and swirl to homogenize the mixture. Add about 1 g of potassium hydroxide pellets (or about 10 pellets) for each 0.1 g of sample. Dry the mixture at 100°C. Fuse the sample in a muffle furnace at 550°C for 15 min. Cool the sample and leach it with small volumes of water. Combine the extracts in a 50 ml volumetric flask. Remove any solids adhering to the crucible with a rubber policeman. Transfer the entire sample to the flask. Add 11 *M* hydrochloric acid to the flask to make the final acid concentration 10% (v/v). Dilute to volume with water and mix. Allow any solids to settle before generating the hydride. Filter any turbid samples.

Acid digestion method

Use this method for organic samples which do not contain appreciable siliceous matter and for the determination of mineral acid-extractable arsenic in other samples. Several samples can be digested simultaneously. Transfer 0.1–0.2 g to a 50 ml digestion tube. The tube must be graduated at 25 and 50 ml volumes. Place the tube in the aluminum block and add 3 ml of 18 *M* sulfuric acid. Heat the block on a hot plate at 350°C. Add 0.5 ml volumes of 30% hydrogen peroxide from time to time until the liquid is clear. Cool the sample and add 5 ml of 11 *M* hydrochloric acid. Dilute to the 50 ml mark with water.

Hydride Generation

Draw a 5 ml volume of sample or standard solution containing 1–500 μg of arsenic into the syringe. Exclude air from the syringe. Dip the needle into 2% sodium borohydride solution and withdraw the plunger in one quick motion to the 50 ml mark. This allows approximately 1.5 ml of sodium borohydride solution to enter the syringe. Invert the syringe, and seal and hold the needle opening with a tissue. Shake the syringe for 20 s and then insert the needle through the plastic tubing at a point 5–8 cm from the end of the glass tube on the atomizer. Slant the syringe to prevent any solution from entering the atomizer. Place a 3.18 cm length of thick-walled rubber tubing (4.85 cm in diameter) on the ribs of the plunger to prevent the plunger from returning to the zero position. In this way, the sample is injected at a uniform rate without concern for solution entering the furnace. Monitor the arsenic peak on the recorder. Remove the needle. Discard the spent solution and rinse the syringe with 10% hydrochloric acid.

Interferences

Potential interferences and the concentrations at which they cause problems are listed in Table 6.19.

D. Determination of Arsenic in Soils, Rocks, Gossans, and Ores by Colorimetry (20)

Comments on the Method

An HF-aqua regia decomposition is used. The arsenic is evolved from solutions containing potential interferents as the covalent hydride. Zinc pellets are used as a reductant in the formation of AsH_3. Arsine is reacted with silver diethyldithiocarbamate in the presence of quinoline to form a color. Between 1 and 800 ppm arsenic can be determined by the procedure. No appreciable interference was noted for 20% Na, K, Mg, Ca, Al, and Fe, 2% Ti, V, Cr, Mn, Co, Ni, Cu, Pb, Ba, Sr, P, Mo, Bi, and U, and 0.2% Ag and Au. The greatest weakness of the method is that no results are reported for actual samples.

Reagents

- Fe/KI solution: 14 g ferrous ammonium sulfate hexahydrate plus 20 g potassium iodide dissolved in 400 ml of approximately 3% sulfuric acid. Add a trace of ascorbic acid to prevent oxidation.
- $SnCl_2$ solution: Prepare a 5% solution of $SnCl_2 \cdot 2H_2O$ in 50% hydrochloric acid by dissolving the salt in hot concentrated hydrochloric acid first and then adding an equal volume of water. Store in presence of tin metal.
- Zinc, Arsenic-free grade pellets, Merck.

Table 6.19
Interferences in As Determination

Cation	Solution concentration of cation (μg/ml)	Concentration equivalent for 1 g sample in 100 ml solution
Ca^{2+}	4000	40%
Al^{3+}	4000	40%
Mg^{2+}	4000	40%
Na^{+}	1000	10%
K^{+}	1000	10%
Li^{+}	1000	10%
Fe^{2+}	750[a]	7.5%
Pb^{2+}	200[a]	2%
Zn^{2+}	200	2%
Mn^{2+}	200	2%
Ba^{2+}	200	2%
Cr^{3+}	100[a]	1%
Co^{2+}	100[a]	1%
Cd^{2+}	100[a]	1%
Cu^{2+}	10[a]	1000 ppm
Ni^{2+}	6[a]	600 ppm
Ag^{+}	5	500 ppm
Ge^{4+}	1[a]	100 ppm
Sb^{3+}	0.06[a]	6 ppm
Sn^{2+}	0.08[a]	80 ppm
Bi^{3+}	0.6	60 ppm
Se^{2+}	0.04[a]	4 ppm
Te^{2+}	0.3[a]	30 ppm

Source: Ref. 19.
[a]Threshold values.

- HCl/EDTA solution: Dissolve 1.5% EDTA disodium salt in hot distilled water and then add hydrochloric acid to bring acidity to 10%. Prepare fresh daily as EDTA crystallizes out on standing.
- Arsine reagent solution: Dissolve, with warming, 1.000 g of silver diethyldithiocarbamate (Merck) in 200 ml chloroform + 100 ml quinoline. Store under refrigeration and keep away from light.

NOTE: The reagent solution should be pale yellow in color; avoid the use of old stocks of quinoline (brown) or silver diethyldithiocarbamate (dark yellow powder).

Silver diethyldithiocarbamate (AgDDC) can be readily prepared in pure form as follows: Dissolve silver nitrate and sodium diethyldithiocarbamate in water in separate beakers. Add an excess of sodium diethyldithiocarbamate to the silver nitrate solution and filter off the creamy yellow precipitate using a Buchner funnel. Wash with distilled water and dry in a desiccator—keep away from strong light.

Purify as follows: Dissolve to make a saturated solution in hot chloroform; then filter through siliconized (phase-separating) filter paper to remove solids and water. Add an excess of ethyl alcohol to the cold filtrate, allow the compound to crystallize out, filter through a Buchner funnel, and suck dry. The product is a pale yellow crystalline solid with a soapy feel and is similar in appearance to the sodium salt. It should be stored in the dark.

Equipment

Any modern spectrophotometer may be used. The wavelength for maximum absorption is 545 nm.

The arsine generator is shown in Fig. 6.4.

Procedure

Heat 0.200 g of finely powdered sample with 4 ml of a solution of 2 parts hydrofluoric acid (40%) and 1 part aqua regia in a Teflon dish for 30 min or until the sample is completely decomposed. Evaporate the solution to dryness. Take up the residue in 10 ml of 10% HCl by warming. Place a 1 to 5 ml aliquot in the arsine generator and proceed as described below.

Add 1 ml of the appropriate As^{5+} standard in 10% HCl to a test tube. Add 2 ml of Fe/KI solution and warm, followed by 2 ml of $SnCl_2$ solution while hot. Allow to cool, and add 5 ml of 1.5% EDTA in 10% HCl. Add four zinc pellets

Figure 6.4. Arsine generator/absorber apparatus (20).

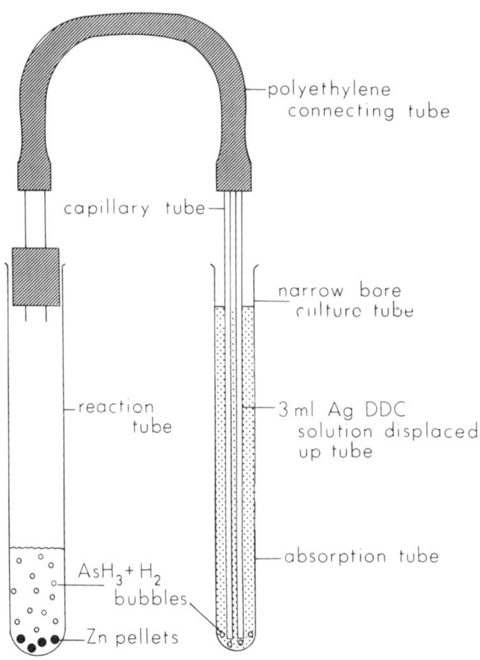

and absorb arsine generated in 3 ml of AgDDC solution (see Fig. 4). Determine absorbance of the solution in reagent tube at 545 nm after 45 min gas evolution. Use standard As^{5+} solutions containing 0–15 μg As for calibration.

For routine determinations on geochemical samples, take 1–5 ml aliquots of the sample digest (×50 dilution factor) previously described and follow with additions of Fe/KI, $SnCl_2$, and EDTA solutions.

Optimum working ranges are 1–160 ppm As in the sample for a 5 ml aliquot and 5–800 ppm for a 1 ml aliquot.

E. Determination of Arsenic, Antimony, and Bismuth in Soils and Sediments by Hydride Generation (ICP–AES) (21, 22)

Comment on the Method

Traces of arsenic, antimony, and bismuth are determined simultaneously by volatile hydride generation and ICP–AES. Samples are attacked with hydrochloric acid at 150°C in sealed tubes. The detection limits for the three elements are about 0.1 μg per gram. The results of analyses of four samples of standard soils were very close to recommended values, but generally slightly lower. Selenium and tellurium could not be dissolved in this procedure. Bismuth was not recovered quantitatively from organic-rich samples.

Equipment and Reagents

- Applied Research Laboratories 29000B quantometer connected to a Radyne R50 plasma generator and a continuous-flow hydride generator. A schematic diagram of the system is shown in Fig. 6.5. Radiation from the torch was focused on the

Figure 6.5. Schematic diagram of the hydride reduction–ICP spectrometry system (21).

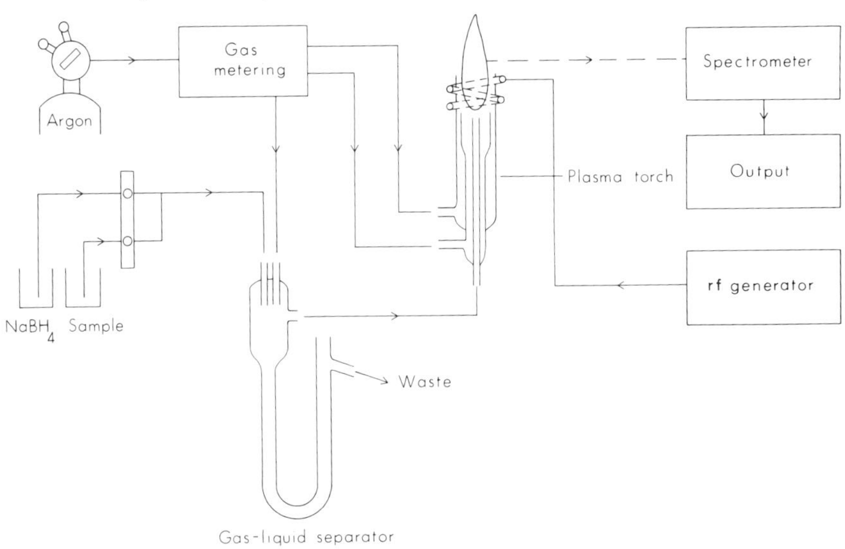

entrance slit by means of a silica lens about 70 mm from the slit. The output of the time-integrated signals was through a digital voltmeter to an IBM typewriter and 545 card punch.

- Watson-Marlow MRHE 200 flow-inducer peristaltic pump. The connections were made with 0.5 and 0.8 mm i.d. tubing. This system produced pulsations at 6.5 Hz, but the resulting flicks of the plasma caused no problem since an integration time of 20 s was used.
- Reaction and separation chamber. This chamber is illustrated in Fig. 6.6.
- Reaction tubes. The test tubes were 160 by 16 mm Pyrex with screw caps. The caps were fitted with liners specially made from a layer of silicone rubber (Esco type SR70, 3.2 mm thickness). The rubber was covered with a fluorocarbon film (Teflon FEP Type A, 0.36 mm thickness).
- Heating block. The aluminum block was drilled to hold the tubes to a depth of about 65 mm. The caps and tops of the tubes were cooled during digestion by means of a current of air from a blower.
- Sodium tetrahydroborate in 0.1 *M* sodium hydroxide, 1% (w/v). The reagent could be pumped with no outgassing of hydrogen. The half-life at 20°C was greater than 100 h.

Procedure

Transfer 0.25 g sample and 5 ml of hydrochloric acid to a test tube. If the sample contains appreciable carbonates, add the acid slowly, and then set the tube aside until the evolution of gas ceases. Cap the tubes, and place them in the heating block at 150°C for 2 h. Remove them, and then cool them rapidly in cold water.

Figure 6.6. Reaction and separation chamber (21).

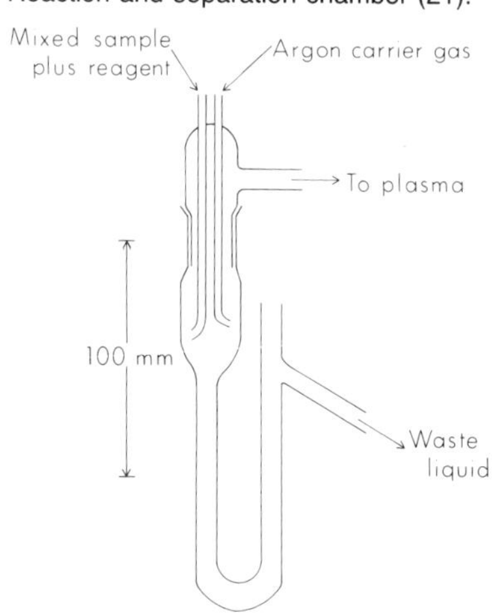

Table 6.20
Operating Conditions

Plasma	
Torch	Fassel type
Power forward	2.7 kW
Viewing height	10 mm above load coil
Plasma support gas flow rate	0.8 liter/min argon
Coolant gas flow rate	17 liters/min argon
Integration time	20 s
Preintegration	15 s
Reduction	
Sample acidity	5 *M* hydrochloric acid
Sample flow rate	9.2 ml/min
Reagent concentration	1% (w/v) sodim tetrahydroborate in 0.1 *M* sodium hydroxide solution
Reagent flow rate	4.5 ml/min
Carrier gas flow rate	0.8 liter/min argon
Arsenic wavelength	228.81 nm
Antimony wavelength	231.15 nm
Bismuth wavelength	306.77 nm

Source: Ref. 21.

Remove the caps. Add 5 ml of 0.2% (w/v) potassium iodide. Mix the contents by shaking the tube. Allow the solids to settle for 4 h. Use the solution for the measurements.

Analyze the sample with the operating conditions shown in Table 6.20. Use a calibrating solution containing 100 ng/ml arsenic(III), antimony(III), and bismuth(III) in 1 : 1 hydrochloric acid. Subtract a blank. The calibrations are linear with concentration up to 800 ng/ml arsenic, 1500 ng/ml antimony, and 500 ng/ml bismuth. Check the calibration after each batch of 10 samples.

Table 6.21
Results for the Canadian Certified Materials Project Standard Soils Compared with the Recommended Values

Sample[a]	Result	As (mg/kg)	Sb (mg/kg)	Bi (mg/kg)
SO-1	Found	1.94	0.12	0.24
	Recommended	1.9 ± 0.3	0.2	0.5
SO-2	Found	0.77	0.1	0.03
	Recommended	1.2 ± 0.2	0.1	0.1
SO-3	Found	2.32	0.22	0.03
	Recommended	2.6 ± 0.1	0.3	0.1
SO-4	Found	6.45 ± 0.07	0.25 ± 0.03	0.19 ± 0.02
	Recommended	7.1 ± 0.7	0.7	0.1

Source: Ref. 22.
[a]SO-1, SO-2, and SO-3 are single determinations; SO-4 is the mean and standard deviation of eight determinations on separate portions of the material.

Table 6.22
Results Obtained by the Method for Some USGS Standard Rocks Compared with Accepted Values

Sample	Result	As (mg/kg)	Sb (mg/kg)	Bi (mg/kg)
W-1	Found	0.89	1.18	<0.04
	Accepted	1.9	1.0	0.046
G-1	Found	0.50	0.55	0.12
	Accepted	0.5	0.31	0.065
G-2	Found	0.22	0.12	0.12
	Accepted	0.25	0.1	0.043
GSP-1	Found	0.12	3.20	0.18
	Accepted	0.09	3.1	0.037

Source: Ref. 22.

Comparisons of the Results of Analyses of Standards

Results of analyses of four standard soil samples from the Canadian Certified Materials Project and four USGS standard rocks are contained in Tables 6.21 and 6.22. The practical detection limit for all three elements is about 0.1 μg/g. Approximately 200 samples can be analyzed by one person in a 2 day cycle.

F. Determination of Antimony in Geological Materials by Hydride Generation (Atomic Absorption) (23)

Comment on the Method

The sample is decomposed in aqua regia. Antimony hydride is generated and measured in a heated quartz tube. The method is free of interference from iron, cobalt, nickel, copper, and silver at levels of 100 times that of antimony. The detection limit is 80 μg/g, and up to 50 samples per day can be analyzed. The results of analyses of 15 standard reference materials were in good agreement with other accepted values. The same method can be used for the determination of arsenic.

Equipment

- Perkin-Elmer model 303 atomic absorption spectrometer with background correction.
- Coleman model 56 strip chart recorder.
- Antimony electrodeless discharge lamp.
- Hydride generation cell and atomizer shown in Fig. 6.7. The generator (cell A) is an 18 × 116 mm plastic test tube. A hole drilled 30 mm above the base accommodates the tip of an Eppendorf pipet containing the sample. Nitrogen sweeps the hydride through a Tygon tubing connection to the quartz cell. The atomizer is mounted 25 mm above a triple-slot air/acetylene burner. The nitrogen

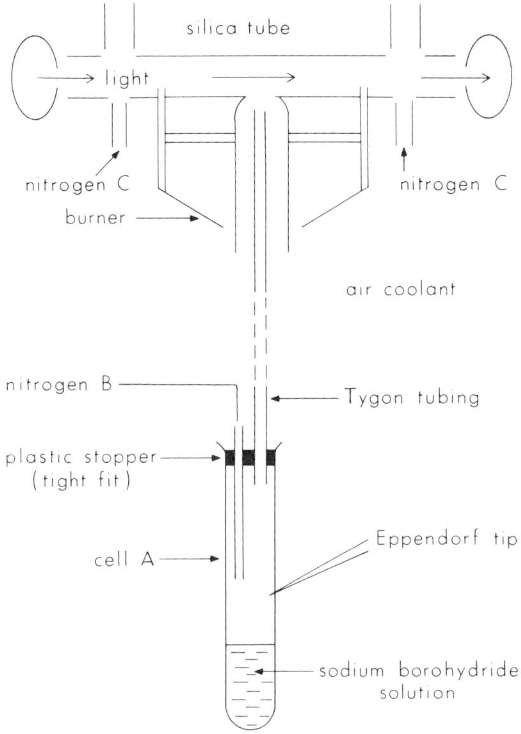

Figure 6.7. Hydride generator and atomizing cell (23).

flowing at either end of the atomizer prevents the combustion of hydrogen. The atomizer is heated by the flame to decompose the hydride.

The instrument conditions include the following: nitrogen flow rate through the generator cell, 2.0 liters/min; wavelength, 217.6 nm; spectral bandpass, 0.2 nm.

Procedure

Weigh a 0.25 g sample into a calibrated test tube. Add 10 ml of aqua regia. Allow the sample to react for 1½ h in the fume hood and then for 2 h in a 90°C water bath. Mix the contents several times during this period. Cool the sample, and make up the volume to 10 ml with aqua regia. Mix the contents, and allow the solids to settle. Transfer an aliquot of 100 μl or less of the solution to a 10 ml volumetric flask. Dilute to volume with 0.5 *M* hydrochloric acid.

Transfer 2 ml of 2% sodium borohydride solution to the generator (the solution must be prepared for each set of measurements). Inject a 500 μl sample (diluted as described above). Use the recorder peak height as a measure of the antimony concentration. Replace the generator cell by another for the next sample. The detection limit is 0.4 μg/ml and the standard deviation 3.1%.

Calibration

Standard solutions containing 0 to 50 μg/ml antimony are used for the construction of a calibration curve. Make the final dilution in 10% aqua regia and 90% 0.5 *M* hydrochloric acid. The calibration curve is linear over this range.

Interference

The signal for antimony is depressed by large amounts of nickel and silver. The interferences can be overcome by addition of EDTA to the final solution. A concentration of 0.01 *M* EDTA is sufficient for 5000 ppb of nickel and 1000 ppb of silver. When large quantities of nickel are present in the sample, a smaller weight, e.g., 0.1 g, should be taken for analyses.

G. Determination of Bismuth in Soils and Rocks by Flameless Atomic Absorption (24)

Comment on the Method

Small samples are fused with sodium hydrogen sulfate. Bismuth in the resulting solution is complexed with APDC and extracted into methyl isobutyl ketone. Quantities as small as 50 ppb of bismuth are determined by means of a graphite furnace. Reference samples of rock and soil were analyzed and satisfactory results were obtained.

Equipment

- Perkin-Elmer model 306 atomic absorption spectrometer with a deuterium background corrector
- Perkin-Elmer model HGA-2100 graphite furnace atomizer
- Perkin-Elmer model 56 strip chart recorder

The instrument conditions are shown in Table 6.23.

Reagents

- Ammonium citrate–EDTA buffer: Dissolve 294 g of sodium citrate dihydrate, 50 g of the disodium salt of EDTA, and 267 ml of concentrated ammonium hydroxide in water. Dilute to 1 liter.

Table 6.23
Instrument Conditions

Wavelength	233.7 nm
Slit width	0.2 nm
Lamp	Perkin-Elmer intensitron bismuth hollow cathode
Furnace	
Dry	125°C, 10 s
Char	400°C, 20 s
Atomize	2000°C, 8 s
Argon flow rate	40 ml/min

- Ammonium 1-pyrrolidinedithiocarbamate (APDC), 1% (w/v) in water.
- Bismuth standard solutions: Prepare standard solutions containing 10, 0.1, and 0.01 μg/ml from a 1000 μg/ml solution. The two most dilute solutions must be prepared daily and the 10 μg/ml solution weekly. They should contain a small amount of nitric acid.

Procedure

Weigh 0.2 g of finely powdered sample into a Pyrex culture tube (18 × 150 mm). Add 1 g of fused sodium hydrogen sulfate. Fuse the sample for 2 or 3 min to evolve dense white fumes. Cool the sample, and then dissolve the melt in 1 ml of hydrochloric acid and 6 ml of water. Warm the sample. Filter the mixture through a No. 41, 9 cm diameter paper. Collect the filtrate in a 16 × 150 mm screw-cap culture tube. Rinse with 2 ml of hot water.

Add 5 ml of EDTA buffer, 1 ml of concentrated ammonium hydroxide, 1 ml of APDC reagent, and then 1 ml of methyl isobutyl ketone (MIBK). Mix after each addition. Cap the tube, and shake it for 5 min on a mechanical shaker. Then centrifuge the tube for 1–2 min to separate the layers. Analyze a 10 μl aliquot (or other suitable volume) in the graphite furnace, using peak heights as a measure of concentration.

Calibration

Prepare a series of calibration standards by transferring the volumes of bismuth standard solutions shown in Table 6.24 to screw-cap culture tubes.

Add 1 ml of concentrated hydrochloric acid and any water required to bring the final volume to 10 ml. Then add, in order, 5 ml of EDTA buffer, 1 ml of concentrated ammonium hydroxide, 1 ml of APDC reagent, and 1 ml of MIBK. Mix after each addition. Shake each tube and centrifuge as in the analysis of samples. Use 10 μl aliquots for the determination of bismuth.

Table 6.24
Bismuth Calibration Standards

Bismuth standard solution (μg/ml)	Volume (ml)	Bismuth content of calibration standard (ng/ml of MIBK)
0.01	0.5	5
0.01	1.0	10
0.01	2.0	20
0.1	0.5	50
0.1	1.0	100
0.1	3.0	300
0.1	5.0	500
0.1	7.0	700
0.1	10.0	1000

H. Determination of Selenium in Geological Materials by Flameless Atomic Absorption (25)

Comments on the Method

This is a rapid, sensitive method for selenium in geological materials. Interferences have been overcome. After the sample has been dissolved in acid, selenium(IV) is extracted into toluene. The toluene solution is analyzed directly for sample selenium levels above 0.2 ppm. For selenium levels between 0.05 and 0.2 ppm, the selenium is reextracted into dilute hydrochloric acid. The results of analyses of standard reference materials GXR-1 to GXR-6 (soil, ore, rock) and glass were in good agreement with accepted values. Relative standard deviations ranged from 4.1 to 8.8%.

Equipment

- Varian AA6 atomic absorption spectrometer with a simultaneous background corrector
- Varian model 63 carbon rod atomizer
- Varian model 53 automatic sample dispenser

The instrument operating conditions are shown in Table 6.25.

Reagents

- Iron solution, 10% (w/v): Dissolve 10 g of Specpure iron powder in 53 ml of concentrated hydrochloric acid while heating on a hot plate; then add 5 ml of 50% hydrogen peroxide. Heat to remove the excess of peroxide and dilute to 100 ml with water. The residual acidity of this solution is approximately 1 *M* as hydrochloric acid.
- Nickel solution, 0.1% (w/v): Dissolve 0.495 g of $Ni(NO_3)_2 \cdot 6H_2O$ in 100 ml of water.
- Nickel–hydrochloric acid solution, 0.1% (w/v) nickel in 0.01 *M* hydrochloric acid: Dissolve 4.95 g of $Ni(NO_3)_2 \cdot 6H_2O$ in 50 ml of water, add 0.83 ml of concentrated hydrochloric acid, and dilute to 1 liter with water.

Table 6.25
Instrument Operating Conditions

Source	Selenium electrodeless discharge lamp
Wavelength	196.0 nm
Power	8 W
Slit width	1.0 nm
Dry	100°C, 30 s
Ash	400°C, 40 s
Atomize	2500°C, 2 s

Source: Ref. 25.

- Selenium stock solution, 1000 μg/ml: Dissolve 2.190 g of Na_2SeO_3 in 50 ml of water, add 10 ml of concentrated hydrochloric acid, and dilute to 1 liter with water.
- Selenium standard solutions, 1 and 10 μg/ml selenium in 1% hydrochloric acid.

Preparation of Sample

Weigh a 0.5 g sample of rock, soil, or stream sediment ground to less than 100 mesh into a 50 ml Teflon beaker. Add 1–2 ml of water and then 10 ml each of nitric, perchloric (70%), and hydrofluoric acids. Heat the beaker on a hot plate set at 150°C and gradually raise the temperature to 220°C over a 1½ h period. When the liquid volume has been reduced to 1–2 ml, remove the beaker from the hot plate. Transfer the sample to a 25 × 150 mm tube by means of two 2.5 ml volumes of 3 *M* hydrochloric acid. Prepare selenium working standards by transferring 0, 0.1, 0.5, 1, 3, 5, and 10 μg of selenium from the standard selenium solutions into similar test tubes. Add to each standard 5 ml of 3 *M* hydrochloric acid. Place sample and standards in an aluminum heating block at 100°C for 5 min. Cool the tubes to room temperature. Add 10 μl of iron solution and mix. Add 10 ml of concentrated hydrochloric acid and mix again.

Extraction of Sample

Add 2 ml of toluene to the test tube. Shake the tube for 5 min and then centrifuge. Pipet 1 ml of the toluene extract into a 16 × 150 mm tube, and cap the tube. Remove the remainder of the toluene with an Eppendorf pipet and discard it. Add 2 ml of toluene to the sample, and repeat the extraction. Again, remove 1 ml of the toluene layer, combine it with the first extract, and mix. Use 5 or 2 μl aliquots for selenium analyses. Pipet a 5 μl volume of nickel solution into the atomizer immediately prior to the injection of the toluene extract. A linear relationship between absorbance and amount of selenium exists for 0–5 μg of selenium in 5 μl aliquots and 3–10 μg of selenium in 2 μl aliquots. These amounts correspond to 0.2–20 ppm in the sample. For selenium levels between 20 and 200 ppm make a 1 : 9 dilution of the organic layer.

Back-Extraction of Samples

For selenium contents between 0.05 and 0.2 ppm, prepare a set of standards to contain 0.025, 0.1, 0.5, 1, 3, and 5 μg of selenium. Treat the sample and standards in the manner described above until after the first centrifugation. Then transfer the entire toluene layer into a 16 × 150 mm screw-capped tube. Add 2 ml of toluene to sample and standards. Shake the tubes for 5 min and centrifuge. Remove the toluene layer and combine it with the first extract. Pipet 1 ml of nickel–hydrochloric acid solution into the combined toluene extract, shake the tube for 5 min, and centrifuge. Use the aqueous phase for the determination of selenium by means of the sample dispenser. Linear relationships exist between absorbance and amount of selenium over the ranges 0.025–1 μg of selenium in 10 μl aliquots and 0.25–5 μg of selenium in 2 μl aliquots. These amounts correspond to 0.05–10 ppm in the sample.

Interferences

There are no interferences from major and trace elements commonly encountered in geological materials. Arsenic in concentrations above 0.10% in the sample interferes by hindering the extraction of selenium into toluene. A smaller sample size (less than 0.50 g) should be used when the sample contains more than 0.1% arsenic. Nitric acid interferes seriously and it must be removed by evaporation.

I. Determination of Trace Concentrations of Selenium in Soils and Sediments by Hydride Generation (ICP–AES) (21, 26)

Comment on the Method

Selenium is extracted with nitric and perchloric acids. The selenium is separated from interferences by coprecipitation with lanthanum hydroxide. The selenium is dissolved in hydrochloric acid and evolved as the hydride for measurement by means of ICP–AES. Calibrations are linear from the detection limit of 1 ng/ml to about 1000 μg/ml. Results of analyses of standard soils and other samples were compared with accepted values and the results obtained by means of other methods. The agreement was good.

Equipment and Reagents

The equipment was mostly the same as that described in the method for arsenic, antimony, and bismuth in soils and sediments; see Section V,E. However, the viewing height was 6 mm above the load coil. See Table 6.26.

- Heating block of aluminum (Scienco-Western Ltd., Cambridge, UK). This block accommodated 252 test tubes (150 by 19 mm) to a depth of 120 mm.
- MSE centrifuge. It accommodated 20 test tubes (120 by 16 mm) and operated at 6000 rpm.
- Potassium bromide, 400 g in 1 liter of water.
- Lanthanum nitrate, 50 g in 1 liter of water.
- Sodium tetrahydroborate, 10 g in 1 liter of 0.1 *M* sodium hydroxide.

Procedure

Grind the air-dried sample to pass a 200 μm sieve. Transfer 0.500 g to a 19 by 150 mm Pyrex test tube. Add 2 ml of nitric acid and some crushed glass. Swirl the contents. Place the tube in the heating block. Raise the block temperature to 100°C. Heat the sample until the evolution of brown fumes ceases. Increase the temperature to 170°C. Remove the sample when it appears to be bleached. Do not dry it completely. Cool the sample.

Add 4 drops of 5 *M* hydrochloric acid. With the help of 5 ml of water, transfer

Table 6.26
Operating Conditions

Plasma	
Torch	Fassel type
Power forward	2.7 kW
Viewing height	6 mm above load coil
Plasma support gas flow rate	0.8 liter/min argon
Coolant gas flow rate	171 liters/min argon
Integration time	20 s
Preintegration	15 s
Reduction	
Sample acidity	5 *M* hydrochloric acid
Sample flow rate	9.2 ml/min
Reagent concentration	1% (w/v) sodium tetrahydroborate in 0.1 *M* sodium hydroxide solution
Reagent flow rate	4.5 ml/min
Carrier gas flow rate	0.8 liter/min argon
Selenium wavelength	196.10 nm

Source: Ref. 21.

the sample into a centrifuge tube. Spin the tube at 4000 rpm for 30 s. Decant the liquid into a graduated 10 ml centrifuge tube.

Add 0.5 ml of lanthanum nitrate solution and 2 ml of ammonia reagent. Swirl to mix. Spin the tube at 4000 rpm for 30 s. Discard the liquid. Wash the solids with 1 ml of ammonia reagent and 3 ml of water. Centrifuge again and discard the liquid. Dissolve the solids in 5 ml of 5 *M* hydrochloric acid. Add 1 ml of potassium bromide solution. Heat in a water bath at 50°C for 50 min. Dilute to 10 ml with 5 *M* hydrochloric acid. Ignore the precipitate of ammonium perchlorate.

Analyze the sample under the operating conditions shown in Table 6.26. Use 5 *M* hydrochloric acid to establish the baseline. Calibrate with a standard containing 0.1 μg/ml selenium. Repeat the calibration after each group of 10 samples.

Comparisons with Standards and Other Analyses

With few exceptions the results compare well with other values. See Tables 6.27 and 6.28.

J. Determination of Tellurium in Silicate Rocks by Hydride Generation (Atomic Absorption) (27)

Comments on the Method

The method is rapid, sensitive, and reasonably accurate and precise. Tellurium hydride is generated in a simple apparatus. The method allows for variations in experimental parameters without adverse effects. The tellurium contents of several

Table 6.27
Comparison of Mean Values (μg/g) Obtained Using Hydride Generation–ICP with Those from Other Laboratories

Sample	ADAS[a]	ADAS[a]	Analyzed by CCRMP[b]	AGRG[c]		
				m	*s*	*n*
SO-1			0.10	0.8	0.01	7
SO-2			0.30	0.36	0.03	6
SO-3			0.05	0.18	0.02	7
SO-4			0.40	0.52	0.03	7
S778		0.20		0.30	0.04	6
S779		0.40		0.43	0.08	7
S783		0.50		0.27	0.03	5
S784		0.90		0.92	0.07	6
S788		1.10		1.08	0.08	6
S789		1.20		1.08	0.09	6
63/65	0.30			0.26	—	1
95/16	0.75			0.63	—	1
63/38	1.52			1.37	—	1

Source: Ref. 26.
[a]Mr. R. J. Hall, Ministry of Agriculture, Fisheries and Food, Newcastle upon Tyne (method: fluorimetry).
[b]Canadian Certified Reference Material Project (method: fluorimetry).
[c]Applied Geochemistry Research Group, Imperial College, London (method: this work). *m*, Arithmetic mean value; *s*, between-batch standard deviation; *n*, number of observations.

USGS rocks were measured. Comparisons with other data were possible only for two standard rocks.

Equipment

Atomic absorption spectrometer and a strip chart recorder. The operating conditions are those recommended by the manufacturer.

Table 6.28
Comparison of Selenium Concentrations (μg/g) in Two Standard Soils Obtained Using Gas–Liquid Chromatography (GLC), High-performance Liquid Chromatography (HPLC), Neutron Activation Analysis (NAA), and an Inductively Coupled Plasma Source (ICP)

Sample	GLC[a]	HPLC[a]	NAA[b]	ICP[c]		
				m	*s*	*n*
Weald loam	0.22	0.17	0.56	0.22	0.04	14
Lower lias	0.59	0.70	0.76	0.05	0.05	14

Source: Ref. 26.
[a]Dr. G. Nickless, Department of Inorganic Chemistry, University of Bristol.
[b]Universities Reactor Centre, Warrington.
[c]Applied Geochemistry Research Group, Imperial College, London (this work).
m, Arithmetic mean value; *s*, between-batch standard deviation; *n*, number of observations.

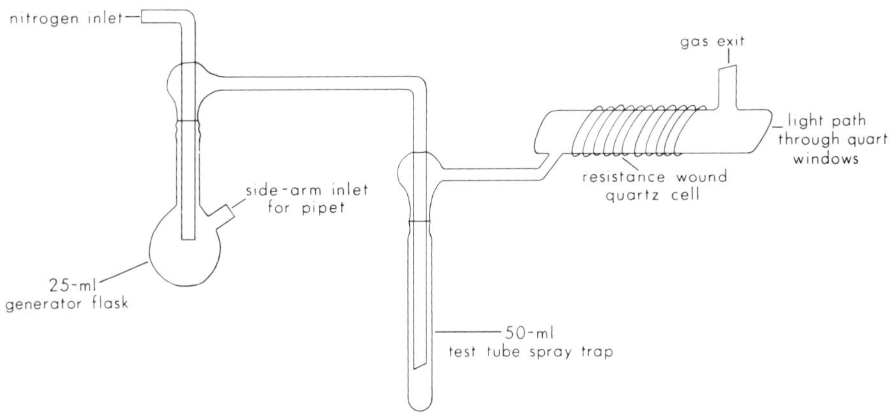

Figure 6.8. Apparatus for the determination of tellurium (27).

The apparatus (Fig. 6.8) consists of a 25 ml generating flask with a sidearm through which the borohydride solution is injected from an automatic pipet. The generated hydrides are swept in the nitrogen stream (4–4.5 liters/min) through a 50 ml test tube serving as a spray trap, to a resistance-heated quartz cell mounted in an atomic absorption spectrometer. The quartz spectrophotometer cell is 2 cm in diameter with a 150 cm light path; it is wrapped with 52 turns of No. 20 Kanthal resistance wire. The cell temperature is controlled with a variac adjusted slightly higher than required for the maximum tellurium sensitivity; this temperature, estimated from the color of the cell, is about 1000°C. The interunit connections are made with plastic tubing and are kept as short as possible. Nitrogen inlet and exit are as shown.

Reagents

- Tellurium standard solution: Prepare a solution containing 100 μg Te/ml from tellurium oxide which is dissolved in a minimum of nitric and hydrochloric acids. Dilute to volume with 6 *M* hydrochloric acid. Just before use, prepare a solution of 100 ng Te/ml by dilution with 6 *M* hydrochloric acid.
- Borohydride solution: Dissolve 5 g of sodium borohydride in 100 ml of water containing two pellets of sodium hydroxide. The solution must be prepared just before use. It is stable for several hours.

Procedure

Weigh a 0.25 g sample into a Teflon beaker, add 5 ml of hydrofluoric acid, and evaporate the contents to dryness at about 180°C. Add 15 ml of hydrofluoric acid, 5 ml of nitric acid, and 0.5 ml of perchloric acid to the beaker, and evaporate to dryness at about 120°C. Add 5 ml of 6 *M* hydrochloric acid and heat the beaker, covered with a watch glass, at about 80°C for 30–60 min to dissolve the residue.

Table 6.29
Tellurium in Standard Rocks

Rock	Tellurium (ng/g)		
	Standard addition[a]	Direct comparison with standards	Other values
BHVO-1, basalt	4.9 ± 1.5	6.3 ± 0.7	
GSP-1, granodiorite	18 ± 2	19 ± 2.9	20, 21[b]
SDS-1, schist	6.8 ± 4.2	5.5 ± 2.7	
BCR-1, basalt	1.1 ± 1.2	0.0 ± 0.4	5.5 ± 2.2[c]

Source: Ref. 27.
[a]Least-squares line fitted to 16 points: 4 replicates of 0, 2, 5, and 10 ng added tellurium. Error is the calculated error of intercept.
[b]Atomic absorption.
[c]Neutron activation.

(Rock solutions prepared in this way are quite viscous but essentially clear.) Decant the entire rock solution to the hydride-generating flask, and seal the side neck of the flask by insertion of an automatic pipet containing 1 ml of the borohydride solution. When the recorder has returned to a stable baseline, inject the borohydride and measure the peak height of the transient tellurium absorbance signal from the recorder trace.

Then disconnect the generating flask from the system at the ground-glass joint and rinse with water before adding the next sample.

Standards, prepared by adding 0.01–0.1 ml aliquots of the dilute standard solution to 5 ml of 6 *M* hydrochloric acid in the generating flask, are mixed at random through the samples. Sample concentrations are obtained by direct comparison with the peak heights of the standards.

Interferences

Up to 100 mg of iron as iron(III) chloride could be tolerated. Over 50 μg of copper as copper(II) chloride suppressed the tellurium absorption.

Accuracy and Precision

The results of analyses of four samples of USGS standard rocks are shown in Table 6.29. Tellurium values were obtained by the method of standard additions as well as by direct comparisons with tellurium standards.

K. Determination of Tin as Cassiterite in Soils and Sediments by Hydride Generation (Atomic Absorption) (28)

Comments on the Method

The following method is suitable for the determination of tin as the mineral cassiterite in soil and sediment samples. Ammonium iodide is used to attack the tin mineral by mixing it with the sample and then heating at 600°C. The tin iodide sublimate thus produced is trapped in a condenser. Tartaric acid solution is used to dissolve the sublimate. The sample solution is treated with sodium borohydride solution to produce volatile tin hydride. This gas is injected into a shielded hydrogen–argon diffusion flame. The precision and detection limits of the method are 12% RSD and 0.2 μg/g, respectively. The 224.6 and 286.3 nm tin lines were tested, and the former gave a 2–3-fold better sensitivity. The method was tested on the GXR series of geochemical standards and on a known synthetic mix of cassiterite and quartz. Satisfactory agreement of results was obtained. Up to 60 samples per hour can be handled.

Reagents

- Ammonium iodide, analytical reagent grade, ground to −200 μm.
- Tartaric acid solution, 1% (w/v): Dissolve 10 g of tartaric acid (analytical reagent grade) in 1 liter of deionized water. Prepare on day of use.
- Sodium tetrahydroborate solution: Dissolve 10 g of powdered reagent (Aldrich Chemical Co.) in 1 liter of 0.1 *M* sodium hydroxide solution. Prepare on day of use.
- Tin(II) chloride stock solution, 1000 μg/ml.

Equipment

- Sample attack tubes (Fig. 6.9), made from Quickfit (14/23) blank cones and sockets.
- Test tubes, borosilicate glass, medium wall.
- Block-bath, operating temperature up to 600°C, drilled to take 16 sample attack tubes (Scienco-Western Ltd.), provided with a rack of cooling coils for the air condenser, made from 0.32 cm outside diameter copper tubing (Fig. 6.9).

Figure 6.9. Sample decomposition apparatus (28).

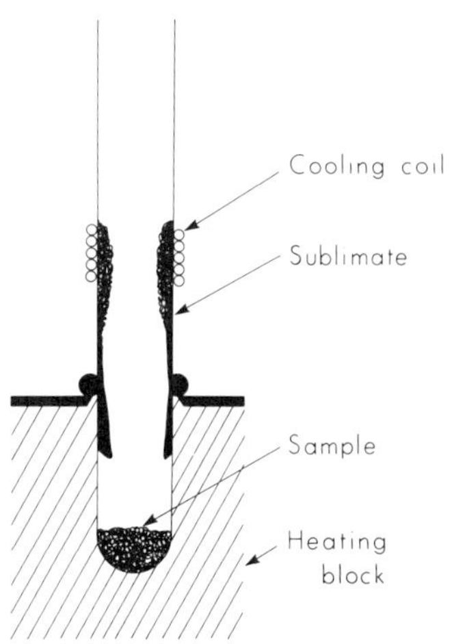

- Hydride generator (Fig. 6.10), comprising a peristaltic pump (Watson-Marlowe MHRE 200), delay coils, and phase separator. The tubing used (silicone rubber) had diameters of 0.8 mm for the test solution (standard or sample) and 0.5 mm for the reducing solution, the respective uptake rates being 9.2 and 4.5 ml/min. The carrier gas (argon) flow was 1.4 liters/min.
- Atomic absorption spectrophotometer, Perkin-Elmer model 403, with the nebulizer capillary closed and the hydride generator connected to the auxiliary oxidant inlet. Gas flows: hydrogen (in the "fuel" inlet), 8 liters/min; argon (in the "oxidant" inlet), 13 liters/min. Wavelength used: 224.6 or 286.3 nm.
- Three-slot burner (Perkin-Elmer), modified by the provision of aluminum shields, for the argon–hydrogen diffusion flame (Fig. 6.11) (Note: The flame is almost invisible).

Procedure

Dissolution of tin

Grind the samples sufficiently to pass a 200 mesh (75 μm) sieve. Weigh each sample (0.20 g) and ammonium iodide (0.30 g) into a sample attack tube and mix them. Fit each tube with a condenser, and heat the tubes in batches of 16 in the block-bath at 600°C for 10 min, the condensers being cooled by a flow of ice-water through the copper coils. Set aside the tubes with the condensers in place to cool in a horizontal position. Detach the condensers and place them in test tubes. To each add tartaric acid solution (10.0 ml) and leach the sublimates at 50°C for 20 min. Remove the condensers and mix the contents of the test tubes.

Determination of tin

Prepare calibration solutions containing tin(II) chloride in the concentration range 0–2 μg/ml in a solution containing tartaric acid (1% w/v) and make a calibration curve in this range. To measure the absorbance of a solution, switch the instrument to "10 average" (which provides updated results at 1 s intervals on the digital

Figure 6.10. Schematic diagram of hydride generator (28).

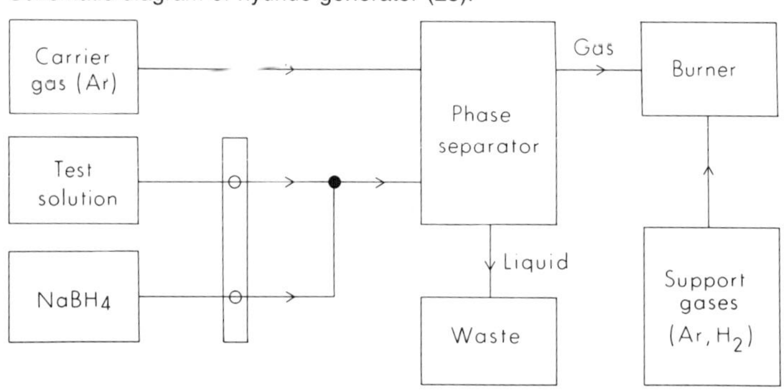

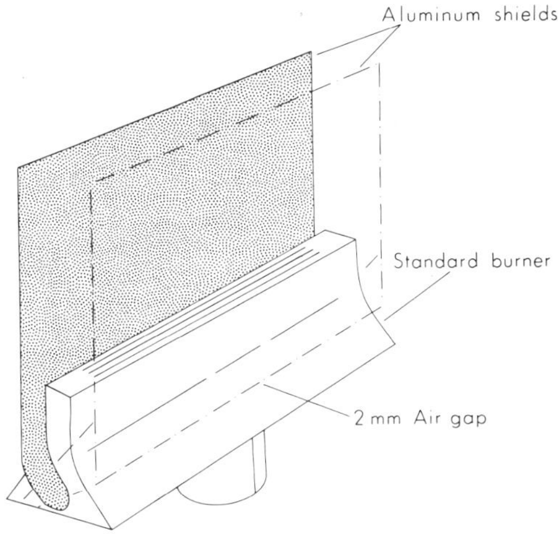

Figure 6.11. Shielded three-slot burner (28).

display), place the uptake tube into the solution, and switch on the pump. When the absorbance has stabilized (after about five 1 s integrations), press the "100 average" button and record the first update result (a 10 s integration). The pump should be left running while the test solutions are being changed. During a run, the autozero should be used after every 10th sample and the absorbance of a standard solution checked after every 20th solution. Solutions with absorbances outside the useful range should be diluted in the tartaric acid and reanalyzed.

VI. SOLVENT EXTRACTION

A. Multielement Extraction System for the Determination of Trace Elements in Geological Samples (Flame and Flameless Atomic Absorption) (29)

Elements Determined

Pd, Pt, Cu, Ag, Au, Zn, Cd, Hg, Ga, Sn, Tl, Sn, Pb, As, Sb, Bi, Se, Te, In

Comment on the Method

The purpose of the multielement extraction procedure is to simplify methods of sample preparation and minimize matrix interferences in atomic absorption spectrometry. The trace elements listed are useful in exploration geochemistry. Halide complexes are extracted into an organic phase consisting of Alamine 336, Aliquot 336, and methyl isobutyl ketone. The elements in the organic phase are determined by means of flame atomic absorption. Although platinum and palladium can be determined in the extract by flameless atomic absorption, the other elements suffer losses during the charring stage. Some methods for stripping the elements from the extracts prior to flameless analysis are outlined (30, 31). The authors state that the extraction procedure is compatible with several methods of sample preparation. Nitric and acetic acids, nitrate and acetate salts, and fluoride ions must be absent.

Equipment

- Perkin-Elmer models 360 and 603 atomic absorption spectrophotometers with deuterium background correction. Single-slot, 10 cm air/acetylene burner for all elements except Sn.
- Universal nitrous oxide/acetylene burner.
- Hollow-cathode lamps for Pd, Pt, Cu, Ag, Au, Zn, Ga, Sn, Pb.
- Electrodeless discharge lamps for Cd, Hg, Tl, Sn, As, Sb, Bi, Se, Te.
- Background correction for As and Se.
- Teflon nebulizer. This is required because the organic extracts attack stainless steel.

Reagents

- Salt solution: Combine 400 g of L-ascorbic acid, 100 g of potassium bromide, 100 g of potassium chloride, and 400 g of potassium iodide in a 2 liter amber glass bottle. Add 1 liter of water, and shake the mixture until all the salts are dissolved. The solution will be stable for several weeks in a cool, dark location. Do not keep it in a refrigerator.
- Extracting solution: Combine 50 ml of Alamine 336 (General Mills Chemicals Division) 100 ml of Aliquot 336 (General Mills), and 100 ml of hexane in a 1 liter flask. Add 500 ml of methyl isobutyl ketone (MIBK), mix, and then dilute to volume with MIBK. Store in an amber glass bottle.
- Standard solutions: The solutions must not contain nitrates or acetates. Aqueous standards are prepared in hydrochloric acid and hydrobromic acid solutions. Combine the following aqueous standards (1000 μg/ml). (a) As, Sb, Bi, Se, Te, Au; (b) Pb, Ag; (c) Ga, Sn; (d) Pd, Pt; (e) Cd, Hg, Tl. Prepare individual 1000 μg/ml aqueous standards for Cu, Sn, and Zn. Prepare 1000 μg/ml organic standards for Sn, Pb, Ga, In, and Hg by dissolving chloride salts in the organic extracting solution. Prepare organic solutions of As, Sb, Se, and Te by dissolving oxides in a small volume of ethanol or hydrochloric acid and diluting to volume with the organic extracting solution.

Procedure

Dissolve samples of approximately 1 g in acid by any of the established procedures. Remove all nitrate, acetate, and fluoride before extracting the samples. Examples of dissolution procedures are given by the author (32).

Transfer the solution to a 16 × 150 mm disposable culture tube. Add enough hydrochloric acid to give a concentration of 2.25 to 4.2 *M* hydrochloric acid in the final solution. All 18 elements except Zn can be quantitatively extracted in this range as shown in Fig. 6.12. (At 2.25 *M*, only a small proportion of zinc is not extracted.) Add 4 ml of salt solution. The total volume of aqueous phase must be 12 ml. Add 5 ml of organic extracting solution. Stopper the tube and shake it for 1 min. Centrifuge, and then allow the tube to stand overnight.

Use the organic phase for the analysis of the trace elements by flame atomic absorption. Prepare standards and blanks in a similar manner.

In order to measure the trace elements by furnace atomic absorption, all except platinum and palladium must be stripped from the organic phase. A sequential procedure for separating groups of the elements is shown in Fig. 6.13. It is useful in eliminating most of the interelement interferences produced by geological materials in the flameless atomic absorption methods.

Transfer 1 ml of the organic extract and 1 ml of 6% (vol) sulfuric acid to a 16 by 100 mm culture tube and extract. Set aside 0.5 ml of the organic phase, and

Figure 6.12. Optimum extraction ranges where at least 95% of metal is in the organic phase. Dashed line indicates that at least 95% is in the acid phase. (Reprinted with permission from Ref. 29. Copyright 1981 American Chemical Society.)

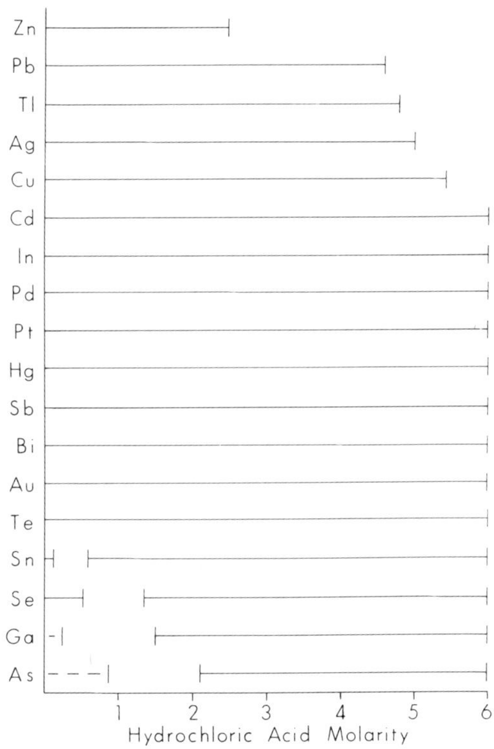

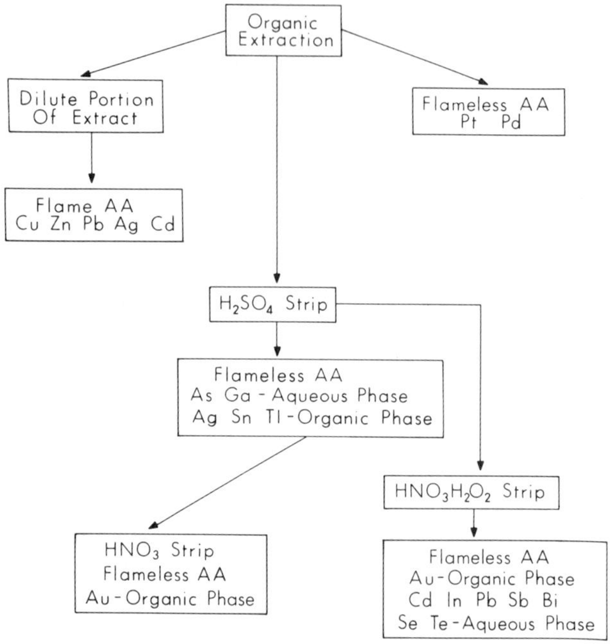

Figure 6.13. Flowchart for processing organic extracts for trace element determinations. (Reprinted with permission from Ref. 29. Copyright 1981 American Chemical Society.)

discard 0.5 ml of the aqueous phase. Add 20 μl of unstabilized hydrogen peroxide (30%), stopper, shake, and let stand overnight. Determine Ag, Sn, and Tl in the organic phase with chromic or ascorbic acid as a matrix modifier. Determine As and Ga in the aqueous phase. Add 75 μl of nitric acid. When the extract has turned pale, measure Au in the organic phase.

To the 0.5 ml volume of organic phase set aside, add 0.5 ml of a solution containing 5% (vol) nitric acid and 5% (vol) acetic acid in 30% hydrogen peroxide. Stopper the tube, extract, and let stand overnight. Measure Cd, In, Sb, Bi, Se, and Te in the aqueous phase.

VII. METHOD FOR WATERS

A. Determination of Metals in Waters by Flame Atomic Absorption (10)

Two types of metal can be determined by the following procedure, "extractable" metal and "total" metal. Extractable and total are placed in quotation marks because the author has used arbitrary definitions as follows. Extractable metal refers to metal, both bound and unbound, which can be extracted using the reagents and conditions of the following procedure. Total metal designates metal which is released by the acid digestion used in the following procedure. Experience suggests that for a wide variety of natural and polluted waters the total metal procedure determines more than 90% of the metal actually present.

Both procedures involve extraction using 1% ammonium 1-pyrrolidinedithiocarbamate (APDC) and methyl isobutyl ketone (MIBK). For extractable metal only a filtration is done with no further sample preparation prior to the extraction. Total metal is determined after the water has been evaporated and digested in the presence of hydrochloric acid and nitric acid.

Atomic absorption spectrometry is used in the determinative step. A flame atomizer was employed.

Reagents

A 1% aqueous solution of APDC is prepared and extracted twice with MIBK. This solution should be prepared fresh weekly.

The buffer solution is made by dissolving 14.7 g citric acid monohydrate and 34.3 g sodium citrate dihydrate in 100 ml water. Purify by extraction with APDC and MIBK.

All acids and chemicals used are reagent grade. Blanks are run at all times to overcome contamination problems.

Equipment

An Instrumentation Laboratories 153 atomic absorption spectrophotometer was employed. A three-slot Boling burner was found best for running solvent samples. The absorption parameters are given in Table 6.30.

Procedure

"Extractable" metal

Filter 200 ml of water through a 0.45 μm filter. Acidify the water to pH 1 with nitric acid. Place the sample in a 250 ml separatory funnel fitted with a Teflon stopcock. Add 4 ml of the buffer. Shake to mix well. The pH should be 4.0 ± 0.1. If this is not the case, add sufficient 20% sodium hydroxide solution to obtain

Table 6.30
Atomic Absorption Parameters

Element	Analytical line (nm)	Oxidant	Slit (μm)
Cd	228.8	Air	80
Pb	283.3	Air	80
Ni	232.0	Air	40
Zn	213.9	Air	80
Cu	324.7	Air	80
Fe	248.3	Air	40

this value. Add 1.5 ml of 1% APDC. Shake briefly. Add 10 to 20 ml (depending on the concentration factor required) of MIBK. Shake vigorously for 60 s. Allow the layers to separate. Retain the MIBK layer in a tightly capped glass bottle until samples have all been made ready for analysis. Prepare standards from a multielement stock solution so that the 200 ml of water extracted contains four concentrations within the following ranges: Fe, 10–200 ppb; Cu, 5–100 ppb; Ni, 5–100 ppb; Cd, 1–20 ppb; Zn, 10–200 ppb; and Pb, 10–200 ppb. In this way a direct concentration relationship exists with samples. Run a reagent blank.

"Total" metal (Cu, Ni, Fe, Pb, Zn, Cd)

Add 1 ml of nitric acid to 200 ml of sample. Evaporate the sample to dryness on medium heat of a hot plate. Add 2 ml of hydrochloric acid plus 1 ml of nitric acid. Evaporate to dryness. Add a drop of hydrochloric acid and dilute to 200 ml. Continue as for "extractable" metal above. Run a blank containing acids.

B. Matrix Interferences in the Analysis of Stream Waters

Fordham (33) encountered serious matrix effects in the analysis of stream waters by furnace atomic absorption. Elements studied were Cd, Pb, Cu, and Zn. Particularly troublesome was the relatively high background absorption obtained in working with these samples. This was reduced by sample treatment with nitric acid directly in the furnace. In addition, other interferences related to the determination of Cd, Pb, and Zn were obviated by the use of higher than normal atomization temperatures.

C. Multielement Analysis of Natural Waters for Hydrogeochemical Prospecting by X-Ray Fluorescence Following Preconcentration and Filter Deposition (34)

Elements Determined

Cu, Pb, Zn, Co, Ni, Cd, Mo, U.

Comments on the Method

Trace metals in water are coprecipitated with ferric ion. The precipitate is collected on a membrane filter. The filter is mounted between layers of x-ray cell film for measurements of metals by means of x-ray fluorescence. Starting with 1 liter of water, the lower limit of detection is 1 ppb. The results of analyses of five standard water samples were in fair agreement with other data.

Equipment

- Siemens SRS x-ray fluorescence spectrometer equipped with LiF 100 and 110 analyzing crystals and with scintillation and flow proportional counters
- Tungsten x-ray tube operated at 50 kV and 40 mA
- Specimen holders—silver cups, 23 mm diameter

- Coleman pH meter
- X-ray cell film, Spex No. 3517
- Filters: 47 mm–0.45 μm, Millipore type HA; 25 mm–3.0 μm, Millipore, type SS
- Plastic curtain rings, 2.54 cm diameter
- Plastic Caplugs EC-16, Protective Closures Co., Buffalo, New York

Chemicals

- Standard solutions of Cu, Pb, Zn, Co, Ni, Cd, or U (1000 μg/ml) in 2 *M* hydrochloric acid: Dissolve 0.6259 g CuO, 0.5486 g PbO, 0.6224 g ZnO, 0.6358 g CoO, 0.6363 g NiO, 0.5712 g CdO, and 0.5896 g U_3O_8 (all of Specpure quality) in 83 ml of concentrated hydrochloric acid and dilute each to 500 ml with deionized water.
- Standard solution (1000 μg/ml) of Mo: Dissolve 0.7502 g of Specpure MoO_3 in 100 ml of 10% sodium hydroxide and dilute to 500 ml.
- Iron solution (1000 μg Fe/ml) in 2 *M* hydrochloric acid: Dissolve 0.715 g of Specpure Fe_2O_3 in 83 ml of hydrochloric acid and dilute to 500 ml.
- Ammonium acetate buffer: Dissolve 170 g of ammonium acetate in 500 ml of deionized water.
- Ammonium 1-pyrrolidinedithiocarbamate (APDC): Powder form and reagent grade (Fisher Scientific Co.).

Procedure

Filter a 1 liter sample of water through a 0.45 μm membrane filter as soon as possible after collection. Then acidify the sample with 2.5 ml hydrochloric acid.

Table 6.31
Instrument Operating Conditions

Metals	2θ (deg)	Crystal	Counter	Target	kV	mA
Cu	45.03	LiF 100	Scintillation	W	50	40
Pb	28.24	LiF 100	Scintillation	W	50	40
Zn	41.80	LiF 100	Flow	W	50	40
Co	52.79	LiF 100	Flow	W	50	40
Ni	48.66	LiF 100	Flow	W	50	40
Cd	21.71	LiF 110	Scintillation	W	50	40
Mo	20.33	LiF 100	Scintillation	W	50	40
U	37.29	LiF 110	Scintillation	W	50	40

Source: Ref. 34.

Pour the sample into a beaker. Add 0.5 ml of iron solution and 10 ml of ammonium acetate buffer. Mix well. Adjust the pH of the sample to 3.0, measured by a pH meter. Use 6 *M* hydrochloric acid, and stir the solution constantly. Then add 0.3 g of APDC powder and mix. Set the sample aside for 1 h.

Filter the sample through a 25 mm diameter, 3.0 μm membrane filter. Stretch a piece of x-ray cell film on a Caplug, and hold it in place with a plastic ring. Place the filter on the film, and then cover it with another piece of x-ray cell film. Hold the film in place with another ring.

Place the sample in a silver specimen holder. Determine the trace metals according to the instrument conditions in Table 6.31. Use a flow-proportional counter for Zn, Co, and Ni and a scintillation counter for the other elements. Count each element for 100 s. Read the quantities of each element from calibration curves prepared as described below.

Calibration

Prepare standards by adding 0, 1, 2, 5, 10, 20, 50, and 100 μg of each trace element to deionized water (approximately 1 liter). Precipitate the metals in the same manner as in the analysis of samples. Count the elements as described above. Prepare

Table 6.32
Comparison of One Determination of Analytical Values Obtained by the X-Ray Method With Those Reported by the U.S. Geological Survey Water Laboratories

Sample No. [a]	Values	Concentration (μg/liter)							
		Cu	Pb	Zn	Co	Ni	Cd	Mo	U
63	X ray	88	—[c]	132	14	6	16	26	6
	Reported[b]	62 ± 8[d]	4.9 ± 3.5[d]	212 ± 14[d]	14.8 ± 3.0[d]	7.4 ± 3.4[d]	14.9 ± 2.1[d]	25.8 ± 4.5[d]	
61	X ray	150	14	42	6	4	1	48	6
	Reported	110 ± 5	10.6 ± 3.9	44 ± 9	4.8 ± 1.3	5.1 ± 2.5	2.4 ± 0.6	43.4 ± 9.5	
57	X ray	358	19	42	8	12	10	40	6
	Reported	320 ± 14	20.0 ± 7.3	36 ± 7	7.7 ± 0.8	11.3 ± 5.8	6.6 ± 2.1	39.7 ± 3.3	
59	X ray	254	24	438	4	4	12	2	2
	Reported	239 ± 17	16.8 ± 4.8	336 ± 16	5.8 ± 0.7	10.2 ± 2.9	4.4 ± 0.9	2.0 ± 0.7	
53	X ray	422	45	888	4	16	20	30	4
	Reported	443 ± 20	41.3 ± 9.7	738 ± 74	9.7 ± 3.2	21.7 ± 7.6	12.4 ± 3.6	38.4 ± 9.3	

Source: Ref. 34.
[a]Standard reference water samples prepared by the U.S. Geological Survey National Water Quality Laboratory (Denver, Colorado).
[b]Reported by the U.S. Geological Survey National Water Quality Laboratory.
[c]Not determined.
[d]Average relative standard deviations: Co, 20.6%; Ni, 41.9%; Cu, 6.7%; Pb, 39.4%; Zn, 12.3%; Cd, 24.1%; Mo, 21.4%.

calibration graphs. The relationship between counts and concentrations of each element is linear over the range 0 to 100 μg.

Results of analyses of standard reference samples of water are shown in Table 6.32.

REFERENCES

1. G. A. Spiers, M. J. Dudas, and L. W. Hodgins, *Clays Clay Miner.* **31,** 397 (1983).
2. P. M. D. Bradshaw, I. Thompson, B. W. Smee, and J. O. Larson, *J. Geochem. Explor.* **3,** 209 (1974).
3. P. J. Lechler, W. R. Roy, and R. K. Leininger, *Soil Sci.* **130,** 238 (1980).
4. H. Agemion and A. S. Y. Chau, *Anal. Chim. Acta* **80,** 61 (1975).
5. J. H. McLaren, S. S. Berman, V. V. Boyko, and D. S. Russell, *Anal. Chem.* **53,** 1802 (1981).
6. M. Odegard, *J. Geochem. Explor.* **14,** 119 (1981).
7. M. Bettinelli, *Anal. Chim. Acta* **148,** 193 (1983).
8. J. A. McKeague, ed., "Manual on Soil Sampling and Methods of the Canada Soil Survey Committee," Can. Soc. Soil Sci., Ottawa, Ontario, Canada, 1978.
9. E. Pruszkowska and P. Barrett, *Spectrochim. Acta, Part B* **39B,** 485 (1984).
10. J. C. Van Loon, internal laboratory method, unpublished.
11. J. Lye, M.Sc. Thesis, Dept. of Geology, University of Toronto (1982).
12. A. Tessier, P. G. C. Campbell, and M. Bisson, *J. Geochem. Explor.* **16,** 77 (1982).
13. S. J. Hoffman and W. K. Fletcher, *J. Geochem. Explor.* **15,** 549 (1981).
14. J. M. Motooka and S. J. Sutley, *Appl. Spectrosc.* **36,** 524 (1982).
15. T. T. Chao and R. F. Sanzalone, *J. Res. U.S. Geol. Surv.* **5,** 409 (1977).
16. D. C. Grégoire, *J. Geochem. Explor.* **23,** 299 (1985).
17. A. M. Ure and C. A. Shand, *Anal. Chim. Acta* **72,** 63 (1974).
18. R. F. Sanzolone and T. T. Chao, *Analyst (London)* **108,** 58 (1983).
19. R. G. Smith, J. C. Van Loon, J. R. Knechtel, V. L. Fraser, A. E. Pitts, and A. E. Hodges, *Anal. Chim. Acta* **93,** 161 (1977).
20. N. T. Marshall, *J. Geochem. Explor.* **10,** 307 (1978).
21. M. Thompson, B. Pahlavanpour, S. J. Walton, and G. F. Kirkbright, *Analyst (London)* **103,** 568 (1978).
22. B. Pahlavanpour, M. Thompson, and L. Thorne, *Analyst (London)* **105,** 756 (1980).
23. G. E. M. Aslin, *J. Geochem. Explor.* **6,** 321 (1976).
24. W. H. Fricklin and F. N. Ward, *J. Res. U.S. Geol. Surv.* **4,** 217 (1976).
25. R. F. Sanzolone and T. T. Chao, *Analyst (London)* **106,** 647 (1981).
26. B. Pahlavanpour, J. H. Pullen, and M. Thompson, *Analyst (London)* **105,** 274 (1980).
27. L. P. Greenland and E. Y. Campbell, *Anal. Chim. Acta* **87,** 323 (1976).
28. D. Gladwell, M. Thompson, and S. J. Wood, *J. Geochem. Explor.* **16,** 41 (1981).
29. J. R. Clark and J. G. Viets, *Anal. Chem.* **53,** 61 (1981).
30. J. R. Clark and J. G. Viets, *Anal. Chem.* **53,** 65 (1981).
31. J. R. Clark and J. G. Viets, *Anal. Chem.* **55,** 166 (1983).
32. J. G. Viets, *Anal. Chem.* **30,** 1097 (1978).
33. A. W. Fordham, *J. Geochem. Explor.* **10,** 41 (1978).
34. A. E. Hubert and T. T. Chao, *Econ. Geol.* **74,** 1669 (1979).

7
Lithogeochemical Analysis

I. DETERMINATION OF MAJOR AND TRACE ELEMENTS IN ROCKS BY ICP–AES (1)

Elements Determined

- Major—Na, K, Mg, Ca, Fe, Al, Ti, P
- Trace—La, Ce, Zr, U, Th, V, Cr, Mn, Co, Ni, Cu, Zn, Mo, Ag, Cd, Sn, Pb, Bi, As, Sb, W

Comment on the Method

The samples are dissolved in acid to yield a final solution for analysis containing 20% hydrochloric acid and 5% perchloric acid (v/v). A matrix-matching calibration is used. A number of trace elements cannot be detected by this procedure unless they occur at anomalous levels. These are Th, U, Mo, W, Ag, Cd, Bi, As, and Sb. The method has been used to analyze 54 geochemical reference materials including coal and fly ash.

Instruments

- ICPQ model QA 137000 and 34000 ICP (Applied Research Laboratories)
- PDP-11 minicomputer

The instrumental operating parameters are given in Table 7.1.

Chemicals and Apparatus

- Hydrochloric acid, 6.2 *M*
- Teflon beakers, 50 ml
- Quartz beakers, 200 ml tall-form

- Steam bath
- Hot plate
- Polypropylene bottles, 125 ml

Procedure

Weigh 1.0 g of powdered sample into a 50 ml FEP beaker. Add about 2 ml of nitric acid, and swirl the contents to wet the solids. Cover the beaker (Teflon cover), and place it on a steam bath for 30 min. Then add 10 ml hydrofluoric acid, 3 ml perchloric acid, and 2 ml nitric acid. Let the mixture react overnight. Remove the cover, and drive off the hydrofluoric acid by heating 4–6 h. Transfer the residue to a 200 ml tall-form quartz beaker. Heat the sample to dryness. Add 5 ml of hydrochloric acid and heat to dryness. Add 40 ml of 6.2 *M* hydrochloric acid and heat to dissolve the solids. Cool. Add enough 6.2 *M* acid to bring the volume to 40 ml. Transfer to a 100 ml volumetric flask. Add water to about 90 ml, and then 5 ml of perchloric acid. Mix, and dilute to 100 ml. Transfer to a 125 ml polypropylene bottle.

If the sample contains appreciable quantities of zircon, beryl, garnet, and spinel, dissolve either the whole sample or the insoluble residue by means of acid in a Teflon bomb.

Dissolve chromite residues in the same quartz beaker with 2 ml of phosphoric acid. Make up the final solution to contain 2% phosphoric acid and 20% hydrochloric acid.

If the sample contains large proportions of carbonate and phosphate, transfer it to a 200 ml tall-form quartz beaker. Cover it with 10 ml of water, and then add dilute hydrochloric acid. Transfer any residue to a Teflon beaker, and dissolve it in the mixed acid (about 5 ml) described above.

Calibration

Use matrix matching to calibrate the instrument. An example is shown in Table 7.2. The proportions of the major elements are chosen to be close to those of the sample. All standards in the table are made up to a total cation content of 2500–3000 ppm (except for #1 and #2). This is equivalent to 1 g of rock diluted to 100 ml. Trace and minor components are measured in the linear portions of the calibration curves. Major element concentrations generally occur in the nonlinear portions. Second- and third-order polynomial equations are required to define the curves.

Interferences

Table 7.3 (2) contains information on spectral interferences for each of the elements. Obtain appropriate correction factors by aspirating samples of standards. Apply the factors as needed in analyses of samples.

Table 7.1
ICP Source Conditions and Operating Parameters

	ICPQ	34000 ICP
rf Generator	Air-cooled, 3 kW continuous-rating generator operated at 27.12 MHz, crystal-controlled.	Air-cooled, 2.5 kW continuous generator operated at 27.12 MHz, crystal-controlled.
Forward power	1600 ± 50 W. Regulated to <1% using a -impedance matching network.	Same
Induction coil	Two-turn, 4.7 mm o.d., silver-plated copper tube coiled on a 28.7 mm diameter form, water-cooled.	Two-turn, 4.7 mm o.d., gold-over-silver-plated copper tube 28.7 mm diameter form, water-cooled.
Plasma torch	Integrated, three-concentric-tube configuration (18 mm o.d.) utilizing 2.0 mm capillary injector tube.	Same
Nebulizer	Permanently aligned glass, coaxial pneumatic nebulizer. All nebulizers used in this study had an uptake rate of 1.8–2.4 ml/min. Water-saturated argon was used as the carrier gas, and the system utilized a tip washer, under computer control, to ensure that there was no salt buildup on the nebulizer.	Same
Spray chamber	"Scott" coaxial spray used in early studies on instrument #1. Conical spray chamber with impact ball used thereafter.	Conical spray chamber with impact ball.
Gas	Welding-grade argon. Coolant, 10.5 liter/min; plasma, 1.5 liter/min; carrier, 1.0 liter/min.	Same Same
Gas flow regulation	Three stages: two gas flow regulators in series followed by capillary restrictors which control the ratio of the gas flow of the coolant to the plasma gases.	Three stages: gas-flow regulators in series, followed by restrictors; flow rates of the gas and plasma gas can be varied independently.
Primary optics	A 4 mm vertical segment of the axial channel of the plasma is focused on the primary slit (12 μm) with 0.7 demagnification. The primary lens is fused quartz with a focal length of 15 cm (at 587 nm).	A 4-mm vertical segment of the channel of the plasma is focused on the primary slit (20 μm) with 1.8 magnification. The primary lens is fused quartz with a focal length of 15 cm (at 587 nm).

	ICPQ	34000 ICP
Height of observation	17 mm above the rf coil. Observation zone is 15–19 mm.	Same
Spectrometer	1.0 m Paschen–Runge mount, *f*/30; air path only.	1.0 m Paschen–Runge mount, *f*/30. A 1/4 in. steel–plate weldment provides vacuum-path capability. Spectrometer temperature maintained at 38° ± 0.2°C in a thermally controlled cabinet.
Gratings	1920 g/mm: interferometrically ruled quartz-blank replica blazed at 270 nm; reciprocal linear dispersion 0.48–0.52 nm/mm, first order; spectral range 180–460 nm. 1080 g/mm: see description under 34000 ICP.	1080 g/mm: interferometrically ruled quartz-blank replica blazed at 600 nm; reciprocal linear dispersion 0.926 nm/mm, first order; 0.463 nm/mm, second order; 0.309 nm/mm, third order; spectral range 175–820 nm.
Secondary optics	50 μm exit slits, concave secondary mirrors collect and refocus light on PMTs. 40 nm bandpass filters (Acton Research) used on all lines below 230 nm.	35 and 50 μm slits, concave secondary mirrors installed at 45° refocus light on the PMTs which are located above (and below) the optical plane. 40 nm bandpass filters (Acton Research) and color filters (Corning) used for order sorting.

Source: Ref. 1.

Accuracy

Major element determinations are accurate to within ±3.6%. Determinations for most minor and trace elements are accurate within 5–10%. Accuracies of determinations of lead and tin are dependent on the concentration of aluminum in the sample.

Alternative Sample Preparation Procedures for Rocks

Major elements: Si, Al, Fe, Mg, Ca, Na, K, Ti, P, Mn

Fusion with lithium metaborate has been used for the preparation of a solution of the major elements. The procedure below requires a 3 : 1 mixture of lithium metaborate : sample (3). Other workers have used 7 : 1 and 1 : 1 mixtures (4, 5). The lower proportions of lithium borate are preferred because of the possibility of solids formation in the nebulizer when concentrated solutions are aspirated. A 1 : 1 mixture did not completely decompose the silicates. However, the products were soluble in dilute (2%) nitric acid and were decomposed in the plasma. The

Table 7.2
Example of a Matrix-Matching Calibration Scheme Used in Determination of the Major, Minor, and Trace Elements in Rocks and Soils

Element	Std. #1 Acid blank	Std. #2 25% HNO_3	Std. #3 20% HCl–5% $HClO_4$	Std. #4 20% HCl–5% $HClO_4$	Std. #5 20% HCl–5% $HClO_4$	Std. #6 20% HCl–5% $HClO_4$	Std. #7 20% HCl–5% $HClO_4$	Std. #8 20% HCl–5% $HClO_4$	Std. #9 20% HCl–5% $HClO_4$	Std. #10 20% HCl–5% $HClO_4$	Std. #11 20% HCl–5% $HClO_4$	Std. #12 20% HCl–5% $HClO_4$	Std. #13 20% HCl–5% $HClO_4$	Std. #14 20% HCl–5% $HClO_4$
Na	0.050		500.0	1500	1000	250.0	50.00	100.0	10.00					
K	0.200		1500.0	1000	500.0	250.0	50.00	100.0	10.00					
Mg	0.020		50.06	10.00			300.0	1000.0		1500			2500	
Ca	0.010		50.01	10.00	500.2	1500	1000	250.0	100.3		2500			
Fe	0.200		50.00	10.00	1000	500.0		250.0	1500	100.0				3600
Al	0.040		50.51	10.10	102.9	251.9	1502		1005		501.0	3000		
Ti	0.000		200.0							326.0				
P	0.000													
Ba	0.000		50.00											
La	0.000		50.16											
Ce	0.000		50.10											
Zr	0.000		50.00											
U	0.000			50.06										
Th	0.000			50.02										
V	0.000		50.16											
Cr	0.000		50.07											
Mn	0.000		50.00	10.00			100.0	1000	250.0	500.0				
Co	0.000		50.32											
Ni	0.000		50.00											
Cu	0.000		50.03											
Zn	0.000		50.01											
Mo	0.000									50.00				
Ag	0.000	10.00												
Cd	0.000			10.00										
Sn	0.000			50.00										
Pb	0.000			50.02										
Bi	0.000	10.00												
As	0.000			50.01										
Sb	0.000			50.02										
Total	—	20.00	2900	2850	3100	2750	3000	2700	2870	(2888)[1a]	3000	3000	2500	3600

[a]The sum includes the K present in the potassium monobasic used as the reagent for P in this standard.

1 : 1 mixture is not suitable for the preparation of solutions for flame atomic absorption analyses because the flame temperatures are too low to decompose the silicates.

Weigh 0.5 g of powdered rock and 1.5 g of lithium metaborate into a platinum crucible. Mix the contents with a platinum rod. Fuse the mixture over a Meker burner for 30 min, and swirl the melt occasionally. Cool the crucible, and then immerse it in a polyethylene beaker containing 165 ml of water, 10 ml of nitric acid, and exactly 20 ml of 1250 μg/ml lanthanum nitrate solution (as an internal standard). Place a Teflon-coated stirring bar inside the crucible, and stir the solution immediately. After the salts have been dissolved (in 1–2 h), dilute the solution to 250 ml. Follow the procedure closely to prevent the formation of insoluble polysilicic acid. The solutions are stable for several months. Use the solutions for ICP–AES analyses of the major components. Consult the publication (3) for details.

Minor and trace elements: Zn, Ni, Co, V, Nb, Cu, Zr, Y, La, Sr, Ce, Cr, Ba, Li, Sc

Acids are used to prepare solutions of samples for trace element work in order to avoid excessive dilutions (3, 4, 6).

Weigh a 0.5 g sample of powdered rock into a plastic beaker. Add 4 ml of perchloric acid and 15 ml of hydrofluoric acid. Evaporate the sample to dryness. Dissolve the residue in 4 ml of perchloric acid and 15 ml of water. Dilute to 50 ml with water. Inspect the solution for the presence of any refractory minerals such as chromite, rutile, cassiterite, and zircon. Use the solution for the determination of minor and trace elements by ICP–AES (3). Major elements may also be determined in this solution.

II. DETERMINATION OF 15 MAJOR AND TRACE ELEMENTS IN ROCK SAMPLES BY X-RAY FLUORESCENCE USING ONE GLASS DISK (7)

Elements Determined

Na, Mg, Al, Si, P, S, K, Ca, Ti, Fe, Mn, Cu, Zn, Rb, Sr, Ba, Pb, Zn

Comment on the Method

Rock samples are fused with lithium borate in the flux-to-sample ratio of 2.2 : 1. Lanthanum oxide is added to reduce variations in mass absorption coefficients. Five rock standards were analyzed to demonstrate the precision and accuracy of the method.

Equipment

- Phillips PW1540 x-ray spectrograph. The operating conditions are given in Table 7.4. The authors noted that a rhodium target tube may be useful for x-ray machines where tube changes are difficult.

Table 7.3
Secondary Arrays, Detection Limits, and Spectral Interference Corrections for the Instruments Used in This Study.

Element	Wavelength (nm)	Spectral order	Detection limit (ppm)		Interferences (ppm interference/ppm concomitant)					
			Measured	Expected[a]	Mg	Ca	Fe	Al	Ti	Other concomitants
Na	588.99	1	0.026	(0.02)						
	589.59	1	0.04	(0.02)						
K	766.49	1	0.15	(0.10)						
Mg	279.08	1	0.024	0.030						
	279.08	2	0.026							
Ca	422.67	1	0.016	0.010						
	317.93	2	0.004	0.010						
Fe	259.94	1	0.003	0.0062						
	259.95	3	0.002							
	263.11	3	0.020	0.015						
Al	309.27	1	0.040	0.023	0.00604					V, 0.087; Be, 0.0165
	309.27	2	0.026							V, 0.083
	308.22	2	0.030	0.045						V, 0.0215
	237.31	3	0.062	0.030	0.0023	0.0004	0.0027			Mn, 0.0049
Ti	334.94	1	0.001	0.0038						Cu, 0.000016
	334.94	2	0.003							
	368.52	2	0.005	0.0011						
	368.52	1	0.008				0.000011			Mn, 0.0008
P	213.62	1	0.095	0.076			0.00025	0.00023		Cu, 0.310; Zn, 0.103; V, 0.00022
	213.62	3	0.083					0.00026		Cu, 0.108
	178.29[b]	3	0.050, 0.035	(0.10)[b]				0.0003		
Li	670.78	1	0.005	(0.003)						
Be	313.04	1	0.00008	0.00027						
Sr	407.77	1	0.00005	0.00041						
Ba	455.40	1	0.00002	0.0013						
	493.41	1	0.001	0.0023						
La	398.85	1	0.005	0.011		0.00015				
	379.48	1	0.011	0.010		0.0001	0.00025	0.000026		V, 0.0033; Ba, 0.0007; Mn, 0.0006
Ce	394.28	1	0.050	0.068		0.011		0.0022		V, 0.0026; Eu, 0.12; Sm, 0.028; Gd, 0.01; Nd, 0.028; Mn, 0.005
	413.77	1	0.066	0.048	0.00036	0.00029	0.00008	0.00011	0.00025	V, 0.0017; Cu, 0.0004; Cr, 0.0005; Ba, 0.0005; Mn, 0.0008
Nd	417.73	1	0.060	0.136				0.00008		Ce, 0.00005; Eu, 0.026; Y, 0.90[c]
Sm	443.43	1	0.020	0.083		0.00003			0.00036	V, 0.001; Ce, 0.013; Nd, 0.018
Eu	381.97	1	0.002	0.0027		0.00003	0.000013			
Gd	376.84	1	0.007	0.025		0.000015	0.00001		0.0001	Mn, 0.00045; V, 0.0005; Nd, 0.0055; Ce, 0.012; Th, 0.015
Yb	369.42	1	0.0006[c]	0.0030			0.000006		0.00003	
Zr	343.82	1	0.003–0.006	0.0071		0.0001				
U	409.01	1	0.40	0.337		0.00005				Mn, 0.003
	385.96	2	0.065	0.250	0.0002	0.0012	0.003	0.0003	0.0024	V, 0.007; Ce, 0.0043; Cr, 0.0037 La, 0.0013

Th	353.96	2	0.110	0.100	0.0004	0.0003	0.0002			Mn, 0.004; V, 0.004; Zr, 0.001; Ce, 0.094
V	311.07	1	0.003	0.010					0.012	Mn, 0.0003; Cr, 0.0001; Zr, 0.004; Cr, 0.0021
	268.80	3	0.004	0.010						
	292.40	2	0.003	0.0075			0.00008		0.0012	Mn, 0.0003; La, 0.0013
Cr	283.56	1	0.007	0.0071	0.00007		0.00125		0.0001	V, 0.00011
	267.72	2	0.004	0.0071						Mn, 0.0002; Mo, 0.0001; U, 0.002
	205.55	2	0.010	0.0061				0.0003		Mn, 0.0006
Mn	257.61	1	0.0007	0.0014						
	257.61	2	0.0009							W, 0.0004
Co	238.89	1	0.007	0.006			0.0026		0.00002	V, 0.00012
	228.62	2	0.006	0.007					0.0018	V, 0.0003; W, 0.001
	228.62	3	0.002				0.00008		0.0016	V, 0.0003
Ni	231.61	1	0.015	0.015						
	231.61	2	0.013, 0.010							
Cu	324.75	1	0.003	0.0054			0.00012		0.00012	Be, 0.0013
	342.75	2	0.005				0.00012			
	224.74	2	0.020	0.0077			0.0017			Pb, 0.0027
Zn	202.55	1	0.002	0.004	0.00014			0.00011		Cu, 0.006
	202.55	2	0.003		0.00012					Cu, 0.005; W, 0.0001
	202.55	3	0.003		0.00007					
	213.86	2	0.005	0.0018	0.00002		0.00002	0.000085		
Mo	313.62	2	0.050	(0.02)			0.0008			Cr, 0.0045; W, 0.002; Zr, 0.002
	202.03	3	0.014	0.0079		0.000012	0.000068	0.00022		As, 0.0001
W	239.70	3	0.030	0.055			0.00069	0.000024		
Ag	328.07	1	0.003	0.007		0.000016	0.000014	0.000017	0.0002	Zr, 0.01
	328.07	2	0.007		0.000022		0.000011	0.000007	0.0002	Zr, 0.004
Au	242.80	1	0.007	0.017			0.00038			Mn, 0.011; V, 0.0001; Sr, 0.00055
	242.80	3	0.014				0.00016			Mn, 0.011
Cd	226.50	1	0.002	0.0034			0.00011		0.00004	Ni, 0.0003
	226.50	2	0.002				0.00008			
	226.50	3	0.003				0.00007	0.000003	0.00014	
Sn	189.99[b]	3	0.027, 0.008	0.025	0.0005	0.000063		0.00026		V, 0.0001; Zr, 0.0008; As, 0.0035
Pb	220.35	1	0.048	0.042			0.0003	0.0014		
	220.35	3	0.027, 0.034		0.000053		0.00017	0.0013	0.0004	V, 0.0006; As, 0.0001; La, 0.0005; Zn, 0.002
Bi	223.06	2	0.260, 0.105	0.034	0.0001	0.0002	0.00039	0.0003	0.0049	V, 0.0014; Cr, 0.0014; La, 0.0013; Ni, 0.002
As	193.76	1	0.130	0.053	0.00026		0.00027	0.0085		V, 0.010; Cr, 0.0007; Zr, 0.005
	193.76	3	0.047					0.005		V, 0.014; Cr, 0.0015
	189.04[b]	3	0.023	(0.136)[b]	0.0001	0.00006	0.00009	0.0006	0.0003	V, 0.0005; Cr, 0.0027; La, 0.043
Sb	206.84	2	0.120				0.00012	0.0024	0.0014	Cr, 0.087; Zn, 0.001
	206.84	3	0.045	0.032	0.00005	0.0004	0.00013	0.0016	0.0012	V, 0.0022; Cr, 0.022
Se	196.09	1	0.095	0.075			0.00035	0.0019		Mn, 0.002; V, 0.0011

[a]Data from Winge *et al.* (329). Data in parentheses from Applies Research Laboratories.
[b]Vacuum path recommended or required.
[c]No Y line was present in the instrument. Therefore, a correction was made on the basis of the measured abundance of Yb and using a Y/Yb ratio of 10.

Table 7.4
Operating Parameters, Count Rates, and Sensitivity[a–d]

Oxide	Crystal	Collimator (μm)	Target	kV	mA	Counting time (s)	Counts (s^{-1} $\%^{-1}$)	Counts for 1% (× 10^4)	Background[e] (%)	Sensitivity[e,f] (lower limit of detection) (ppm)
Na_2O	TlAP	480	Cr	44	22	25	16	0.04	0.60	775
MgO	TlAP	480	Cr	44	22	50	40	0.2	0.30	245
Al_2O_3	PE	480	Cr	44	20	16	126	0.2	0.15	170
SiO_2	PE	480	Cr	44	20	15	134	0.2	0.12	155
P_2O_5	Ge	480	Cr	44	22	25	320	0.8	0.04	45
SO_3	Ge	480	Cr	44	22	25	320	0.8	0.04	45
K_2O	PE	480	Cr	44	20	2.7	7,500	2	0.022	20
CaO	LiF200	480	Cr	44	20	1.6	12,500	2	0.035	25
TiO_2	LiF200	480	Cr	44	20	1.5	13,300	2	0.066	35
MnO	LiF200	480	W	50	20	10	2,000	2	0.060	35
Fe_2O_3	LiF200	480	W	50	20	8	2,500	2	0.084	40
CuO	LiF200	160	Au	50	20	17	2,350	4	0.028	17
ZnO	LiF200	160	Au	50	20	15	2,670	4	0.028	15
Rb_2O	LiF200	160	Mo	50	20	25	16,000	40	0.032	6
SrO	LiF220	160	Mo	50	20	24	4,150	10	0.038	12
BaO	LiF200	160	Cr	44	20	10	2,000	2	0.042	30
PbO	LiF200	160	Mo	50	20	10	4,000	4	0.115	35

Source: Ref. 7.
[a] *K*α lines for all elements except Ba–*L*α and Pb–β.
[b] Vacuum used when necessary to increase count rate.
[c] Flow proportional counter with 1 μm polypropylene window was used for all elements except Pb, Sr, and Rb, where a scintillation counter was used.
[d] Pulse-height selector used for all elements with a 5–20 V window.
[e] Percent and ppm refer to oxide content of original sample.
[f] Sensitivity is the amount that is significantly above background at 2 S.D. level (95% confidence).

- Muffle furnace.
- Glass disks formed in graphite molds, 31.5 mm in diameter, heated on a hot plate at 220°C. The molten glass is pressed with an aluminum plunger as shown in Fig. 7.1 (8).

Chemicals

Prefused lithium borate ($3Li_2O \cdot 2B_2O_3$). Mix 38.0 g lithium tetraborate (anhydrous), 29.6 g lithium carbonate, and 13.2 g lanthanum oxide. The lithium salts must be ignited at 500°C prior to mixing and the lanthanam oxide at 900°C. Melt the mixture in a graphite or platinum crucible (with a cover) at 1000°C. Pour the melt onto a large sheet of polished aluminum, 3 mm in thickness. Combine several melts and grind to a coarse powder. (Each melt yields 63 g of product.) Store in an airtight container. Heat the borate at 550°C before mixing it with the sample in order to

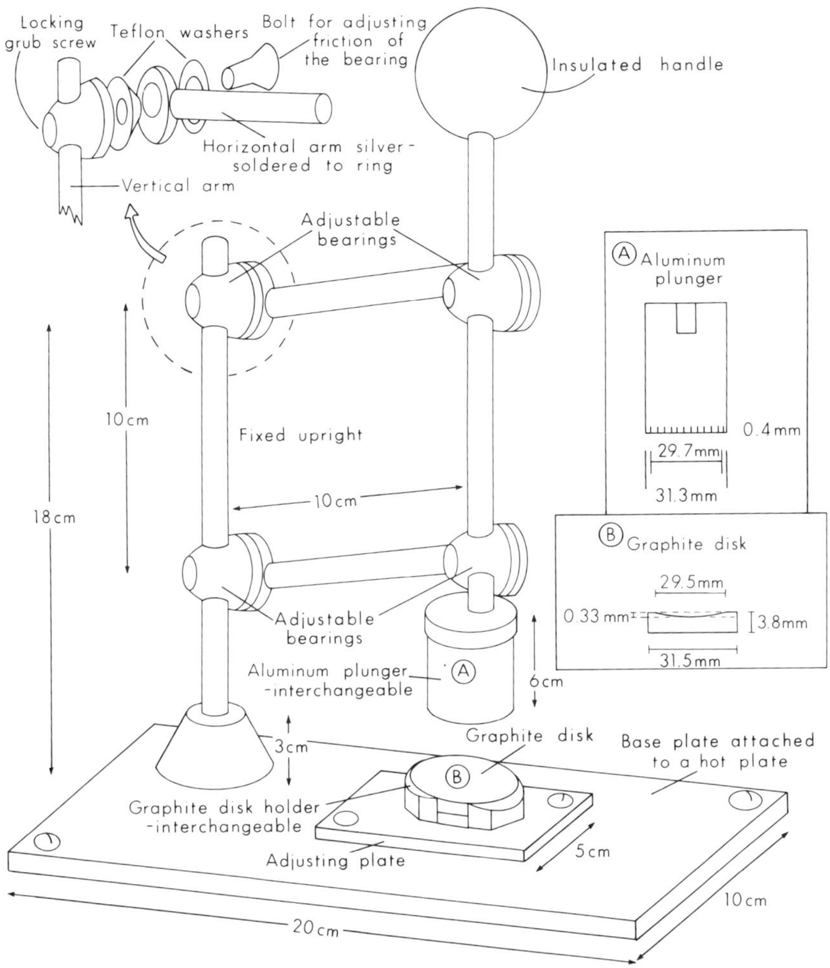

Figure 7.1. Equipment for preparation of glass disks (8).

drive off moisture. An alternative source is Spectroflux 105, Johnson Matthey Ltd. The flux must be dried at 550°C.

Procedure

Grind a sample of coarsely broken rock in a 100 ml ring and puck mill for 3 min. For samples containing less than 1% sulfur, the fusion formula is 1.22 g prefused lithium borate, 0.02 g lithium nitrate, and 0.56 g powdered rock sample. Fuse the mixture in a Pt–Au crucible in a muffle furnace at 1000°C for 10–15 min. Withdraw the crucible and heat it to 1050°C in an oxygen–propane flame. Pour the fusion onto a heated graphite disk at 220°C. Press and chill the disk in the equipment described earlier.

Analyze the disk according to the parameters in Table 7.4.

Table 7.5
Recommended Coefficients for Matrix Corrections[a]

Element	*M* flux	Fe_2O_3	MnO	TiO_2	CaO	K_2O	SO_3	*X* P_2O_5	SiO_2	Al_2O_3	MgO	Na_2O	Loss
Fe	1.046	−0.027	−0.031	0.146	0.134	0.126	−0.060	−0.060	−0.065	−0.074	−0.090	−0.110	−0.163
Mn	1.045	−0.044	−0.044	0.146	0.135	0.130	−0.037	−0.063	−0.063	−0.074	−0.078	−0.100	−0.163
Ti	0.851	0.081	0.077	0.179	0.647	0.644	0.194	0.181	0.110	0.078	0.069	0.051	−0.132
Ca	0.865	0.090	0.092	0.065	0.130	0.723	0.201	0.182	0.128	0.105	0.068	0.051	−0.134
K	0.897	0.098	0.086	0.017	000	0.069	0.182	0.179	0.119	0.101	0.080	0.057	−0.139
S	0.894	0.086	0.074	0.002	−0.023	−0.037	−0.053	0.167	0.131	0.112	0.087	0.063	−0.139
P	0.896	0.108	0.094	−0.020	−0.037	−0.047	−0.059	−0.063	0.127	0.110	0.094	0.046	−0.139
Si	1.014	0.082	0.086	−0.034	−0.042	−0.055	−0.057	−0.061	−0.061	0.122	0.093	0.063	−0.158
Al	1.056	0.112	0.116	−0.032	−0.037	−0.048	−0.056	−0.060	−0.088	−0.072	0.116	0.058	−0.164
Mg	1.050	0.136	0.126	0.010	−0.021	−0.043	−0.046	−0.016	−0.070	−0.078	−0.084	0.080	−0.163

[a]The coefficients are for use with weight fractions of the oxides, i.e., percent/100.

Matrix Corrections

A detailed discussion of matrix corrections for the analysis of silicate rocks has been published (9).

The general equation for calculating the matrix correction applied to iron, for example, is

$$\mathrm{Fe} = \mathrm{Fe}'(M + \sum_{i}^{n} X_i E_i) - B$$

where Fe is the nominal concentration of Fe_2O_3 obtained from a calibration graph, M = Am/As = mass absorption coefficient of flux/mass absorption coefficient of sample, X_i is the coefficient for matrix correction of the ith element (see Table 7.4) and E_i is the nominal concentration of the ith element divided by 100. Table 7.5 contains the coefficients for matrix corrections. Additional factors for elements heavier than Fe are not listed in the original papers. They were calculated from published mass absorption coefficients and normalized to the average rock (10).

Calibration

Prepare calibration samples by using known weights of the pure chemicals. For elements heavier than iron, use a mixture of 10% oxide and 90% silica. Correct the observed counts for matrix and subtract the background. Use the net values to prepare calibration graphs.

Table 7.6
Comparison of Results for Standards BCR, AGV, GSP, GH, and Sy-1 with Published Values

Oxide	Range (%)	Published value less determined value (mean of 5)	Maximum relative error (%)
Na_2O	2.8–4.3	0.11	4
MgO	0.03–4.2	−0.10	10
Al_2O_3	9.6–17.2	0.32	4
SiO_2	54.5–75.8	0.80	1.5
P_2O_5	0.01–0.49	−0.03	10
K_2O	1.7–5.5	0.01	1
CaO	0.7–10.5	−0.01	4
TiO_2	0.1–2.2	0.02	6
MnO	0.04–0.4	0.00	10
Fe_2O_3	1.3–13.4	−0.06	2
ZnO	0.01–0.028	−0.0007	10
Rb_2O	0.005–0.042	0.0006	10
SrO	0.001–0.078	0.0007	10
BaO	0.002–0.15	0.0015	4
ZrO_2	0.02–0.08	−0.0004	4

Background

Obtain background values for each element by measuring line intensities in a sample free from the elements, and correct for matrix. This represents the background due to impurities in the borate and scatter in the fused disk. In this manner, backgrounds for each sample need not be measured.

The authors state that some additional background arises with samples having large amounts of elements with atomic numbers greater than 18. They recommend that the method described by Norrish and Hutton (11) be used. It is also described in Chapter 9. The calculated background corrections must be subtracted from the nominal concentrations before matrix corrections are applied.

Results for Standard Samples

Table 7.6 contains comparisons of the x-ray results with published data for five standards.

III. DETERMINATION OF MAJOR AND TRACE ELEMENTS IN SILICATE ROCKS BY ENERGY-DISPERSIVE X-RAY FLUORESCENCE (12)

Elements Determined

- Major—Si, Ti, Ba, Al, Fe, Mn, Mg, Ca, Na, K, P, V, Cr, Ni
- Trace—Rb, Sr, Y, Zr, Nb, Pb, Th, Cu, Zn, Ga

Comments on the Method

Glass beads are used for analyses of major elements, and trace elements are measured in powder pellets. A high degree of accuracy and precision in analyses of silicate rocks was achieved. For many elements, 2σ standard deviations are better than 2% relative, and limits of determination of many trace elements are in the range 4–10 ppm. Results of analyses of seven standard rocks were tabulated.

Equipment

Mecca 10-44 ED–XRF spectrometer and analysis system manufactured by Link Systems Ltd. It consisted of a low-power (49 W) silver side window x-ray tube and an Si(Li) detector.

Detector characteristics were: resolution, 165 eV at 5.9 keV; active area, 30 mm^2; Be entrance window thickness, 6 μm.

Sample mounting was in a horizontal 20-position turntable. The detector viewed approximately 200 mm^2 of the fluorescent area of the sample through a coarse collimator. The sample chamber was evacuated to a pressure of 0.5 torr.

Pulse amplification and shaping were performed on a Harwell processor with a time constant of 20 μs. It had a linear response in the range up to and above 20,000 counts/s.

The instrument was operated in the pulsed mode. This allowed operations at output rates of 10,000 counts/s at no more than 50% dead time. Spectra were accumulated in 1024 channels with a capacity of 2^{20} for each channel. Pulse gains of 10 and 20 eV per channel were employed for the 0–10 and 0–20 keV regions, respectively.

Chemicals and Standards

- Flux. Spectroflux 100B, Johnson Matthey Chemicals Ltd. This is a 4 : 1 lithium metaborate : tetraborate mixture.
- Polyvinyl alcohol, 5% (saturated aqueous solution).

Procedure

Sample preparation—major elements

Dry finely ground sample overnight at 110°C. Carry out a trial fusion of flux at 1100°C, and determine the weight loss. Correct for any loss with additional flux so that the ratio of flux to sample is 6 : 1. Mix the flux and sample thoroughly in a Pt/Au crucible (5% Au). The charge should be enough to make a 32 mm diameter glass bead. Fuse the mixture in a furnace at 1100°C for 20 min.

Pour the melt into a Duralumin platen preheated to 225°C on a hot plate. Apply pressure for a few seconds with a preheated plunger. Transfer the platen to a 200°C hot plate for 10 min, and then cool to room temperature. Determine the ignition loss at 1100°C for each sample.

Sample preparation—trace elements

Make a powder pellet weighing 6 g or more by mixing finely ground sample with a polyvinyl alcohol binder. Transfer the moist powder to a 32 mm diameter die, and press it at 10 ton/in.^{2} Bake the pellet at 110°C.

Analysis

Count disks for major elements for 500 s at 10 kV and 0.2 mA with no primary beam filter. Count disks for minor elements for 800 s at 45 kV and 0.3 mA with a primary beam filter of 127 μm of silver. Normalize peak intensities to the total spectrum counts. Determine peak areas using a computer program. (The authors used a Link Systems FLS program, which applies a "top hat" digital filter to recorded spectra to suppress statistical noise and continuum components.) Determine the net peak area of a specified element by using a linear least-squares procedure which fits the element x-ray profile (filtered with the same top hat function) to the filtered sample spectrum. In order to do this, it is necessary to construct a library of individual x-ray profiles to account for all interferences due to overlaps. *K* and *L* line profiles for major and trace elements in silicate rocks are shown in Figs. 7.2 and 7.3.

Figure 7.2 shows that x-ray line profiles of V *(K)*, Cr *(K)*, and Ba*(L)* lines must be routinely fitted if these elements are present in detectable amounts. Also, Fig.

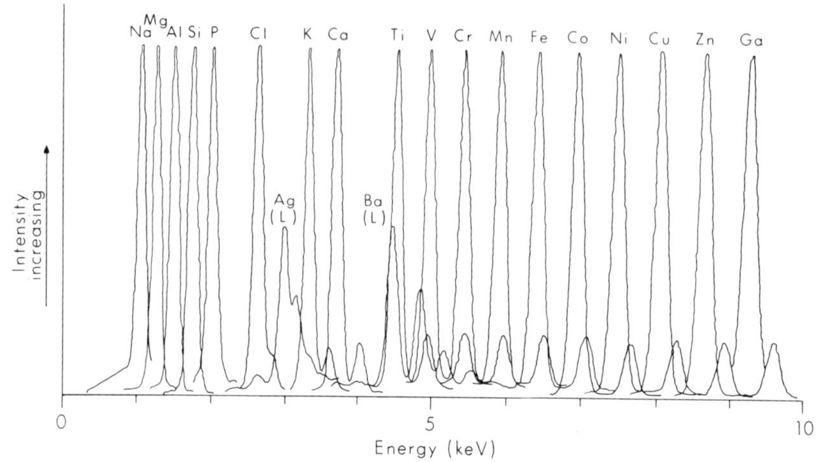

Figure 7.2. *K* line profiles of the analytically important elements in the 0–10 keV region of the x-ray spectrum with the *L* lines of Ag (scattered from the silver x-ray tube) and Ba superimposed. The profiles of Ag and Ba have been arbitrarily plotted at half the relative intensity for clarity. (Reproduced from Ref. 12 by permission of John Wiley and Sons Limited.)

7.3 shows that *L* lines of Pb, Th, and U and *K* lines of As and Mo must be fitted to spectra.

Prepare element profiles by spiking lithium borate glasses with a known amount of an element and high-purity silica to promote glass formation. Do not add silica to disks spiked with Al_2O_3, SiO_2, and KH_2PO_4. Since Na_2CO_3 and MgO (both without SiO_2) will not form stable glasses, grind the fused and quenched mixture

Figure 7.3. X-ray profiles of the *K* lines of As, Rb, Sr, Y, Nb, Zr, and Mo (solid lines) and the *L* lines of Pb, Th, and U (marked areas at half relative intensity) superimposed to indicate the degree of overlap in trace element spectra between 10 and 20 keV. (Reproduced from Ref. 12 by permission of John Wiley and Sons Limited.)

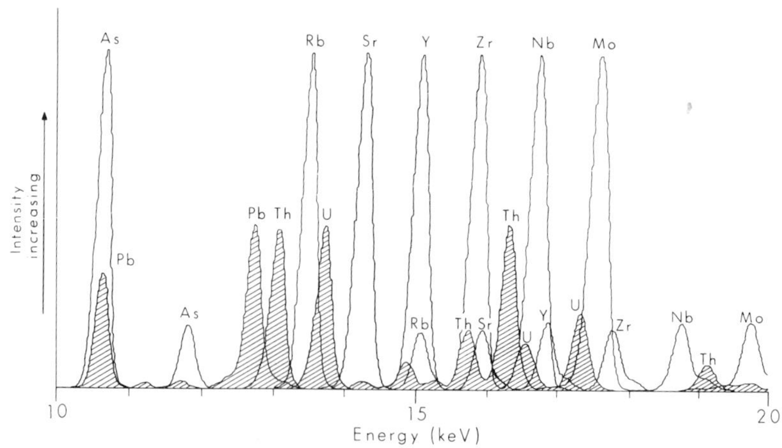

and use pressed pellets. Obtain profiles of trace elements from the pure element oxides in the form of pellets or as powders held on Mylar film.

Place a 50 μm aluminum filter over the primary x-ray beam to prevent interference from silver tube *L* lines when profiles of *K* and Ca *K* lines are recorded. The silver *L* line profile can be obtained from a silica glass disk. Restrict count rates to those giving a maximum dead time of 50% in pulsed-tube operation.

Apply the matrix correction procedure described in the Calibration section (in an inverse manner). This is applied iteratively. For trace elements, obtain counts for the major elements needed for matrix corrections by recording an additional spectrum for each pellet. The instrument conditions are 10 kV and 0.1 mA with no filter.

Calibration

Prepare calibration graphs by means of a selection of the reference materials listed in Table 7.7. Use the same instrument conditions for major and trace elements as for analyses of samples. For major elements count each disk twice, and for minor elements count both sides of each pellet.

Obtain calibration lines by computation of linear regression of net peak intensities plotted against expected rock composites. See Table 7.8. Data on expected compositions are from Abbey (13, 14) and Govindaraju (15).

For major elements convert expected concentrations to apparent fluorescent values according to the method of Norrish and Hutton (9). See also Section IX, Chapter 9 of this book.

For trace elements, correct for mass absorption effects. The authors used the correction program Link Systems XRF2 based on Ref. 16.

Results

The results recorded in Table 7.8 illustrate the values obtained for standard samples. The limits of determination are illustrated in Tables 7.9 and 7.10.

Table 7.7
Reference Materials Used for Calibration

USGS (U.S.A.)	CRPG/ANRT (France)	NIM (South Africa)	BAS (UK)[a]	GSJ (Japan)
AGV-1 andesite	BR basalt	NIM-D dunite	BCS-267 silica brick[b]	JB-1
BCR-1 basalt	GA granite	NIM-G granite	BCS-269 firebrick[b]	JG-1
G-2 granite	GH granite	NIM-L Lujavrite[b]	BCS-375 soda feldspar[b]	
GSP-1 granodiorite	DR-N diorite	NIM-N norite	BCS-376 potash feldspar[b]	
PCC-1 peridotite[b]	MICA-Fe biotite[c]	NIM-P pyroxenite		
	MICA-Mg phlogopite[b]	NIM-S syenite		
	UB-N serpentine[b]			

Source: Ref. 12. Reproduced by permission of John Wiley and Sons Limited.
[a]BAS (Bureau of Analysed Samples) samples are certified reference materials.
[b]Reference materials used for major element calibration only.
[c]MICA-Fe was used for the trace element calibration only.

Table 7.8
Precision and Accuracy in Routine Analysis[a]

	BCR-1			G-2			AN-G		BE-N		MA-N	
Oxide	Expected	Mean analyzed	2s	Expected	Mean analyzed	2s	Expected	Analyzed	Expected	Analyzed	Expected	Analyzed
SiO_2	54.53	54.84	0.38	69.22	68.73	0.60	46.33	46.83	38.45	38.40	66.64	66.87
TiO_2	2.26	2.24	0.02	0.48	0.50	0.02	0.22	0.22	2.64	2.58	0.03	0.02
Al_2O_3	13.72	13.59	0.18	15.40	15.22	0.02	29.79	29.38	10.10	10.39	17.68	17.59
Fe_2O_3	13.41	13.31	0.09	2.69	2.70	0.04	3.36	3.25	12.87	12.61	0.48	0.25
MnO	0.18	0.19	0.01	0.03	0.03	—	0.04	0.04	0.20	0.19	0.04	0.04
MgO	3.48	3.56	0.12	0.75	0.73	0.18	1.82	1.97	13.25	13.11	0.07	0.26
CaO	6.97	7.05	0.06	1.96	1.92	0.04	15.89	15.94	13.95	13.86	0.60	0.61
Na_2O	3.30	3.36	0.30	4.06	4.09	0.36	1.65	1.56	3.19	3.10	5.84	6.05
K_2O	1.70	1.66	0.04	4.46	4.50	0.06	0.14	0.06	1.40	1.34	3.21	3.17
P_2O_5	0.36	0.36	0.02	0.13	0.13	0.04	0.03	0.07	1.05	1.14	1.38	1.52
V_2O_5	0.07	0.07	0.01	0.01	0.01	—	0.012	0.014	0.042	0.046	0.001	0
Cr_2O_3	—	—	—	—	—	—	0.007	0	0.053	0.042	0.001	0
NiO	—	—	—	—	—	—	0.004	0.01	0.034	0.031	0.001	0
BaO	0.08	0.07	0.02	0.21	0.22	0.02	0.004	0.03	0.114	0.14	0.005	0.003
Total	100.06	100.30		99.40	98.78		99.30	99.37	97.34	96.98	95.98	96.41

	GSP-1			AGV-1			AN-G		BE-N		MA-N	
Element	Expected	Mean analyzed	2s	Expected	Mean analyzed	2s	Expected	Analyzed	Expected	Analyzed	Expected	Analyzed
Rb	250	249.5	5.0	67	67.8	1.8	1	2	47	48	3600	3500
Sr	240	236.3	3.6	660	675	9.8	76	79	1370	1385	84	82
Y	29	29.7	4.0	19	21.3	1.4	8	8	30	31	1	(40)
Zr	500	495.4	4.2	230	233	3.4	15	21	265	276	27	39
Nb	23	25.3	1.6	16	14.8	0.8	2	3	100	103	173	152
Pb	54	54.1	10.2	33	35.8	2.8	2	7	4	6	29	21
Th	105	103.7	1.4	6.4	6.9	2.0	—	1	11	9	1	1
Cu	33	31.1	5.6	59	58.1	9.6	19	21	72	70	140	145
Zn	105	105.1	6.4	86	88.9	6.0	20	21	120	114	220	216
Ga	23	21.0	3.4	19	20.4	2.8	18	20	17	16	59	70

Source: Ref. 12. Reproduced by permission of John Wiley and Sons Limited.

[a]Units: major elements, percent oxides; trace elements, parts per million. Expected compositions for BCR-1, G-2, GSP-1, and AGV-1 taken from Abbey (13,14) and for CRPG standards from Govindaraju (15). Analyzed data are averages of the following numbers of determinations: BCR-1, 12; G-2, 12; GSP-1, 20; AGV-1, 13; CRPG standards, 2.

Table 7.9
Calibration Data for Major Elements Determined on Glass Disks (1 : 6 Sample-to-Flux Ratio)[a]

Oxide	Composition range of reference materials (% oxide) Low	High	Counting sensitivity[b] (counts per 100 s per 1% oxide)	Limit of determination[c] (500 s count time) (% oxide)
SiO_2	38.30	95.90	3,450	0.21
TiO_2	0.01	2.61	24,700	0.03
Al_2O_3	0.32	33.90	1,580	0.19
Fe_2O_3	0.10	16.98	28,900	0.025
MnO	0.01	0.77	35,400	0.015
MgO	0.03	43.51	600	0.33
CaO	0.08	13.87	22,900	0.025
Na_2O	0.01	10.40	170	0.96
K_2O	0.01	15.35	20,300	0.04
P_2O_5	0.01	1.05	5,700	0.05
Cr_2O_3	0.01	3.50	33,500	0.015
NiO	0.01	0.31	20,500	0.03
BaO	0.01	0.47	20,200	0.05
V_2O_5	0.01	0.08	34,000	0.015

Source: Ref. 12. Reproduced by permission of John Wiley and Sons Limited.

[a]Samples were fluoresced with a silver x-ray tube operated at 10 kV, 0.2 mA and counted for 500 live seconds with no primary beam filter.

[b]Counting sensitivity was calculated from a spectrum of CRPG-BR fluoresced under routine operating conditions (500 s count time).

[c]Limit of determination is three times the fitting error expressed as percent oxide and is therefore dependent on mass absorption effects due to the matrix composition of CRPG-BR.

Table 7.10
Calibration Data for Trace Elements Determined on Powder Pellets

Element	Composition range of reference materials (ppm) Low	High	Counting sensitivity[a] (counts per 100 s per ppm)	Limit of determination[b] (800 s count time) (ppm)
Rb	5	2200	35	5.6
Sr	3	1300	43	3.5
Y	3	145	44	3.5
Zr	10	800	61	14.8
Nb	2	270	69	4.4
Pb	4	65	22	7.7
Th	0.6	150	31	8.6
Cu	4	72	5	9.9
Zn	10	1300	7	11.5
Ga	8	95	9	7.8

Source: Ref. 12. Reproduced by permission of John Wiley and Sons Limited.

[a]Counting sensitivity calculated from a spectrum of CRPG-BR.

[b]Reference materials were fluoresced with a silver x-ray tube operated at 45 kV, 0.3 mA using a 127 μm silver foil primary beam filter and counted for 800 live seconds.

IV. DETERMINATION OF 14 TRACE ELEMENTS IN ROCK SAMPLES BY STEPWISE SOLVENT EXTRACTION AND FLAME ATOMIC ABSORPTION SPECTROMETRY (17)

Comments on the Method

The trace elements are separated from the sample matrix in chloride solution by stepwise solvent extraction with *n*-butyl acetate, diethyldithiocarbamate, and 8-hydroxyquinoline. The procedure is summarized in Fig. 7.4. Rock samples are first dissolved in hydrofluoric, nitric, sulfuric, and perchloric acids and then taken up in hydrochloric acid containing free chlorine. Gold, Tl, Sb, Ga, Mo, Cu, Ni, Co, Ag, Bi, Cd, Pb, Zn, and Mn can be determined. These elements were determined in NIM and USGS standard rocks. Generally good agreement with the "accepted" values was obtained. Achievable detection limits are shown in Table 7.11.

Reagents

The reagents used were "Analar" grade with the exception of MIBK, which was reagent grade. All water was distilled and deionized.

Solutions

- Sodium diethydithiocarbamate (NaDDC): Dissolve 3 g NaDDC in 100 ml of distilled and deionized (DDI) water; filter the resultant solution and extract three times with 15 ml portions of methyl isobutyl ketone (MIBK).
- 8-Hydroxyquinoline: Dissolve 1 g 8-hydroxyquinoline in 100 ml of MIBK.
- Buffer: Dissolve 102 g potassium biphthalate in 500 ml DDI water.
- Sodium tartrate: Dissolve 40 g sodium tartrate in 200 ml DDI water; extract with 1% (w/v) 8-hydroxyquinoline until all trace metals are removed.
- EDTA: Dissolve 37.2 g analar grade disodium ethylenediaminetetraacetate dihydrate in 1000 ml DDI water.

Standard solutions

For stock standard solutions, dissolve appropriate metal or stoichiometric salt (99.9% purity or analar) to prepare stock standard solutions at a concentration of 1000 ppm.

Equipment

A Techtron model AA-3 with Varian Techtron 1M-6D Indicator Module was used for all AAS determinations. The pH adjustment and measurement were done using a Radiometer type TTT-1 titrator–pH meter.

Procedure

Figure 7.4 shows the procedure for this determination.

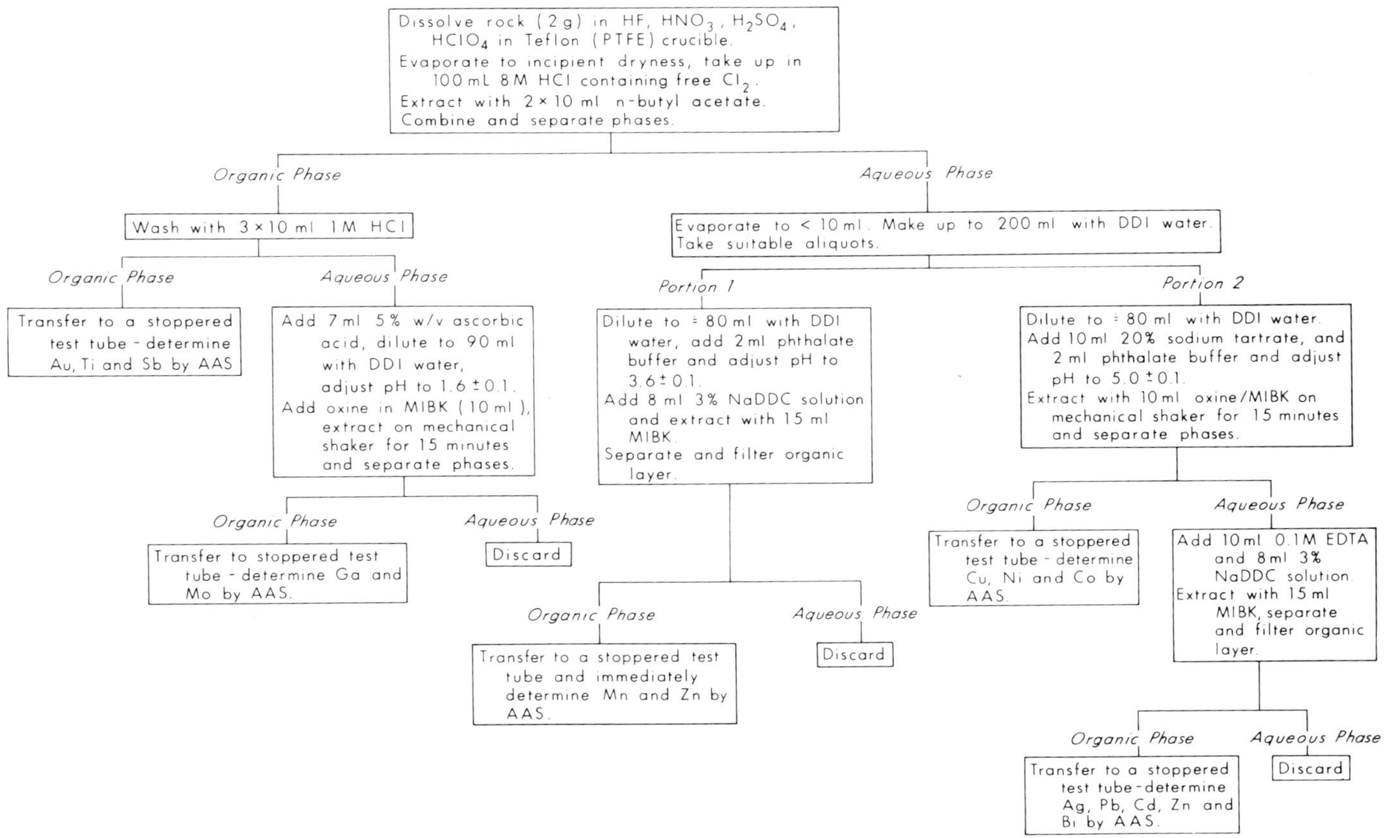

Figure 7.4. Summary of procedure (17).

Table 7.11
Elemental Detection Limits (Using 2 g of Sample)

Element	Detection limit (ppm)
Au	0.06
Tl	0.08
Sb	0.3
Ga	1.0
Mo	0.9
Ag	0.09
Bi	0.3
Cd	0.01
Co	0.04
Cu	0.03
Mn	0.02
Ni	0.04
Pb	0.15

Source: Ref. 17.

V. DETERMINATION OF TRACE ELEMENTS IN ROCKS AND MINERALS BY FLAME ATOMIC ABSORPTION (18)

Elements Determined

Ag, Cd, Co, Cr, Cu, Fe, Mn, Ni, Pb, Zn

Comment on the Method

The method is rapid and straightforward for the elements listed when their concentrations exceed about 5 ppm in the sample. When lead is to be analyzed in the sample, sulfuric acid is replaced by perchloric acid. Hydrochloric acid should not be used when silver is being analyzed. The method has yielded satisfactory results for the analysis of USGS standard rocks.

Equipment

A flame atomic absorption spectrometer equipped with a background corrector and an air/acetylene burner was used.

Chemicals and Apparatus

- Teflon dishes, 150 ml
- Hot plate

Procedure

Weigh 1.0 g of finely ground sample into a 150 ml Teflon dish. Moisten the solids with water, and then add 25 ml hydrofluoric acid and 1 ml sulfuric acid. If lead is being determined, substitute perchloric acid for sulfuric acid. Heat the sample to dryness on a hot plate set at medium heat. Add 1 ml of hydrochloric acid (or nitric acid if silver is being determined) and heat again for 5 min at medium heat. Dilute with 10 ml of water. Filter the solution into a 25 ml volumetric flask. Wash the residue and filter with 1% hydrochloric or nitric acid. Then dilute the solution to volume with water. If additional dilutions of the sample solution are required, make sure that the diluted solutions contain 1% total acid.

Prepare dilute standard solutions of the elements so that the solutions contain 1% total acid. Multielement standard solutions may also be used.

Analyze the samples using the instrument conditions recommended by the manufacturer. All the elements can be measured with an air/acetylene flame. Lean flames are needed for chromium and nickel. Use background correction for all elements.

VI. DETERMINATION OF TRACE ELEMENTS IN ROCKS BY FLAMELESS ATOMIC ABSORPTION (19)

Elements Determined

Cd, Co, Cu, Ni, Pb, Zn

Comment on the Method

Nonflame methods for the determination of trace elements in heavy-matrix solutions are subject to severe spectral and chemical interferences. In the recommended method, instructions for overcoming the interferences are provided. However, for the metals listed, nonflame techniques offer little advantage over flame techniques. Analyses of two standard rocks by flameless and flame methods yielded satisfactory results.

Equipment

- Perkin-Elmer model 303 atomic absorption spectrometer equipped with a PE model 2000 graphite furnace
- Perkin-Elmer model 56 strip chart recorder

The instrument parameters are given in Table 7.12.

Preparation of Samples

Weigh a 1 g sample of powdered rock into a Teflon dish. Add 25 ml of hydrofluoric acid and 2 ml of sulfuric acid. Evaporate the liquid over a medium heat until fumes appear. Cool the sample, and then add 7 ml of nitric acid and 21 ml of hydrochloric acid. Evaporate again over low heat to dryness. Remove the excess hydrochloric acid by adding nitric acid until the evolution of brown fumes ceases. Add about 15 ml of water, and filter the mixture into a 25 ml volumetric flask. Dilute to volume with water. (This solution can be used directly for flame methods.) Dilute the solution for analysis (usually 1 : 250).

Table 7.12
Instrumental Parameters for Perkin-Elmer Model 303 Atomic Absorption Spectrophotometer and HGA-2000[a]

Parameter	Cd	Co	Cu	Ni	Pb	Zn
Wavelength, nm						
Absorbing	228.9	240.9	324.8	232.0	217.0	213.9
Background	226.8	239.3	296.1	231.6	220.4	210.4
Spectral slit width, nm	0.65	0.20	0.20	0.065	0.65	0.65
HGA temperature, °C						
Dry	125	125	125	125	125	125
Ash	500	1100	1100	1100	750	500
Atomize	1800	2500	2500	2500	2300	1800
Time, s						
Dry	30	30	30	30	30	30
Ash	60	60	60	60	60	60
Atomize	20	15	15	15	15	20

Source: Ref. 19.
[a]Purge gas: nitrogen, 2.5 liters/min.

Make background corrections by means of the nonabsorbing line method using the wavelengths listed in Table 7.12. Because the absorbance curves are nonlinear beyond 50–60% of full scale, absorbance values must be converted to concentrations before they are subtracted from uncorrected absorbance readings.

Background corrections by means of a deuterium lamp are not recommended because the lamp did not compensate fully for the large background levels generated by these samples.

Calibration

Prepare multielement standard solutions from the individual element standards. Obtain calibration curves by means of standard additions using peak heights. The

Table 7.13
Comparison of Flame and Nonflame Detection Limits in Heavy Matrix Solution

	Detection limit (ppb)	
Element	Flame	Flameless
Cd	10	2
Co	10	20
Cu	30	12
Ni	40	80
Pb	30	10
Zn	10	3

Source: Ref. 19.

curves will be linear up to concentrations giving absorptions of 50–60% of full scale regardless of the element. (Zinc values were consistently high in spite of all precautions.)

Detection Limits

A comparison of detection limits for flame and nonflame methods is given in Table 7.13.

VII. DETERMINATION OF SCANDIUM, YTTRIUM, AND RARE EARTH ELEMENTS IN ROCKS BY ICP–AES (20)

Elements Determined

Sc, Y, La, Ce, Nd, Sm, Eu, Gd, Dy, Yb

Comments on the Method

Samples are dissolved in mixed acids. High-level concentrations of rare earths are analyzed directly. Low (or chondrite) levels are separated by ion exchange, and fractions are analyzed. The methods were used in the analyses of 12 standard reference materials.

Equipment

- JY48P 36-channel direct reading spectrometer (Instruments S.A.)
- Plasma-Therm ICP torch
- Digital 11/23 computer
- Decwritter 111 terminal and Tektronix 4006-3 video terminal
- Concentric glass (Meinhard) nebulizer

Chemicals and Apparatus

- Mixed acid (1 : 1 : 3 HNO_3 : $HClO_4$: HF)
- Hydrochloric acid, 2 *M* and 6 *M*
- Amberlite cation-exchange resin, 100–200 mesh
- Teflon beakers, 50 ml
- Hot plate

Sample Preparation

High-level concentrations

Transfer 0.5 g of powdered rock and 5 ml of mixed acid to a 50 ml PTFE beaker. Heat on a hot plate to drive off all perchloric acid. Cool, and dissolve residue in 1 ml hydrochloric acid. Dilute to a volume of 50 ml. Use this solution for analysis.

Low-level concentrations (ion-exchange separation)

Transfer 1.0 g powdered rock to a 50 ml PTFE beaker. Wet the solids with water, and decompose carbonates with 1–2 ml 2 *M* hydrochloric acid and then 1–2 ml 6 *M* hydrochloric acid. Add 5 ml perchloric acid and 20 ml hydrofluoric acid. Heat the sample to dryness on a hot plate. Dissolve the residue in 15 ml 2 *M* hydrochloric acid and 2 ml perchloric acid. Heat to dryness. Cool, and dissolve the residue in 10 ml 2 *M* hydrochloric acid.

Procedure

Prepare a column containing 18 g of Amberlite cation-exchange resin. The column diameter is 15 mm, and the height of resin is 16–20 cm. Equilibrate the resin in 2 *M* hydrochloric acid. Calibrate the column by eluting a sample of strontium solution with 2 *M* hydrochloric acid. This is the volume required to remove the major–minor element fraction preceding the rare earth elements and will be about 160 ml. Begin collection of the rare earth fraction after the addition of 140 ml of 2 *M* acid. Elute the rare earths with 20 ml of 2 *M* acid and 160 ml of 6 *M* hydrochloric acid. Evaporate the fraction to dryness. Dissolve the residue in 2 ml of concentrated hydrochloric acid. Dilute to 15 ml with water. Use this solution for analyses.

Calibration

Prepare calibration curves from the nebulization of single- and mixed-element standard solutions. Instrument conditions are contained in Table 7.14. Determine interference correction coefficients for the elements shown in Table 7.15. Do not

Table 7.14
Instrument Conditions and Wavelengths of Elements[a]

Element	Wavelength (nm)	Order
Gadolinium	342.246	1
Scandium	255.237	1
Ytterbium	328.937	1
Dysprosium	353.170	1
Samarium	359.260	1
Yttrium	371.030	1
Lanthanum	379.478	1
Europium	381.967	1
Neodymium	401.609	1
Cerium	413.765	1

[a]Instrument conditions: plasma rf power, 1100 W; observation height, 15–16 mm; sample feed rate, 1–2 ml/min; integration time, 10 s.

Table 7.15
Interelement Correction Coefficients for Rare Earth Element Determinations

Interfered element	Interfering element[a]	
Gd	Ni(0.0043)	Cr(0.014)
	Fe(0.00048)	Ce(0.02)
	Dy(0.005)	Sm(0.002)
Sc	Mn(0.00027)	Fe(0.000083)
Yb	V(0.0037)	Fe(0.000001)
Dy	Mn(0.0043)	Ti(0.00004)
	Sm(0.002)	Nd(0.003)
Sm	W(0.014)	Ni(0.01)
	Mn(0.000024)	Fe(0.0001)
	V(0.0038)	Ca(0.000067)
	Ti(0.00013)	Nd(0.033)
	Gd(0.11)	
Y	Ca(0.000007)	Ti(0.000027)
La	Fe(0.00091)	V(0.0032)
	Ti(0.000059)	Ce(0.01)
	Ca(0.000045)	Nd(0.0067)
Eu	Fe(0.000015)	Nd(0.003)
	Gd(0.0067)	
Nd	Ni(0.017)	Fe(0.00009)
	Ca(0.0006)	Ti(0.00016)
	Sr(0.016)	Sm(0.025)
Ce	Ca(0.00036)	Ti(0.00077)
	Sr(0.025)	

[a]Corrections are obtained from analysis of individual element spectroscopic standards of known concentrations. Units of coefficients are defined as (μg/ml interfered)/(μg/ml interfering).

apply off-peak corrections since no significant improvement in the results will be obtained. Neodymium is an exception for which a single-point off-peak correction can be applied.

Analyses

Analyze the sample solutions under the same instrumental conditions as the standards. Obtain the average intensities after three 10 s integrations. Correct for interelement interferences.

Detection Limits

The limits are expressed in terms of lowest quantitative determinable concentrations (LQDs) in Table 7.16. The LQD was calculated as five times greater than the experimentally determined detection limit. The detection limit is twice the standard deviation of the background.

Table 7.16
Lowest Quantitatively Determinable Concentrations (LQDs) for Sc, Y, and the Rare Earth Elements

Element	Digestate[a]	Ion exchange[b]
Gd	5	1
Sc	1	0.2
Y	2	0.2
Yb	1	0.1
Dy	4	0.5
Sm	15	1
La	4	0.5
Eu	1	0.1
Nd	30	2
Ce	35	6

[a]Values given in micrograms per gram of rock based on LQD solution × 100 (dilution).
[b]Values given in micrograms per gram of rock based on LQD solution × 15 (dilution).

VIII. DETERMINATION OF RARE EARTH ELEMENTS IN MINERALS BY FLAME ATOMIC ABSORPTION (21)

Elements Determined

Y, Eu, Tb, Dy, Ho, Er, Th, Yb, Lu

Comment on the Method

The more sensitive rare earths are determined after they have been separated from solution as hydroxides. Lanthanum is added to overcome interferences from iron and other elements. The results of analyses of zirconium silicate and calcium rare earth silicate were compared with x-ray fluorescence results and proved to be satisfactory.

Equipment

- Perkin-Elmer model 303 atomic absorption spectrophotometer equipped with a 5 cm nitrous oxide/acetylene burner
- Hollow-cathode lamps for Dy, Ho, Er, Yb
- Perkin-Elmer Intensitron lamps for Y, Eu, Th, Lu

Standard Solutions of Rare Earths

Dissolve weighed amounts of the rare earth oxides (spectroscopically pure, A. D. McKay Co.) in 25 ml of 1 : 1 nitric acid by gently heating. Transfer the solutions to 1 liter volumetric flasks, and dilute to the mark with water. These solutions should contain 1000 ppm of the element.

Prepare working standards by diluting aliquots of the stock solutions with sufficient 10% lanthanum solution to yield a final lanthanum concentration of 1%. Add sufficient nitric acid to correspond to the acid concentration of the samples.

Prepare a 10% lanthanum nitrate solution from lanthanum oxide and nitric acid. The final acid concentration must be 1%. Test the solution to ensure that rare earths are absent.

Preparation of Samples

Zirconium rare earth silicates

Weigh a 0.5 g sample into a plastic beaker, and add concentrated hydrofluoric acid. Warm the mixture to decompose the silicates and to precipitate the insoluble fluorides. Filter the sample to remove soluble zirconium fluoride. Ignite the precipitate in a platinum crucible with the aid of a few drops of sulfuric acid. Fuse the oxides with potassium pyrosulfate, and then dissolve the cooled salts in dilute acid. Precipitate the hydroxides with dilute ammonia solution. Filter the precipitate, and treat it as described below.

Calcium rare earth silicates

Weigh a 0.5 g sample into a platinum crucible. Fuse it with sodium carbonate. Dissolve the salts in dilute acid. Precipitate the hydroxides with dilute ammonia solution. Filter the precipitate, and treat it as described below.

Other rare earth silicates

Dissolve a sample by an established procedure. Precipitate the rare earths by means of fluoride, hydroxide, or oxalate to free them from large salt contents. Obtain a hydroxide precipitate as a final step, and treat it as described below.

Procedure

Dissolve the precipitate in hot, dilute nitric acid in a 50 ml beaker. Wash the paper or filter crucible thoroughly with hot 1 + 50 nitric acid. Add sufficient lanthanum solution to give a final lanthanum concentration of 1% in the chosen volume. Evaporate the solution to 2 ml. Transfer the sample to a volumetric flask (5 or 10 ml) and dilute to volume with water.

Analyze the sample and standards for each element using the instrument settings shown in Table 7.17. Each standard solution must contain the same quantities of

Table 7.17
Instrument Parameters

Parameter	Eu	Tb	Dy	Ho	Er	Th	Yb	Lu	Y
Wavelength, nm	459.4	432.6	421.2	410.4	400.8	409.4	398.8	331.2	410.2
Current, mA	35	30	15	30	25	30	20	30	30
Slit width, nm	0.3	0.3	0.3	0.3	0.3	0.3	0.3	0.1	0.1

Source: Ref. 21.

Table 7.18
Detection Limits and Sensitivities

Element	Sensitivity (ppm per 1% absorption)	Detection limit (ppm)
Terbium	10	2
Dysprosium	1.0	0.5
Holmium	2.0	0.3
Erbium	0.8	0.4
Thulium	1.0	0.3
Ytterbium	0.3	0.1
Lutetium	30	10
Europium	0.7	0.3
Yttrium	1.8	—

Source: Ref. 21.

lanthanum and nitric acid as the sample. Adjust the flame conditions to give maximum absorption in a slightly reducing flame. Under these conditions, interferences due to silica (twofold excess) and Fe, K, Mg, Na, Ca, Ti, and other rare earths (eightfold excess) were reduced to negligible levels.

In the presence of aluminum, use a burner-to-beam height of 2 cm. In the presence of zirconium, use a 2 cm burner-to-beam height. Also adjust the flame conditions until equal absorbances are obtained from zirconium-bearing and zirconium-free solutions.

Detection limits and sensitivities are given in Table 7.18.

IX. DETERMINATION OF THE RARE EARTH ELEMENTS, YTTRIUM, AND SCANDIUM IN SILICATE ROCKS BY FLAMELESS ATOMIC ABSORPTION (22–24)

Comment on the Method

The rock is decomposed by acids, and the lanthanides are separated from most of the matrix by an ion-exchange column. The elements are determined by electrothermal atomization from a pyrolytically coated graphite surface. A tantalum foil-lined graphite furnace tube yields greater sensitivity (10–40-fold) for La, Ce, Pr, Gd, Tb, and Lu. Satisfactory results were obtained for five international reference rock samples. Four new Canadian iron-formation reference materials, FeR-1 to FeR-4 were also analyzed.

Apparatus

- Varian AA-475 spectrometer
- Varian GTA-95 graphite tube atomizer equipped with a programmable sample dispenser, and memory storage for 8 programs

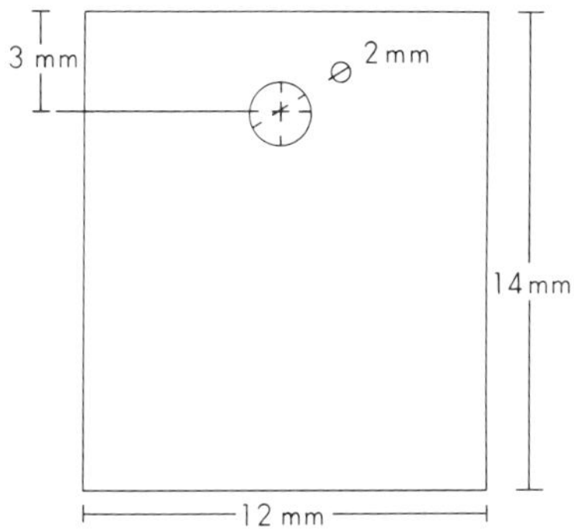

Figure 7.5. Shape and dimensions of the tantalum foil. (Reprinted with permission from Ref. 24. Copyright 1984, Pergamon Press, Ltd.)

- Epson MX-82 type 11 printer
- Pyrolytically coated tubes
- Tantalum foil, 99.5% pure, 0.05 mm thick

Preparation of Tantalum Foil-Lined Furnace Tube

Drill a 2 mm hole in a 12 × 14 mm piece of foil as shown in Fig. 7.5. Wrap the foil around a glass rod 4 mm in diameter with the short side parallel to the rod. Insert the foil into a new pyrolytically coated graphite tube so that the holes coincide. Insert a metal rod (3 mm diameter) into the tube, and roll it around firmly to press the metal foil firmly to the graphite tube.

Anneal the coated tube by placing it in the GTA-95 and heating it in a current of argon. Heat the tube to 1000°C in 10 s, hold for 10 s, and the cool it down in 16 s. Repeat this operation for temperatures of 1500 and 2000°C. Perform a blank run according to the program in Table 7.19.

Table 7.19
Blank Run for Ta-Lined Furnace

Step	Temperature (°C)	Time (s)	Gas flow, (liter/min)	Read[a]
1	75	15	3	
2	90	60	3	
3	120	60	3	
4	850	10	3	
5	1800	10	3	
6	1800	2	0	
7	2600	1.3	0	*
8	2600	2	0	*
9	2600	1	3	

Reprinted with permission from Ref. 24. Copyright 1984, Pergamon Press, Ltd.

[a]*, Read absorbance.

The foil may shift inside the graphite tube after several firings. Check the alignment of the holes periodically.

Check the condition of the foil by firing a standard after every 5–10 samples. Change the foil when deterioration in performance is noted (usually after 40 firings). Discard the foil and graphite tube after 80 firings.

Preparation of Ion-Exchange Column

Use a Pyrex glass column as illustrated in Fig. 7.6. Pack a 30 cm length with Dowex 50W-X8 cation-exchange resin. Wash the column with 6 *M* hydrochloric acid until the effluent is free of Fe^{3+}. Then wash the column with demineralized water until the effluent is neutral to blue litmus paper.

Instrumental Operation Conditions

Tables 7.20–7.22 give the operating parameters of the instruments used in this determination.

Figure 7.6. Ion-exchange column. (Reprinted with permission from Ref. 23. Copyright 1984, Pergamon Press, Ltd.)

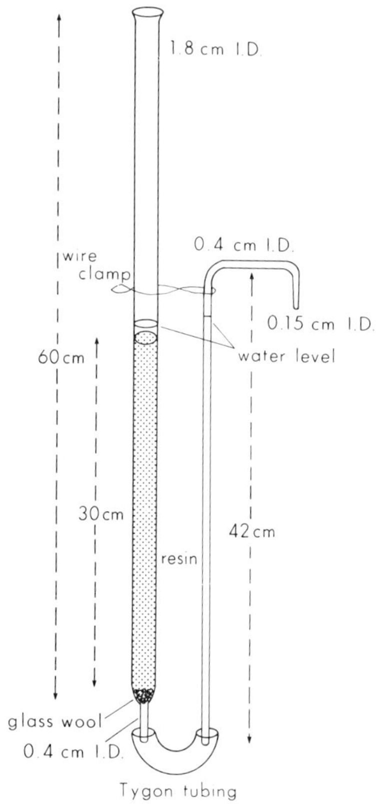

Table 7.20
Instrumental Parameters

Element	Wavelength (nm)	Hollow-cathode lamp	Current (mA)	Spectral bandpass (nm)
Sc	391.18	Westinghouse	5	0.2
Y	410.2	Westinghouse	5	0.5
La	550.13	Varian-Techtron	10	R[a]
Ce	567.0	Westinghouse	10	R
Pr	495.1	Westinghouse	10	R
Nd	492.45	Westinghouse	10	0.2 or R
Sm	429.67	Westinghouse	10	0.2 or R
Eu	459.4	Westinghouse	5	0.2
Gd	368.4	Westinghouse	10	0.2 or R
Tb	432.7	Cathodeon	10	0.2 of R
Dy	421.2	Westinghouse	10	0.2
Ho	410.4	Westinghouse	10	0.2
Er	400.8	Westinghouse	5	0.2
Tm	371.8	Varian	10	0.2
Yb	398.8	Westinghouse	5	0.5
Lu	336.0	Varian-Techtron	10	R

Reprinted with permission from Ref. 22. Copyright 1985, Pergamon Press, Ltd.
[a]R, Restricted 0.5 nm slit width.

Table 7.21
Operating Parameters for Pyrolytically Coated Graphite Furnace

Element	Step number	Temperature (°C)	Time (s)	Argon gas flow (liter/min)	Read[a]
Sc, Y, Nd,	1	75	15	3	
Sm, Eu, Dy,	2	90	60	3	
Ho, Er, Tm,	3	120	10	3	
and Yb	4	850	10	3	
	5	1800[b]	10[c]	0	
	6	1800[b]	2	0	
	7	2700[d]	1.3	0	*
	8	2700[d]	2	0	*
	9	2800[d]	5	3	

Reprinted with permission from Ref. 22. Copyright 1985, Pergamon Press, Ltd.
[a]*, Read absorbance.
[b]1400°C for Eu and Yb; 2200°C for Sm, Dy, Er, and Tm; 2400°C for Ho.
[c]15 s for Nd and Dy; 20 s for Ho.
[d]2200°C for Yb.

Table 7.22
Operating Parameters for Tantalum Foil-Lined Graphite Furnace

Element	Step number	Temperature (°C)	Time (s)	Argon gas flow (liter/min)	Read[a]
La, Ce, Pr,	1	75	15	3	
Gd, Tb, and	2	90	60	3	
Lu	3	120	60	3	
	4	850	10	3	
	5	1800[b]	10	3	
	6	1800[b]	2	0	
	7	2600	1.3	0	*
	8	2600	2	0	*
	9	2600	1	3	

[a]*, Read absorbance.
[b]2000°C for Gd; 2200°C for Tb and Lu.

Reagents

Standard solutions

Prepare stock solutions (1000 ppm) of scandium, yttrium, and the lanthanides by dissolving the appropriate quantities of the pure oxides in hot, dilute nitric acid. Remove excess acid by evaporation. Dilute each to a known volume with 1 *M* nitric acid, and store the solutions in plastic bottles.

Standardize each solution by transferring 4 ml to a 10 ml platinum crucible previously ignited to constant weight at 1000°C. Evaporate to dryness on a steam bath. Heat the crucible gently with a low flame of a burner until no more fumes are evolved. Heat the crucible strongly. Ignite in a muffle furnace at 1000°C for 30 min. Cool in a desiccator for 20 min and weigh.

Dilute the stock solutions with 0.1 *M* nitric acid to obtain working standards.

Standard mixtures

Prepare synthetic standard mixtures of reference materials GA (granite) and SY-2 (syenite) by mixing the standard solutions in appropriate amounts as shown in Table 7.23.

Calibration solutions

Prepare three solutions with concentrations corresponding to 0.05, 0.1, and 0.2 g of sample per milliliter by diluting known volumes of GA and SY-2 standard mixtures to known volumes with 0.1 *M* nitric acid. Use the standard mixture that corresponds to the type of sample for analysis.

Table 7.23
Concentrations of Sc, Y, Th, and Rare Earth Elements in Synthetic Mixtures of GA and SY-2[a]

Element	GA (granite) (ppm)	SY-2 (syenite) (ppm)
Sc	7	7
Y	17	120
La	38	88
Ce	70	210
Pr	7	16
Nd	20	75
Sm	5.2	17
Eu	0.83	2.3
Gd	3	11
Tb	0.5	2.7
Dy	2.7	21
Ho	0.5	7
Er	1.5	13
Tm	0.21	2.5
Yb	1.4	16
Lu	0.3	3
Th	17.6	370

Reprinted with permission from Ref. 24. Copyright 1984, Pergamon Press, Ltd.

[a]These values were derived from the author's own published and unpublished works as well as those found in the literature.

Procedure

Preparation of sample solution (solution A)

Depending on the rock type, transfer a 1–2 g sample (accurately weighed) to a 50 ml platinum dish. Moisten it with water. Cover the dish with a platinum lid and gradually, while stirring with a platinum rod, add 10–20 ml of concentrated hydrofluoric acid. Heat on a steam bath for 2–3 h; then remove the lid and rinse it with water, and evaporate the solution to dryness. Add 10–20 ml of concentrated nitric acid to the residue, stir to break up any lumps, and evaporate to dryness. For shale samples of high carbon content (2–8%), place the dish on a silica triangle supported on a tripod, cover, and heat gently with a Bunsen burner until all acid fumes are expelled and all organic material is destroyed. Heat briefly to red heat, cool, then proceed as described.

Add 15–20 ml of concentrated perchloric acid and stir the mixture until a thin slurry is produced; then cover the dish and heat gently on a sand bath until fumes of perchloric acid start to appear. Remove the lid, rinse with water, stir to mix, then evaporate the solution to a syrupy consistency.

Cool the dish, add 25 ml of 10% nitric acid–5% hydrogen peroxide solution, cover, and, with occasional stirring, heat on a steam bath to dissolve the salts. Transfer the solution to a 400 ml beaker, rinse the dish thoroughly with water, cover the beaker with a watch glass, and boil vigorously to decompose the excess of hydrogen peroxide. If insoluble material is present or a precipitate forms on cooling, filter the solution through a Whatman No. 40 paper, wash with hot 10% nitric acid–5% hydrogen peroxide solution, then transfer the paper and residue to a platinum crucible and burn off the paper at ~400°C in a muffle furnace. Ignite the residue at 800°C for 15 min, cool the crucible, and, depending on the quantity of residue, add 0.5–1 g of potassium pyrosulfate. Mix thoroughly; then fuse the mixture over an open flame and keep the melt at red heat for a short time to ensure complete decomposition. Cool the crucible, add 15 ml of water, loosen the cake with a glass rod, and transfer to a 50 ml beaker. Heat on a hot plate and add 50% sulfuric acid dropwise, if necessary, until a clear solution is obtained. Add this solution to the main solution.

Ion-exchange separation

Using a pH meter, adjust the pH of the solution to 1.2 ± 0.2 by dilution with demineralized water; then transfer the solution to a 250 ml separatory funnel clamped at the top of the cation-exchange resin column. Pass the solution through the column at about 2.5 ml/min. Wash the beaker three times with 10 ml portions of acidulated water (pH 1.2 ± 0.2) and add the washings to the column.

Elute the common ions from the column with 250–300 ml of 1.85 *M* hydrochloric acid (at 2.5 ml/min) until the effluent is free from Fe^{3+} (test with ammonium thiocyanate solution). Discard the effluent.

Elute the rare earths, scandium, and yttrium by passing 500 ml of 6 *M* hydrochloric acid through the column at 2.5 ml/min. Collect the eluate in a 600 ml beaker and evaporate it to about 5 ml. Transfer the solution to a 20 ml beaker, rinsing the original beaker with hot 6 *M* hydrochloric acid. Evaporate the solution to a syrupy liquid on a hot plate and finally to dryness on a steam bath. Add 5 ml of concentrated nitric acid and 1 ml of 30% hydrogen peroxide, cover the beaker with a watch glass, and heat on a steam bath to decompose any resin and convert the chlorides to the nitrates. When the reaction has subsided, raise the watch glass by means of glass hooks and evaporate the solution to dryness. Repeat the nitric acid–peroxide treatment and subsequent evaporation twice more to ensure complete decomposition of the organic material and conversion of the salts into nitrates. Dissolve the final residue in 5 ml of 0.1 *M* nitric acid by stirring and warming briefly on a steam bath; then transfer to a 10 ml standard flask and dilute to the mark with 0.1 *M* nitric acid. If necessary, filter the 5 ml of solution through a 5.5 cm glass fiber paper in a 58° funnel 35 mm in diameter, rinsing the beaker and

Table 7.24
Sensitivities for Sc, Y, and Lanthanides in Pyrolytically Coated and Tantalum Foil-Lined Graphite Furnaces

		Sensitivity[a]		
Element	Atomization temperature (°C)	Pyrolytically coated furnace (PCF)	Ta foil-lined furnace (TaF)	Sensitivity enhancement factor (PCF/TaF) at 2600°C
Sc	2600	1.3×10^{-10}	1.2×10^{-11}	10.8
	2700	1.0×10^{-10}		
Y	2600	2.6×10^{-9}	2.8×10^{-10}	9.3
	2700	1.3×10^{-9}		
La	2600	4×10^{-8}	2×10^{-9}	20
	2700	2×10^{-8}		
Ce	2600	2×10^{-7}	5×10^{-9}	40
	2700	0.8×10^{-7}		
Pr	2600	1.6×10^{-8}	1.5×10^{-9}	10.7
	2700	1×10^{-8}		
Nd	2600	3×10^{-9}	2×10^{-10}	15
	2700	1.5×10^{-9}		
Sm	2600	1×10^{-9}	7×10^{-9}	14.3
	2700	0.8×10^{-9}		
Eu	2600	5.5×10^{-11}	4×10^{-12}	13.8
	2700	3.4×10^{-11}		
Gd	2600	2.1×10^{-8}	7.3×10^{-10}	28.8
	2700	1.4×10^{-8}		
Tb	2600	6.4×10^{-9}	3.2×10^{-10}	20
	2700	4.1×10^{-9}		
Dy	2600	2×10^{-10}	2×10^{-11}	10
	2700	1.4×10^{-10}		
Ho	2600	5.5×10^{-10}	3.1×10^{-11}	17.7
	2700	2.8×10^{-10}		
Er	2600	4.4×10^{-10}	1.3×10^{-11}	33.8
	2700	2.8×10^{-10}		
Tm	2600	3.7×10^{-11}	4×10^{-12}	9.3
	2700	1.8×10^{-11}		
Yb	2600	3.3×10^{-12}	5.4×10^{-13}	6.1
	2700	2.5×10^{-12}		
Lu	2600	4.9×10^{-9}	1.4×10^{-10}	35
	2700	3.5×10^{-9}		

Reprinted with permission from Ref. 22. Copyright 1985, Pergamon Press, Ltd.

[a]Defined as the weight of the element in grams which produces a change, compared with a pure solvent or blank, of 0.0044 absorbance unit.

washing the paper with 0.1 *M* nitric acid and collecting the filtrate and washings in a 10 ml standard flask. This 10 ml of solution is solution A.

Measurement

Transfer calibration and sample solutions to the vials in the sample dispenser. Depending on the sensitivity (see Table 7.24), use sample solutions containing about 0.1 mg of sample per milliliter and a scale expansion of up to 10×. Use a sample volume of 20 μl for all the elements listed except cerium. Calibrate with 5, 10, 15, and 20 μl of either GA or SY-2 calibration solutions, plus sufficient 0.1 *M* nitric acid to yield total volumes of 20 μl. The instrumental operating conditions are listed in Tables 7.20–7.22. Obtain the concentrations of elements from either the instrument readouts or calibration curves. As a recommended practice, two reference rocks for which "usable values" for the rare earth elements have been reported should be run with every batch of samples to check the recovery.

Clean the sample dispenser capillary tip periodically with a tissue. After about 50 firings of a new pyrolytically coated graphite tube, absorbances decrease. Check the absorbances of standards and apply correction factors if necessary.

Minor and trace amounts of some common elements which are retained on the ion-exchange resin and coeluted with the rare earths interfere seriously with the determination of cerium. Determine cerium in solution A by the oxalate–hydrous ferric oxide procedure described by the author (24, 25).

X. DETERMINATION OF RARE EARTH ELEMENTS IN ROCKS BY THIN-FILM X-RAY FLUORESCENCE SPECTROMETRY (26)

Elements Determined

La, Ce, Pr, Nd, Sm, Eu, Gd, Tb, Dy, Ho, Er, Yb

Comment on the Method

A sample is decomposed by fusion with sodium peroxide and hydroxide. Rare earth elements (REE) in the leachate are coprecipitated with hydrous ferric oxide. The precipitate is dissolved in acid, and REE are separated from most other elements on a cation-exchange resin. The REE in the eluate are concentrated again with hydrous ferric oxide. A thin film of this precipitate on a Millipore filter is used for x-ray measurements. The results of analyses of five standard reference rocks were in good agreement with those obtained by ICP–AES and with other values reported in the literature.

Apparatus and Reagents

A Siemens sequential SRS-1 x-ray fluorescence spectrometer with a 10-position sample changer was used. Extra apertures were installed between the primary and secondary beams to mask all radiation but that from the 16 mm diameter sample surface. The instrument parameters are listed in Table 7.25.

An ion-exchange column of borosilicate glass was 170 mm in length with a bore of 18 mm. A reservoir with a volume of about 330 ml (173 mm in length and 57 mm in diameter) was attached to the top of each column.

The column was charged with about 22 g of Bio-Rad AG 50 W-X8 polystyrene cation-exchange resin, 200–400 mesh. The flow rate was controlled at 1–2 ml/min by constricting the plastic outlet tube.

Stock solutions of REE (1000 ppm) were prepared from high-purity oxides (Johnson Matthey Ltd). All oxides except cerium were dissolved in 20% (vol) nitric acid. Cerium oxide was dissolved in 30% (vol) sulfuric acid.

Mixed standard solutions containing 10 ppm Lu and 4 ppm other rare earths were prepared from the above stock solutions and 10% (vol) hydrochloric acid.

Millipore filters of types GS, 0.22 μm pore size, and GA-6, 0.45 μm pore size, were used.

Analytical grade acids were used to prepare the following eluents: 1 and 5.5 *M* hydrochloric acid, 2 *M* nitric acid, 1 *M* sulfuric acid.

All reagents were examined for REE impurities by applying the procedure to pure silica.

Procedure

Sample decomposition

Transfer 1 g sample to a nickel crucible. Mix with 3 g sodium hydroxide and 3 g sodium peroxide. Fuse the mixture at 600°C for 100 min over a Meker burner. Cool the fusion for 1–2 min.

Place the crucible and lid in a 150 ml beaker containing 2 ml water. Leach the

Table 7.25
XRF Instrument Parameters

X-ray spectrometer	Siemens SRS-1
X-ray generator	Crystalloflex 4 type H4
X-ray tube target	Au
Power	40 kV; 60 mA
Collimator	0.15° divergence
Analyzing crystal	LiF (220)
Detector	F.P.C. with P 10 gas[a]
Counting time	Fixed time, 60 s
Pulse height selection	8.5/7 V
Sample changer	10-position, under vacuum

Reprinted with permission from Ref. 26. Copyright 1985, Pergamon Press, Ltd.

[a]F.P.C., flow proportional counter; P10, 10% methane and 90% argon.

fused salt for 10 min. Transfer the contents to a 100 ml plastic centrifuge tube, and centrifuge at 2000 rpm for 5 min.

Decant the supernatant liquid into another similar centrifuge tube. Add 1 ml of 1% iron(III) chloride solution to coprecipitate any dissolved REE. Centrifuge again. Discard the supernatant liquid.

Dissolve the precipitate in 10 ml of water and 5 ml of 6 *M* hydrochloric acid by heating on a steam bath. (Keep the acid concentration to a minimum of 2 *M* to avoid precipitation of any silica.)

Ion-exchange separation

Transfer the solution prepared above to an ion-exchange column prepared as outlined. After the sample solution has passed through the column, wash the column successively with 150 ml of 1 *M* hydrochloric acid, 170 ml of 2 *M* nitric acid, 120 ml of 1 *M* sulfuric acid, 120 ml of 1 *M* hydrochloric acid, and 300 ml of 5.5 *M* hydrochloric acid. Wash the column with 15 ml of water between acid washes. Collect the final eluate (300 ml of 5.5 *M* hydrochloric acid), which contains the REE, in a 400 ml beaker. Evaporate the sample to dryness. Dissolve the residue in a little water and 0.5 ml of 10 *M* nitric acid. Dilute to about 100 ml with water.

Preparation of thin film

Add 300 μg of iron(III) and 50 μg of lutetium(III) (as an internal standard) to the above solution. Add 10 ml of ammonia solution to raise the pH to above 12 and to coprecipitate the REE. Filter the sample through a Millipore filter to obtain a thin film for XRF measurements.

Prepare a reference disk from a 5 ml volume of the mixed standard solution (which contains 50 μg of lutetium and 20 μg of each of the other REE). Transfer the solution directly to the column as described above and obtain the REE fraction.

Prepare a blank disk containing only lutetium and iron by applying the procedure to 1 g of pure silica. (Disks prepared from individual REE can be used for identification of overlapping spectral lines.)

Construction of the calibration line

Make all XRF measurements by counting each line and background for 60 s in sequence for each of the REE. Use the line and background positions listed in Table 7.26. This table also contains lower limits of detection.

Measure REE peak and background intensities for a blank disk containing only lutetium and iron. Calculate the corresponding peak/background ratios. Multiply the measured sample and standard backgrounds by these ratios to obtain the background intensities under the peaks.

Line overlap interferences for which corrections are required are listed in Table 7.27. Make corrections by measuring the intensity of the interfering line at the appropriate analyte wavelength on a disk containing a known quantity of an interfering element. Calculate the counts per unit quantity of interfering element. Subtract the count resulting from the interfering element from the analyte count for the multielement standard disk.

Table 7.26
Measurement Position and Detection Limits by XRF

Analyte	Line	Background − peak $2\theta^a$ (deg)	Lower limit of detection[b] (ppm)
La	$L\alpha_1$	−2.590	0.06
Ce	$L\beta_1$	−1.378	0.14
Pr	$L\beta_1$	+1.008	0.16
Nd	$L\alpha_1$	−2.398	0.08
Sm	$L\beta_1$	+0.816	0.21
Eu	$L\alpha_1$	+1.784	0.15
Gd	$L\alpha_1$	−1.938	0.11
Tb	$L\beta_2$	+0.720	0.42
Dy	$L\beta_1$	−0.682	0.27
Ho	$L\beta_1$	−0.406	0.29
Er	$L\beta_1$	+0.821	0.31
Yb	$L\alpha_1$	+1.250	0.14
Lu	$L\alpha_1$	−0.730	—

[a]Background position relative to measured peak, negative if background 2θ lower than peak 2θ.

[b]Peak and background counting time is 1 min and the 95% confidence limit is 2σ. Lower limit of detection from counting statistics = standard concentration $\times$ 3 $\times$ $N_b\sqrt{(N_p - N_b)}$, where N_p and N_b are total counts for peak position and background.

As an alternative, calculate and subtract the amount of analyte equivalent to the concentration of the interfering element as listed in Table 7.27.

Express the final net counts obtained for the blank and the multielement standard as ratios to the internal standard count. Use these values to calculate calibration factors on the assumption that the XRF count is a linear function of analyte concentration. In this manner, the calibration can be used indefinitely. For each new batch of chemicals, prepare and measure a new blank disk.

Table 7.27
Line Overlap Interferences and Required Corrections

Analyte	Interferent (100 ppm)	Analyte apparent increase (ppm)
Eu	Pr	12.4
Tb	Yb	44
Gd	Ce	5.8
Gd	Nd	0.19

Table 7.28
Results for International Reference Samples

Element	NIM-G (ppm)			BCR-1 (ppm)				BHVO-1 (ppm)				GSP-1 (ppm)			AGV-1 (ppm)			
	This work		Other values	This work		Other values		This work		Other values		This work	Other values		This work		Other values	
	XRF	ICP		XRF	ICP			XRF	ICP			XRF			XRF	ICP		
La	114.4	118.4	105?[a]	28.5	26.8	26.6[b]	27[a]	20.2	16.7	16.2[b]	16.7±0.8[c]	192.4	188[b]	195[a]	43.5	41.0	39[b]	36[a]
Ce	184.2	203.7	200[a]	51.6	54.6	53.8[b]	53[a]	42.2	39.6	38.2[b]	41±4[c]	405.2	432[b]	360[a]	66.2	68.4	69[b]	71[a]
Pr	19.7		19.4[d]	6.9		7.29[b]	7?[a]	6.4		5.66[b]	5.6	51.7	56[b]	50?[e]	8.4		8.5[b]	7?[e]
Nd	78.3		68?[a]	32.8		29.7[b]	26[a]	31.9		25.4[b]	24±6[c]	229.0	206[b]	190?[a]	36.3		32[b]	37?[a]
Sm	13.0	16.8	16?	6.9	7.3	6.7[b]	6.5[a]	6.9	6.0	6.4[b]	6.1±0.7[c]	23.6	27.4[b]	25?[a]	5.6	10.8	5.8[b]	5.9[a]
Eu	—	0.49	0.4?[a]	1.8	1.9	1.98[b]	2[a]	2.0	2.0	2.18[b]	2.0±0.4[c]	1.4	2.16[b]	2.4[a]	1.4	1.7	1.6[b]	1.6[a]
Gd	13.6	14.3	11?[a]	6.6	6.7	6.9[b]	6.6[e]	6.6	5.7	7.0[b]	7±2[c]	11.5	12.4[b]	15?[e]	4.5	7.4	4.6[b]	5.5[e]
Tb	3.0		3?[a]	1.7		1.0[b]	1.0[a]	1.6		0.86[b]	1.0±0.2[c]	1.6	0.4[b]	1.47[a]	0.8		0.5[b]	0.7?[a]
Dy	17.5		16[e]	6.5		6.72[b]	7?[a]	5.4		5.59[b]	4.8±0.2[c]	5.1	5.8[b]	5.7?[a]	3.3		3.7[b]	3.5[a]
Ho	3.5		3?[e]	1.3		1.4[b]	1.2?[a]	1.1		1.06[b]	0.94	1.1	1.0[b]		0.5		0.7[b]	0.6[e]
Er	12.7		10[e]	3.7		3.8[b]	3.5?[a]	2.0		2.63[b]	2.0±0.3[c]	2.5	1.7[b]	3?[e]	1.7		1.8[b]	1.2[e]
Yb	13.4	12.9	14[a]	3.3	3.0	3.7[b]	3.4[a]	2.0	1.9	2.19[b]	2.1±0.5[c]	1.6	1.4[b]	1.9[a]	1.6	1.6	1.72[b]	1.9[a]

[a] "Usable value" of Abbey (28). Question mark (?) indicates uncertainty because it represents the median of only 5–9 available results (28).
[b] ICP-AES value of Crock and Lichte (31)
[c] Reported by Gladney and Goode (27)
[d] Steele *et al.* (29)
[e] "Usable value" of Abbey (30)

Comparisons of Results

All the REE were determined except thulium and lutetium, which are usually below the detection limits specified here (see Table 7.26). In Table 7.28 (27–31), the XRF results for five reference materials are listed together with ICP–AES results obtained by the same authors and with other accepted values. The agreement is good.

XI. DETERMINATION OF GOLD, INDIUM, TELLURIUM, AND THALLIUM IN ROCKS, SOILS, AND SEDIMENTS BY SOLVENT EXTRACTION AND FLAME ATOMIC ABSORPTION (32)

Comment on the Method

Gold, indium, and tellurium in rocks, soils, and sediments are measured in concentrations much higher than crustal abundance, and thallium is measured at both crustal and anomalous levels in the same sample digest. The sample is decomposed in acids, and the four elements are separated from the sample digest and concentrated in a two-step extraction with 4-methyl-2-pentanone (methyl isobutyl ketone, MIBK). Six geochemical reference samples were analyzed, and values for gold, thallium, and tellurium were in good agreement with literature values for these elements. There were differences in the two sets of indium results, but recoveries of added indium were very good. For a 2 g sample, detection limits were 0.1 ppm for gold and tellurium and 0.2 ppm for thallium and indium. The authors stated that analyses of the extracts by electrothermal atomization will lower the detection limits for gold, indium, and tellurium.

Equipment

- Perkin-Elmer 603 atomic absorption spectrometer equipped with a variable nebulizer, a 10 cm three-slot burner, and a background corrector
- Electrodeless-discharge lamp power supply
- Tellurium and thallium electrodeless-discharge lamps
- Gold and indium hollow-cathode lamps

The instrumental and operating conditions were those recommended by the manufacturer for air/acetylene flames.

Chemicals

All chemicals were reagent grade quality.

- Hydrobromic acid–10% bromine solution: Dilute 10 ml of bromine to 100 ml with hydrobromic acid (48%).

- Iron solution, 4% in 3 *M* hydrobromic acid: Dissolve 4 g of pure iron powder in hydrobromic acid (48%). Add 1 ml of bromine and evaporate to dryness. Add 34 ml of hydrobromic acid and warm to dissolve the residue. Dilute to 100 ml.
- Metal stock solutions, 100 μg/ml: These solutions are approximately 9 *M* in hydrobromic acid. Dissolve the weight of metal or oxide listed below in 20 ml of hydrobromic acid–10% bromine solution. Expell excess bromine by gentle heating, and make up to volume in 1 liter flasks with hydrobromic acid. Weights: gold, indium, or thallium, 1.000 g metal; tellurium, 1.250 g TeO_2.

Working Standards in MIBK

- Gold and thallium: Prepare a series of 20 ml aqueous solutions containing 0, 0.5, 1.0, 2.5, 5.0, and 10.0 μg of metal from the required volumes of the diluted stock solutions. Add enough hydrobromic acid in the dilution to make the final acidity 0.1 *M*. Shake each of the final 20 ml solutions with 5 ml of MIBK for 5 min. Separate the MIBK phases by means of a centrifuge.
- Indium and tellurium: Prepare these solutions to contain the same quantities of metals in 20 ml as those listed above. Make the final solutions 3 *M* in hydrobromic acid and add 0.1 ml of the iron solution to each 20 ml standard. Add 1 g of ascorbic acid. (The iron is reduced and the solutions are light yellow in color.) Extract each standard with 5 ml of MIBK and centrifuge.

NOTE: The concentrations of the individual metals in the standards prepared above correspond to 0, 0.25, 0.50, 1.25, 2.50, and 5.00 ppm metal in a 2 g sample.

Procedure

Weigh a 2.00 g sample (<100 mesh) of rock, soil, or sediment into a 50 ml Teflon beaker. Add 1 ml of water to wet the sample. Then add slowly 10 ml of aqua regia. Add 10 ml of hydrofluoric acid. Place the beaker on an oscillating hot plate at room temperature, and allow the sample to dissolve (about 2 h). Evaporate the solution to dryness (100–110°C). Cool the sample, and add 0.22 ml of hydrobromic acid–10% bromine solution and 5 ml of water. Warm and stir to dissolve the residue. Wash the solution into a 25 × 200 mm screw-cap test tube with water. Make up the volume to 20 ml. Add 5 ml of MIBK. Cap the tube and shake it for 5 min. Centrifuge to separate the MIBK layer.

Use the MIBK extract for the determination of gold and thallium. Use the aqueous layer for the determination of indium and tellurium.

Remove any residual MIBK layer. Add 10 ml of hydrobromic acid, and mix. Add 1 g or more of ascorbic acid to reduce the iron and to change the color of the sample from brown to light yellow. (The reduction overcomes the interference of iron in the indium and tellurium determinations.) Extract the sample with 5 ml of MIBK as described above. Use the MIBK layer for the determination of indium and tellurium.

Measure the metals in the MIBK extracts (peak absorbances). Calibrate by means of the standards prepared above.

Table 7.29
Electrothermal Atomic Absorption Analyses of MIBK extracts

Metal	Lower limit of determination in 2 g sample (ppm)	Reference
Au	0.001	33
In	0.025	34
Te	0.004	35

Sensitivity

In analyses of 2 g samples, sensitivities were 0.1 ppm Au, 0.2 ppm Te (0.1 ppm if 2.5 ml MIBK was used), 0.2 ppm Tl, and 0.2 ppm In.

The authors stated that the MIBK extracts are suitable for electrothermal atomization, and they provided the references in Table 7.29 (33–35). See Section XIII for the determination of gold.

Analyses of Reference Samples

The results of analyses of six U.S. Geological Survey reference samples are shown in Table 7.30 (34–36).

Table 7.30
Replicate Analyses ($n = 6$) for Au, Tl, In, and Te in Six Geochemical Reference Samples

Element	Value	GXR-1	GXR-2	GXR-3	GXR-4	GXR-5	GXR-6
Au	Mean, ppm	3.1	[a]	[a]	0.43	[a]	[a]
	Literature value[f], ppm	3.0			0.35		
	RSD, %	4.3			11.9		
	Recovery[b], %					100,98	96, 98
Tl	Mean, ppm	[a]	1.2	3.9	3.5	0.32	2.5
	Literature value[c], ppm		1.2	3.8	3.5	0.35	2.5
	RSD, %		4.2	1.6	2.4	12.5	2.0
	Recovery[b], %					96, 100	90, 102
In	Mean, ppm	0.53	[a]	[a]	[a]	0.32	[a]
	Literature value[d], ppm	0.39				0.11	
	RSD, %	9.7				12.9	
	Recovery[b], %					102,98	96, 96
Te	Mean, ppm	9.3	0.83	[a]	0.92	[a]	[a]
	Literature value[e], ppm	8.7	0.88		1.0		
	RSD, %	7.3	6.2		4.5		
	Recovery[b], %					96,90	96, 95

[a]Below detection limit, 0.1 ppm for Au and Te, 0.2 ppm for Tl and In.
[b]For 0.5, 1.0 μg of the element, added to the sample.
[c]Ref. 36.
[d]Ref. 34.
[e]Ref. 35.
[f]Ref. 33.

XII. DETERMINATION OF TIN IN ROCKS BY ICP–AES (37)

Comments on the Method

This method is useful for the determination of tin as the mineral cassiterite in rocks. Decomposition of cassiterite is effected by heating at 600°C with ammonium iodide. The sublimate of tin iodide thus produced is condensed and subsequently dissolved with tartaric acid. Depending on the concentration of tin in the sample, the tartaric acid solution is nebulized or treated by hydride generation. The lower concentrations of the working range are 10 and 0.02 mg/kg for nebulization and hydride generation, respectively. A 0.1% tartaric acid solution is used in the hydride generation because the tartrate masks the iron interference prevalent in the more commonly used hydrochloric acid solutions. The method was tested on USGS standard rocks. This method is similar to an atomic absorption procedure by the same authors.

Equipment

The equipment is shown in Fig. 7.7, which illustrates the two alternative methods of sample introduction. The plasma generator was a Radyne R50 cavity controlled oscillator, fitted with a plasma torch of 20 mm outside diameter of the design described by Scott *et al.* (38). Light from the tail flame of the plasma was focused on the slit of an Applied Research Laboratories 29000B quantometer, set up with

Figure 7.7. Schematic diagram of the analytical instrumentation, showing alternative means of introducing the sample solution (37).

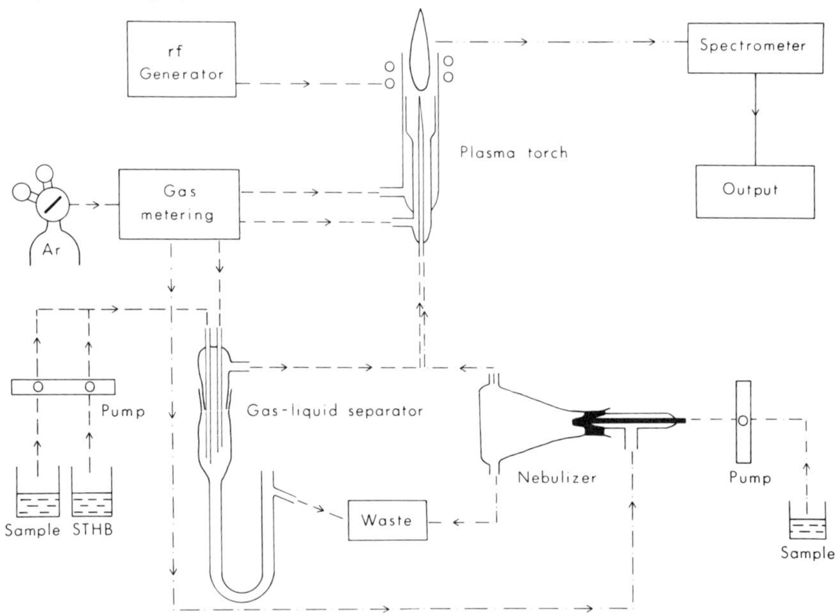

lines for about 40 elements including tin. The conventional sample introduction technique utilizes a glass pneumatic nebulizer (J. E. Meinhard, type T-230-A3). The performance of this system is generally broadly similar to that of specifically designed commercial equipment.

The hydride generator consisted of a peristaltic flow inducer (Watson-Marlow type MHRE 200) to pump and mix the reagent solution and the sample solution and a cell designed to bring about the separation of the hydrides and hydrogen from the spent solution. The hydrides were transferred to the injector tube of the plasma torch by argon carrier gas.

Operating conditions are given in Table 7.31.

Reagents

- Sodium tetrahydroborate powder was obtained from Aldrich Chemical Co. Analytical grade acids and ammonium iodide, Specpure grade metallic compounds, and demineralized water were used throughout.
- Tartaric acid solution: Dissolve 10.0 g of analytical grade reagent in 1 liter of purified water. This solution should be freshly made each day.
- Ammonium iodide, analytical grade, ground to pass a 0.2 mm sieve.
- Sodium tetrahydroborate solution: Dissolve 10 g of the reagent in 1 liter of 0.1 *M* sodium hydroxide.

Procedure

Grind the sample to pass a 200 mesh sieve. Weigh the sample (0.200 ± 0.001 g) and mix with ammonium iodide (0.3 ± 0.01 g) in the bottom of a heating tube.

Table 7.31
Operational Parameters

Parameter	Sampling method: Nebulization	Sampling method: Hydride generation
Forward power, kW	2.4	2.7
Viewing height, mm	16	11
Plasma support gas flow rate,[a] liter/min	0–1	0–1
Coolant gas flow rate, liter/min	10	17
Injector flow rate, liter/min	1	1
Sample uptake rate, ml/min	3	9.2
Integration time, s	20	20
Preintegration, s	10	15
Reagent flow rate, ml/min	—	4.5
Reagent concentration	—	1% (w/v) sodium tetrahydroborate in 0.1 *M* sodium hydroxide solution

Source: Ref. 37.
[a]Must be optimized for each individual torch.

Heat the tube in the hot block at 600°C for 10 min with the condenser in place and a glass sphere on the condenser. When cool, detach the condenser and place it in a test tube. Add 20.0 ± 0.05 ml of tartaric acid solution and heat at 50°C for 20–25 min. When cool, the solution is ready for analysis by hydride generation or by nebulization. Make the calibration solutions with tin(II) chloride of suitable concentration in a solution containing tartaric acid (10.0 g/liter) and ammonium iodide (15.0 g/liter).

XIII. DETERMINATION OF GOLD IN ROCKS AND SOILS BY FLAMELESS ATOMIC ABSORPTION (33)

Comments on the Method

A variety of geochemical (mainly soils) and rock samples can be analyzed by this procedure. Standard GXR samples including jasperoid, soils, Fe–Mn-rich deposits, and copper mill heads and several USGS rocks were analyzed by the proposed procedure. Generally good agreement was obtained with the reported values.

Gold is dissolved using a hydrobromic acid and bromine solution. Subsequently, the gold–bromide complex is extracted into MIBK. An aliquot is then injected into the graphite furnace for atomic absorption determination. Samples containing organic carbon or sulfide must be roasted at 700°C for 1 h. Approximately 0.4 ng/ml Au gave an absorbance of 0.0044.

Reagents

- Hydrobromic acid, concentrated, reagent grade.
- Hydrobromic acid, 0.1 *M*, MIBK-saturated: Dilute 23 ml of concentrated hydrobromic acid to 2 liters with water; add an excess of methyl isobutyl ketone (approximately 5 ml), shake thoroughly, allow the phases to separate, and discard the MIBK layer.
- Methyl isobutyl ketone (MIBK), reagent grade.
- Bromine, reagent grade.
- Hydrobromic acid–bromine solution, 0.5% bromine: Dissolve 10 ml of bromine in 2 liters of concentrated hydrobromic acid.
- Gold stock solution, 1000 μg/ml: Dissolve exactly 1.00 g of gold in 100 ml HBr–Br_2 solution. Heat the solution gently to expel excess bromine. Cool and dilute to 1000 ml with concentrated hydrobromic acid.
- Prepare solutions containing 0.05, 0.02, 0.01, 0.005, and 0.002 μg/ml gold by serial dilution of the 1000 μg/ml gold stock solution with concentrated hydrobromic acid. Transfer 10.0 ml of each of these solutions and a blank solution to a series of 25 by 150 mm screw-capped tubes. Add 10.0 ml of MIBK and 10 ml of water to each tube and shake for 5 min. Transfer the MIBK layer, with an automatic pipet, to another tube containing 40 ml of MIBK-saturated 0.1 *M*

hydrobromic acid. Shake the tubes for 2 min and use the MIBK layer for calibration of the atomic absorption spectrophotometer.

Equipment

- Muffle furnace, electric, Blue M model 8630C-1
- Shaking machine, equipoise, heavy duty, Precision Scientific model 65855
- Centrifuge, size 2, model V, International Equipment, accommodates 25 by 150 mm tubes
- Hot plate, 37 by 47 cm, Fisher Scientific model 11-492-11
- Test tube rack, aluminum, approximately 20 cm wide, 30 cm long, and 10 cm deep, drilled to accommodate fifty 25 by 150 mm test tubes
- Vortex mixer
- Culture tubes, 25 by 150 mm, screw-cap
- Evaporating dishes, porcelain, size 0000
- Automatic pipet, 5 ml
- Atomic absorption spectrometer, Perkin-Elmer model 360
- Graphite furnace atomizer, Perkin-Elmer model 2200
- Automatic sampler, Perkin-Elmer model AS-1

Procedure

Transfer 10 g of pulverized sample to a 0000 porcelain evaporating dish. Roast the sample in a muffle furnace at 700°C for 1 h. When the sample is cool, transfer it to a 25 by 150 mm screw-cap test tube containing 10 ml of $HBr{-}Br_2$ solution. Mix sample and acid thoroughly using a vortex mixer. Heat the tubes by placing the metal rack on a hot plate set at its maximum setting until the acid begins to boil at the bottom of the tube. Bumping will occur, causing the tubes to chatter in the rack. At that point remove the tubes from the hot plate and allow them to cool. Add 10 ml of water and 10 ml of MIBK to each tube. Cap the tubes and shake on a shaking machine for 5 min. Centrifuge the samples to separate the MIBK layer from the acid layer. Using a 5 ml automatic pipet, transfer the MIBK layer to another 25 by 150 mm screw-cap tube containing 40 ml of MIBK-saturated 0.1 *M* hydrobromic acid. Cap the tubes and shake on the shaking machine for 2 min. Allow the phases to separate and estimate the gold in the MIBK layer using an atomic absorption spectrometer equipped with a graphite furnace.

Pipet a 20 μl aliquot of the MIBK layer into the graphite furnace using either a micropipet or an automatic sampler. Observe the peak absorbance and compare with absorbance values for standards prepared and atomized using the same conditions to determine the gold concentration of the sample. The following atomic absorption conditions are used.

- Wavelength, 242.8 nm
- Slit, 0.7 mm, alternate (alternate setting reduces slit height for furnace use)

- Mode, absorbance (readout is in unexpanded absorbance units)
- Signal, peak (peak height is measured and displayed on readout)
- Lamp current, 10 mA
- Purge gas, nitrogen
- Flow time, normal, 3 s (internal gas flow through the graphite tube is reduced to approximately 35 cm^3/min during the first 3 s of atomization)
- Drying time, 20 s
- Ramp drying time, 15 s
- Drying temperature, 120°C
- Charring time, 20 s
- Charring temperature, 500°C
- Atomizing time, 8 s
- Atomizing temperature, 2700°C
- Graphite tube, pyrolitically coated

XIV. DETERMINATION OF PLATINUM IN ROCKS, SOILS, AND SEDIMENTS BY FLAME ATOMIC ABSORPTION (39)

Comment on the Method

Platinum is recovered from a 25 g sample of rock, soil, or sediment using bromine in hydrobromic acid. The solution containing Pt(IV) is reacted with stannous chloride and then extracted twice using methyl isobutyl ketone (MIBK). The platinum is determined in the MIBK phase using flame atomic absorption spectrometry.

Samples of Canada Center for Mineral and Energy Technology reference materials PTA-1, PTC-1, and PTM-1 were analyzed. The results obtained by the proposed procedure agreed acceptably with the certified values.

Reagents

- Hydrobromic acid, approximately 49% (w/w).
- Bromine.
- Bromine solution: Mix 20 ml of bromine with 1 liter of hydrobromic acid.
- Methyl isobutyl ketone (MIBK).
- Hydrochloric acid, approximately 36% (w/w).
- Stannous chloride, dihydrate.
- Stannous chloride solution: Dissolve 500 g of stannous chloride in hydrochloric acid and dilute to 1 liter with this acid.
- Platinum, sponge.

- Standard platinum solutions: (a) 100 μg Pt/ml: Dissolve 50 mg of sponge in 10 ml of bromine solution and dilute to 500 ml with 2 *M* hydrobromic acid. (b) 5 μg Pt/ml: Dilute 5 ml of the 100 μg/ml solution to 100 ml with 2 *M* hydrobromic acid. This solution is stable for 1 day.

Equipment

A Pye-Unicam SP90A Series II atomic absorption spectrometer was used with the following experimental conditions:

- Wavelength: 267 nm
- Slit width: 0.08 mm
- Burner height: 0.7 cm
- Coarse gain: 4
- Air/acetylene: 5 liters/450 ml per minute

Procedure

Weigh 25 g of sample into a beaker (400 ml) and add 100 ml of bromine solution. Mix well and leave to stand overnight, agitating the mixture at intervals during the day. Boil to expel the excess of bromine, cool the solution, and dilute it to 250 ml; leave until the undissolved matter has settled, overnight being convenient. Take 100 ml of the clear solution and extract it with 15 ml and then 10 ml of MIBK, shaking the solution for 2 min. The ketone phase is suitable for the determination of gold and/or tellurium by atomic absorption spectrometry; otherwise, discard it. To the aqueous phase add 25 ml of stannous chloride solution and extract with 15 ml and then 10 ml of MIBK. Wash the combined ketone phases twice by extraction with 25 ml volumes of 2 *M* hydrobromic acid, dilute to 25 ml with the ketone, and aspirate the ketone extract into the flame of the instrument. Interpolate the platinum concentration from a calibration graph prepared by extraction of platinum in the range 0–10 μg from 2 *M* hydrobromic acid solution.

REFERENCES

1. S. E. Church, *Geostand. Newsl.* **5,** 133 (1981).
2. R. K. Winge, V. J. Peterson, and V. A. Fassel, *Appl. Spectrosc.* **33,** 206 (1979).
3. J. N. Walsh and R. A. Howie, *Mineral. Mag.* **43,** 967 (1980).
4. J. Burman and K. Bostrom, *Anal. Chem.* **51,** 516 (1979).
5. J. Burman, C. Ponter, and K. Bostrom, *Anal. Chem.* **50,** 679 (1978).
6. J. N. Walsh, *Spectrochim. Acta, Part B* **35B,** 107 (1980).
7. J. T. Hutton and S. M. Elliott, *Chem. Geol.* **29,** 1 (1980).
8. G. C. Kerrigan, *J. Phys. E* **4,** 544 (1971).
9. K. Norrish and J. T. Hutton, *Geochim. Cosmochim. Acta* **33,** 431 (1969).
10. K. F. J. Heinrich, *in* "The Electron Microprobe" (T. D. McKinley *et al.*, eds.), p. 296. Wiley, New York, 1966.
11. K. Norrish and J. T. Hutton, *X-Ray Spectrom.* **6,** 6 (1977).

12. P. J. Potts, P. C. Webb, and J. S. Watson, *X-Ray Spectrom.* **13,** 2 (1984).
13. S. Abbey, *Geostand. Newsl.* **4,** 163 (1980).
14. S. Abbey, *Geostand. Newsl.* **5,** 103 (1981).
15. K. Govindaraju, *Geostand. Newsl.* **4,** 49 (1980).
16. G. C. Brown, D. J. Hughes, and J. Esson, *Chem. Geol.* **11,** 223 (1973).
17. P. Hannaker and T. C. Hughes, *J. Geochem. Explor.* **10,** 169 (1978).
18. J. C. Van Loon, internal laboratory method, unpublished.
19. R. B. Cruz and J. C. Van Loon, *Anal. Chim. Acta* **72,** 231 (1974).
20. A. Bolton, J. Huang, and A. Vander Voet, *Spectrochim. Acta, Part B* **38B,** 165 (1983).
21. J. C. Van Loon, J. H. Galbraith, and H. M. Aarden, *Analyst (London)* **96,** 47 (1971).
22. J. G. Sen Gupta, *Talanta* **32,** 1 (1985).
23. J. G. Sen Gupta, *Talanta* **31,** 1045 (1984).
24. J. G. Sen Gupta, *Talanta* **31,** 1053 (1984).
25. J. G. Sen Gupta, *Talanta* **28,** 31 (1981).
26. Hou, Q.-L., T. C. Hughes, M. Haukka, and P. Hannaker, *Talanta* **32,** 495 (1985).
27. E. S. Gladney and W. E. Goode, *Geostand. Newsl.* **5,** 31 (1981).
28. S. Abbey, *Geol. Surv. Pap. (Geol. Surv. Can.)* **83–15** (1983).
29. T. W. Steele, A. Wilson, R. Goudois, P. J. Ellis, and A. J. Redford, *Geostand. Newsl.* **2,** 71 (1978).
30. S. Abbey, *Geol. Surv. Pap. (Geol. Surv. Can.)* **80–14** (1980).
31. J. G. Crock and F. E. Lichte, *Anal. Chem.* **54,** 1329 (1982).
32. A. E. Hubert and T. T. Chao, *Talanta* **32,** 568 (1985).
33. A. L. Meier, *J. Geochem. Explor.* **13,** 77 (1980).
34. L. Zhou, T. T. Chao, and A. L. Meier, *Anal. Chim. Acta* **161,** 369 (1984).
35. T. T. Chao, R. F. Sanzolone, and A. E. Hubert, *Anal. Chim. Acta* **96,** 251 (1978).
36. G. H. Allcott and H. W. Lakin, *J. Geochem. Explor.* **8,** 659 (1975).
37. B. Pahlavanpour, M. Thompson, and S. J. Walton, *J. Geochem. Explor.* **12,** 45 (1979).
38. R. H. Scott, V. A. Fassel, R. N. Kniseley, and D. E. Nixon, *Anal. Chem.* **46,** 75 (1974).
39. S. E. Stanton and S. Ranankutti, *J. Geochem. Explor.* **7,** 73 (1977).

8

Determination of the Platinum Group Metals and Gold

I. INTRODUCTION

Although advances in analytical instrumentation have been great in recent times, the basic approaches to the accurate determination of the precious metals remain largely unchanged. In particular, the fire assay remains an essential pretreatment of a large variety of samples prior to the determinative step. Likewise, separational procedures (mainly solvent extraction, ion exchange, distillations, and precipitation) are still often necessary to minimize interference effects.

The choice of a determinative method is usually dictated by the levels of precious metal to be handled, nature of the sample matrix, and availability of equipment. (It is to be hoped that the latter factor is not preeminent.) For trace levels (to the parts per million range) flame atomic absorption and inductively coupled plasma (ICP) emission spectrometry can be used. ICP emission is favored because of ease of multielement operation. When levels are sub-parts per million, furnace atomic absorption and nuclear techniques (mainly neutron activation) are favored. Minor and percent concentrations are best handled by x-ray fluorescence, gravimetric, and titrimetric approaches. Solution spectrophotometry, although diminished in importance, can be extremely valuable for solution composition studies and for determination of osmium and ruthenium subsequent to distillations. Other techniques such as electrochemical methods and catalytic approaches have important but isolated applications.

Sampling and sample decomposition are important aspects of the accurate determination of precious metals. Fire assay, oxidizing fusions, mixed acid attack, and chlorinations are all important sample decomposition methods.

The use of standard reference samples and internal and external laboratory control schemes is essential to the accurate determination of precious metals. Fortunately, a variety of standard reference samples are now appearing.

The noble metals present the ultimate challenge to the analytical chemistry researcher. Difficulties are encountered in every analytical step from sampling through to the final determination. Despite volumes of research, stimulated by the high monetary value of these metals, many analytical problems remain unsolved.

Most noble metal analytical work has centered around the fire assay. The first comprehensive treatment of the subject of gold and silver assaying was by Lazarus Ercker in his book "Treatise on Ores and Assaying," published in 1574. This work is a detailed account of procedures used in the assay laboratory of the Holy Roman Empire. It is surprising to note the attention paid to the running of blanks and duplicates, the latter being repeated if satisfactory agreement was not obtained.

With the increasing economic interest in the platinum metals at the beginning of this century, there was a corresponding increase in research activity. F. E. Beamish records the frustration he experienced in entering the field in the 1930s because of secretive research practices of the few rival industrial groups engaged in noble metal production (1). This problem persists in a number of companies even today.

II. FIRE ASSAY

A. Classical Lead Assay

Without any doubt the classical lead assay has proved to be the most important procedure for the concentration and isolation of the noble metals. In 1972 Beamish stated that during 40 years of research in this field he had not experienced a single example of failure of the classical assay to find a paying ore. The continued popularity of the assay is borne out by a 1975 South African ore certification study (2). This work showed that 92% of all platinum analyses received were accomplished by fire assay, the largest number of these being done by the classical lead method.

The reasons for the success and popularity of the classical lead assay are not clearly understood. In fact, the mechanism by which lead quantitatively extracts and collects the noble metals is not clear, particularly in the case of ruthenium, iridium, osmium, and rhodium, which are not very soluble in the lead collector. Noble metals are often present in samples at submicrogram levels. In addition, particularly in the case of gold, the distribution of the metals can be very inhomogeneous. These two factors favor the fire assay approach, where a large sample size may be used and the noble metals are concentrated into a small bead or prill. An additional benefit is achieved because the metals are extracted from a complex matrix into a relatively simple metal alloy. The latter effect is particularly important in view of the complex interference problems encountered in noble metal analytical chemistry. Another attractive feature of the fire assay is its wide applicability to ores, concentrates, rocks, and many industrial products.

Although the fire assay has been proved to work well, achieving success with the method is very much dependent on the experience and skill of the assayer. It is not uncommon, therefore, for problems to be encountered.

Losses of noble metals during assays have been, and still are, the subject of much research. Whether losses occur depends on the particular metal, the mineralogy, and the skill and experience of the assayer.

By the latter half of the 19th century a large number of publications existed on assay work. However, T. Wilm stated in 1885 that in his opinion no satisfactory procedure had yet been recorded (3). As recently as 1940 F. E. Lathe (4) recorded a similar statement. These comments refer to the quantitative capabilities of the procedures and not their ability to find an ore. Frequently, problems can be related to the nature of the slag. An experienced fire assayer, faced with a new sample type, will try a number of flux compositions. A reassay of pots, surface fragments, and slags must be done each time to check for losses. This trial-and-error process is necessary because the acidity or basicity of the slag can be calculated only approximately.

A number of studies have been done in an attempt to better outline steps in the assay which are prone to losses. A good example of such a study was reported by Wall and Chow (5). In this research the effects of fusion temperature, lead button weight, flux composition, cupellation temperature, and use of a cover during fusion on losses of gold were reported. These studies were based on gold recoveries and on use of radioactive gold tracer (^{195}Au) added to the sample. (The latter approach is a particularly powerful method of studying fire assay losses.) The results of the work showed that a total loss of gold of less than 1% occurred under ideal conditions.

In general, gold, platinum, palladium, and rhodium are readily collected by lead and appear quantitatively in a silver cupellation bead. Achieving efficient collections of these elements, however, may require extensive experimentation with flux composition and assay conditions. A reassay of the slag may be necessary for best results.

Ruthenium, iridium, and osmium collection is more seriously affected by flux composition and assay conditions. Ruthenium losses are not closely related to the degree of slag acidity. Collection of osmium and ruthenium is favored by high temperature. Appreciable losses of osmium are experienced with basic fluxes, and a minimum button size of 25 g must be maintained. Basic or subsilicate slags cause severe iridium losses. A reassay of slags is usually necessary for quantitative recovery of iridium because of its low affinity for lead.

Using cupellation involving a silver bead, losses to the cupel may be high for iridium and ruthenium. Gold beads are often employed for iridium collection. No cupellation is possible when osmium is to be analyzed because of its extreme volatility as the tetroxide.

For both osmium and ruthenium the cupellation step is invalidated. However, well-proven procedures exist for the separation and recovery of these elements by perchloric acid decomposition of the lead button directly, followed by distillation of the volatile tetroxides.

One of the important areas for continued fire assay research is the development of improved procedures for the separation and recovery of the platinum metals and gold obtained from the dissolution of the lead button.

A recent example of such work is the paper by Diamantatos (6). A conventional lead assay was performed and the lead button thus produced was then treated according to the outline shown in Fig. 8.1. As can be seen, such a scheme is relatively complex. If accurate results are to be obtained, a very highly skilled analyst must be used.

Figure 8.1. Flow sheet of the integrated analytical scheme for Pt, Pd, Rh, Ir, Os, and Au. (Reprinted from Ref. 6.)

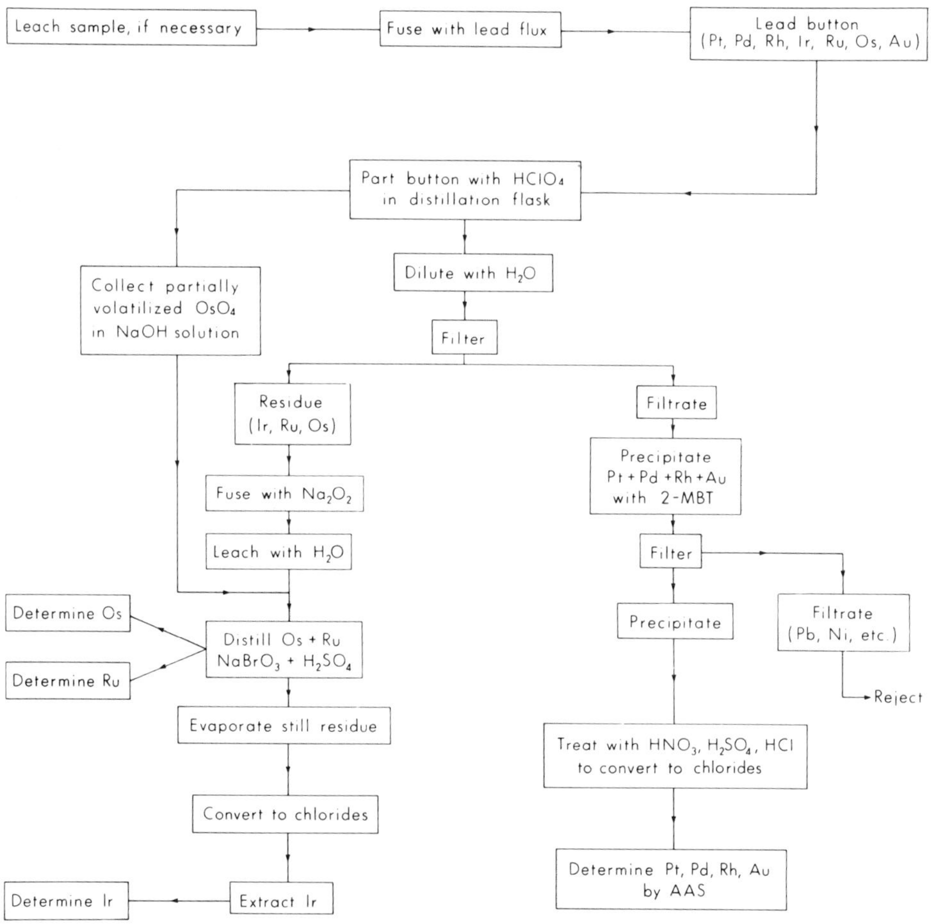

B. Other Assay Procedures

A number of alternative fire assay procedures employing other collectors have been proposed. The main alternative collectors are iron, copper, and nickel used together (or copper by itself) (7), tin (8), and nickel sulfide (9). These are particularly useful for the determination of osmium, iridium, and ruthenium, but in general there is little advantage to these approaches for the other precious metals compared to the lead method. The nickel sulfide procedure has been found to be very important when used prior to neutron activation analysis. In the nickel sulfide procedure (9) the residue left after a hydrochloric acid decomposition of the nickel sulfide button contains all the platinum group metals and gold. This residue on a filter paper is analyzed by neutron activation directly.

C. Disadvantages of the Assay

Fire assay furnace equipment is expensive and often bulky. A highly skilled assayer is required. The technique is dirty and prevents the use of the space for other purposes. Large quantities of chemicals are used, often resulting in substantial contamination of the charge with silver and sometimes gold. This problem can usually be obviated through the use of blanks.

A good deal more work must be done to quantify losses of noble metals during the fire assay procedure. This work is now being assisted by the use of radioactive tracers. At present, success with the assay depends to a large extent on the experience of the assayer. It would be of great benefit to have the assay on a less subjective footing.

The present authors believe that the popularity of the classical lead assay will be maintained in the near future. Other assay procedures will gain in popularity, particularly for the analysis of osmium, iridium, and ruthenium.

III. OTHER METHODS OF DECOMPOSITION

The massive noble metals platinum, palladium, gold, rhodium, and iridium, when properly annealed, are little attacked by single mineral acids. In fact, some samples of platinum wire have proved to be resistant to quantitative attack by even aqua regia. However, it has also been shown that small amounts of palladium, present as a finely divided metallic residue, can be completely dissolved using hydrochloric acid alone in the presence of oxygen.

Beads resulting from a fire assay are often dissolved using aqua regia when platinum, palladium, and gold are to be determined. For beads obtained from gold ores it is often desirable to preferentially leach the silver from the matrix, leaving a gold residue, which can be annealed and subsequently weighed. For this purpose nitric acid is usually employed. This decomposition will also attack finely divided palladium.

Fusions are sometimes used for decomposing noble metals. Most generally effective is a mixture of sodium hydroxide, sodium peroxide, and potassium nitrate.

In dealing with complex samples of ores, rocks, minerals, wastes, plating liquids, industrial products, and organic samples it is often preferable to utilize the fire assay as a first step in the decomposition procedure. This will usually allow the analyst to proceed with a greater degree of certainty with the remainder of the sample decomposition steps.

Dry chlorination at elevated temperatures is a very effective way to corrode precious metal samples. It has been found especially useful for decomposition of inert substances such as iridium metal. The present authors have discovered that chlorination is to be recommended for a variety of complex commercial samples for which a fire assay would not be practical (10). An example of such samples is capacitor waste materials, which consist of precious metal-coated leads on broken pieces of alumina capacitors. There are many instances where such an attack would be useful. However, when chlorination is suggested, many workers are unwilling to consider such an approach because of the need for special equipment for handling such a corrosive and poisonous material. Chlorinations have been done frequently in the present authors' laboratory for a period of 20 years without any problems.

IV. EVAPORATIONS

Evaporations are often necessary during sample preparation to remove large volumes of liquid obtained during the decomposition steps. The following points are important. Osmium and to some extent ruthenium may be lost by volatilization during this step. Osmium is particularly susceptible to this problem. Evaporations of osmium solutions should be done in the absence of oxidizing agents such as oxides of nitrogen, even in trace amounts. Prolonged heating of the evaporated residue should be avoided. Insoluble residues, including trichlorides of iridium and rhodium and metallic gold, may be produced. In evaporating chloride solutions it is useful to add a small amount of sodium chloride to minimize the latter problem.

One of the more difficult industrial problems is the analysis of cyanide mill and plating solutions for gold. Although a variety of analytical schemes have been proposed for sample preparation, a critical evaluation of the effectiveness of some of these methods, particularly when they call for a precipitation recovery, is essential. Evaporation of cyanide solutions in a lead vessel, followed by cupellation or fire assay, is a commonly employed procedure.

More recently atomic absorption spectroscopy has been used to analyze cyanide solutions. This determination is usually done after a solvent extraction separation of gold from the cyanide solution has been accomplished (11).

V. SEPARATION TECHNIQUES

Historically, the hydrolytic precipitation procedures developed during the early

1900s for the isolation of platinum from palladium, rhodium, and iridium were important. These formed the basis of the comprehensive schemes of separation and determination of the noble metals developed in the 1920s and 1930s at the National Bureau of Standards in Washington, D.C. by R. Gilchrist (12) and at the University of Toronto by Beamish and Russell (13).

It is of special interest to note the preeminent positions occupied by distillation procedures for the separation of osmium and ruthenium, as the tetroxides, from complex solution and from each other. The approach, developed shortly after the discovery of these elements, is still the method of choice. Failure to recognize the ease with which particularly osmium tetroxide is produced has, in the present authors' opinion, led to serious losses of osmium during industrial processing and analytical procedures (e.g., roasting in air). The greatest impediment to the application of other separational techniques to the noble metals is lack of knowledge about the solution chemistry and reaction mechanisms of these metals in the complex analytical solutions. Volumes of data exist describing noble metal species in simple solution. However, while it may be possible to assume that a certain precious metal species exists, for example, in a rather pure chloride solution, it is not logical to suppose that the same species would be present in chloride solutions containing a variety of other anions and cations.

Little is known about the effect of reaction kinetics on the performance of separational procedures. For example, it is well known that while quantitative cation-exchange separation of ruthenium from base metal chloride solution can be obtained using freshly prepared solutions, solutions left to stand yield low recoveries of ruthenium (14).

In a biogeochemical investigation, Samiullah (15) studied losses of gold, platinum, and silver on filter paper and borosilicate glass. He found that gold solutions containing 0.1% nitric acid and in the 1–10 ng/ml metal concentration range must not be stored for more than 48 h in borosilicate glass. Losses of up to 21% were recorded. Filtration through paper filters has resulted in losses of up to 90% for gold and silver and 50% for platinum. The magnitude of the loss was generally greatest for fine-porosity filter paper. The presence of calcium (2000 mg/ml) together with gold in the solution reduced the gold losses by about 30%.

A. Chromatography

Despite these problems, much progress has been made in applying the very powerful approach of chromatography to separational problems. Of widest applicability is the technique of ion-exchange chromatography. Cation exchange is used for separating the noble metals from other metals and anion exchange for separating the individual noble metals.

Among the other chromatographic methods, paper chromatography is of greatest importance. A number of paper chromatographic procedures exist for separations of individual noble metals from one another.

B. Solvent Extractions

Solvent extraction procedures have played a major role in noble metal separations. In the past these methods have been widely used with spectrophotometry.

At present solvent extraction methods are playing an important role in the widespread acceptance of atomic absorption spectroscopy for noble metal determinations. In this application not only does solvent extraction achieve a separation and concentration of noble metal constituents from large volumes of complex solution but also the resultant organic extract often yields an enhanced atomic absorption signal compared to aqueous solution. Hence solvent extraction has retained, if not gained in, popularity as a separational technique.

With the increasing popularity of ICP emission spectrometry, solvent extraction can have a new and very important role to play. Because detection limits by this plasma technique are relatively poor (e.g., compared to furnace atomic absorption) a preconcentration step is often necessary. It is very important therefore to develop solvent extraction schemes which will extract the precious metals as a group, thus taking advantage of the inherent multielement capability of plasma instrumentation.

Many of the existing wet separational methods for isolating noble metals from ore wastes and industrial products are complex, tedious, and time-consuming. A great deal of effort must be expended to improve this situation. The aqua regia leaching of gold from a large sample of finely ground ore, followed by direct extraction of the chloroaurate or bromoaurate into methyl isobutyl ketone for atomic absorption determination, is an example of a simple, very effective procedure for a number of siliceous ore types.

VI. ATOMIC ABSORPTION SPECTROSCOPY

One of the major landmarks in noble metal analytical chemistry was the development of analytical atomic absorption spectroscopy by A. Walsh in 1955 (16). Although not enthusiastically received in North America until 1964, the technique was immediately investigated for noble metal analyses particularly by Australian and South African researchers.

At present atomic absorption spectroscopy is the most widely used determinative tool for these metals. In the 1975 South African ore certification study (17), 77% of all results quoted were obtained by atomic absorption, 12% by emission spectroscopy, and 9% by spectrophotometry.

As might have been expected, the interferences experienced in the atomic absorption determination of noble metals are more complex than those for most other elements. This fact is not always appreciated by the analyst and has led to some serious errors in routine work. Fortunately, extensive interference studies have now been completed and a variety of remedies proposed. Much of the work has been done with lower temperature flames, mainly air/acetylene. A complex pattern of both specific and nonspecific interferences has been found. A large variety of releasing agents and buffers have been proposed for overcoming interferences. These include vanadium for Pt, Pd, Rh, Ru, and Au (18); vanadium or uranium for Os and Ir (19); a combination of copper and sodium for Ir (20); lanthanum for Pt, Pd, Au, and Rh (21); and a combination of cadmium and copper for Ru (22).

Separations are widely used for interference problems. This approach, although more time-consuming, has the added advantage of effecting a preconcentration of the noble metals. The latter consideration can be of great importance. Noble metal levels in many samples are well below the detection limit of flame atomic absorption, the atomic absorption sensitivity of iridium, ruthenium, and osmium being relatively poor in aqueous solution.

Using solvent extraction in which the solvent chosen is readily combustible often results in signal enhancement, in addition to effecting the desired separation. In this regard, methyl isobutyl ketone is most commonly employed.

Frequently, a fire assay step precedes determination by flame atomic absorption. Using fire assay not only provides separation of the precious metals from many of the harmful matrix constituents but also allows use of a large sample size.

Platinum, palladium, and gold can be determined in an acid solution of the silver bead using an air/acetylene flame. In this case a releasing agent such as lanthanum must be employed (23). Chemical interference in this determination can also be largely overcome with the use of a hotter nitrous oxide/acetylene flame (24). However, noise levels associated with this flame often negate its use when highest precision and accuracy and best detection limits are required. Since iridium and rhodium are not soluble to any appreciable degree in the silver bead, a gold bead is employed for collection of these elements during cupellation (25, 26).

Atomic absorption is particularly well suited to the analysis of Au, Pd, Rh, and possibly Pt. Relatively poor detection limits are obtained for Ir, Ru, and Os. Because of this, it is surprising to note the increase in numbers of atomic absorption determinations of the latter elements in ores and rocks. The authors believe that for the trace analysis of Ir, Ru, and Os, solution spectrophotometry, neutron activation, or inductively coupled argon plasma emission should be employed.

In recognizing that the detection limits for the precious metals are poorer than are required for many applications, much recent research has centered on the behavior of these metals, using an electrothermal atomizer with atomic absorption (27). It must always be emphasized that work with electrothermal atomizers is much more difficult and much slower than work with flame atomizers. The worker is thus cautioned to use flame atomic absorption in preference to electrothermal atomic absorption whenever possible.

Sample types currently analyzed for noble metals by atomic absorption span the field. With the advent of this technique many companies were, for the first time, able to undertake reliable noble metal analyses in their own laboratories. Applications include analysis of rocks, minerals, ores, waters, waste solutions and solids (including nuclear wastes), cyanide baths and other plating liquids, alloys, and biological, clinical, and environmental samples. In addition, atomic absorption has found wide application in determining the impurities in high-purity noble metal alloys.

Because atomic absorption is relatively inexpensive and relatively easy to operate, it has come within the reach of most industries working with noble metals.

VII. EMISSION SPECTROSCOPY

This technique has the distinct advantage over atomic absorption spectroscopy of having simultaneous multielement capability and a much greater linear dynamic working range. In addition, in contrast to conventional atomic absorption, emission spectroscopy can easily be applied to conducting solid samples. In spite of these favorable points, the simplicity, low cost, and relative freedom from interference exhibited by atomic absorption spectroscopy until recently have resulted in its widespread adoption at the expense of emission spectroscopy.

In the middle 1970s inductively coupled and direct-current (dc) argon plasma source emission spectrometers became commercially available. Both simultaneous multielement and sequential instruments can be obtained. This equipment yields detection limits comparable to those of flame atomic absorption and in addition is capable of yielding results with better precision while showing greater freedom from chemical interference. On the negative side, because of the extremely high temperatures of plasmas, very rich line spectra and background shifts are obtained, which mean that a greater potential for spectral and background interference exists. Also negative in comparison with atomic absorption are the facts that the initial cost of good plasma emission spectrometers is very high and their operation requires the attention of a well-trained spectroscopist. Compared to electrothermal atomic absorption, detection limits by plasma emission spectrometry are poorer by up to two orders of magnitude.

Choice of a dc or an inductively coupled plasma must be made. Recent research makes it clear that there are fewer chemical interferences with the inductively coupled plasma. In addition, with the inductively coupled plasma it seems that the atoms "see" a relatively higher temperature than background species, perhaps because of the absence of thermodynamic equilibrium. This anomaly improves the signal-to-noise ratio considerably compared to a dc plasma, which does not have the same analyte-to-background signal enhancement.

In summary, plasma emission spectrometry can be recommended when the precious metals (28) are present in solution at microgram per milliliter levels, when simultaneous multielement analysis is an advantage, and when best precision is essential. Also, iridium, osmium, and ruthenium are better done by this approach than by flame atomic absorption.

It is important to emphasize that plasma emission spectrometry is, as it is applied to the analysis of precious metal samples, in its infancy. Thus extensive research is still necessary to evaluate its potential for application to a particular precious metal analysis problem.

Conventional arc/spark optical emission spectroscopy remains preeminent for a variety of applications. For the determination of individual noble metals in precipitates and residues the technique has no peers. In addition, it is widely used for the detection and determination of impurities in high-purity noble metals. Because emission spectroscopy is applicable to solids, it is often employed for the direct analysis of assay beads.

Preparing standards for arc/spark emission analysis is a difficult problem. In general these should be in a form and in a matrix similar to those encountered in the samples. When a fire assay is employed this problem is somewhat simplified.

A renewed interest is evident in the Grimm-type glow discharge tube for use in emission analysis of alloys (29). This source has been used with conventional prism and grating monochromators and with resonance spectrometers. In the case of the latter, the flame resonance spectrometer is particularly attractive because of the ease with which a multielement capability may be achieved. Glow discharge emission techniques, like cathodic sputtering absorption cells, should find useful application in the analysis of noble metals in alloys in the near future.

VIII. NEUTRON ACTIVATION SPECTROSCOPY AND OTHER NUCLEAR TECHNIQUES

Neutron activation spectroscopy occupies a unique position in the analytical chemistry of the noble metals. Sensitivities, under ideal conditions, are several orders of magnitude lower than those obtainable by other approaches. Unfortunately, these sensitivities are not readily achieved on real samples because of the complexity of interferences.

As a result of interference problems it is often necessary to employ separational procedures. To compensate for losses, the chemical yield approach is often used. There was a tendency, particularly in early work, for researchers to employ analytical separations and other manipulations which are of questionable usefulness on the assumption that chemical yield would correct for errors encountered. For example, it is not uncommon to find the isolation of a final precipitate accomplished using a zinc or magnesium reductant. This practice, even controlled by a chemical yield determination, is unacceptable.

It is important in applying chemical yield methodology to add the carrier in a form similar to that of the element in the sample. Complete isotopic exchange must occur between sample element and carrier. Failure to recognize this pitfall can lead to serious errors.

Neutron activation analysis usually requires use of a relatively small sample (1 g or less). This can be a serious problem in dealing with noble metals, particularly gold, which is notorious for its inhomogeneous distribution in many sample types. Much care is necessary in sampling the material for analysis. For the purpose a relatively large sample, e.g., 30 g of finely ground sample, can be spread out in a thin layer on a cellophane sheet. The sample for analysis is obtained by scooping a small amount from 1.3 cm^2 segments marked over the surface. A number of replicate samples must be taken to obtain the mean composition.

Osmium and ruthenium are the two elements most often neglected in the analysis of the noble metals in samples. In the 1975 certification study of a South African ore (17), out of 38 participating laboratories only 6 and 10 reported results for osmium and ruthenium, respectively. Fortunately, research activity is on a relatively high level for the analysis of these elements by neutron activation analysis. To date, most researchers developing procedures have wisely employed the well-accepted distillation separation procedures prior to the determinative step.

Neutron activation analysis has been applied to a fairly broad range of noble metal samples. Particular note should be made of the prevalent use of this technique to determine trace amounts of noble metal impurities in high-purity noble metal samples (30). The determination of the noble metals at background levels in soils, rocks, and other geological samples is also noteworthy (31). It is hoped that the use of neutron activation analysis will become more widespread, particularly in the applications mentioned above. This will undoubtedly occur as the number of neutron sources increases.

Some of the noble metals have analytically interesting radionuclides of very short half-life (e.g., ^{105}Rh, half-life 4.41 min). Recently, reactors of somewhat lower flux intensity, but requiring much less capital investment and personnel commitment, have been developed which are ideal for work with such isotopes. This may bring neutron activation analysis within the reach of a broader segment of noble metal analysts.

Neutron activation analysis is, at present, uniquely suited to the determination of very small concentrations of the noble metals. Much work is essential in perfecting analytical schemes which can accomplish this purpose. The present authors believe that this area of research must be pursued with vigor. However, in our opinion, electrothermal atomizers used in conjunction with atomic absorption and perhaps atomic fluorescence spectroscopy may be competitive in this field.

IX. RADIOACTIVE TRACERS

Tracer techniques are particularly useful for examining the efficiency of analytical treatments of noble metals and for determining the losses and distribution of the noble metals in the various stages of industrial methods for recovery and purification. In numerous cases this technique has been applied to the study of losses during fire assay. Results, in general, substantiate the findings obtained by more laborious previous work.

X. X-RAY FLUORESCENCE SPECTROSCOPY

This approach is often favored, when applicable, because of the speed of analysis. Unfortunately, sensitivities achieved by the technique are relatively poor, being limited to samples with milligram or higher concentrations.

A number of procedures have been developed for the x-ray fluorescence analysis of noble metals, after separation and preconcentration of these elements. Predominant among these is the fire assay concentration of noble metals into a lead button, followed by cupellation to produce a gold or silver bead. The latter alloy is flattened and annealed and then placed in a sample holder for analysis (32). Resin-impregnated filter paper has also been used for the preconcentration of noble metals for x-ray analysis. In this application the filter paper circle is placed directly into the sample holder.

Other approaches to sample presentation involve analysis of solutions of dissolved assay beads. This is especially important when the bead constituents are not homogeneously distributed even after the annealing process. Energy-dispersive x-ray fluorescence equipment is leading to more extensive application of the x-ray spectroscopy technique in precious metal analysis. The present authors look forward to the expanded use of this equipment in the field.

XI. ELECTROCHEMICAL METHODS

Prior to the development of atomic absorption spectrometry methods there was considerable activity, particularly in Czechoslovakia and the USSR, in applying polarographic methods to the analysis of noble metals. However, the greater inherent simplicity of the former technique caused a sharp decline in the use of electrochemical methods for quantitative work.

Recently, the revolution in electronics has led to a new generation of electrochemical equipment. As a result, techniques such as differential pulse polarography and anodic stripping voltametry can be performed with relatively inexpensive equipment, compared to atomic absorption. These techniques, together with twin cell and alternating-current (ac) polarography, are being investigated with renewed interest. However, it is the present authors' opinion that despite certain advantages, such as greater sensitivity compared to atomic absorption, these techniques will make few inroads on the widespread use of the former method for quantitative work. Even with the vastly improved electrochemical equipment, interference problems are of a severity and complexity which will discourage the routine use of these methods.

Electrochemical methods have been, and still are, important in the study of noble metal solution chemistry. Again, however, the complexity of interferences negates the possibility of using this approach to study many complex, real, analytical solutions.

XII. GRAVIMETRY AND TITRIMETRY

Historically, these methods have been of great importance. While retaining a preeminence for standardizing solutions they have, in modern times, often been supplanted by instrumental techniques. Of the two approaches, gravimetry is the most employed.

A large number of precipitants have been proposed for the individual metals. Some produce a direct weighing form, thus obviating the need for reducing the precipitate to the metal.

In spite of the substantial number of gravimetric reagents which have been proposed over the past two decades, few have gained widespread acceptance over, for example, dimethylglyoxime for palladium, hydroquinone for gold, and thionalide for osmium and ruthenium (33). This is despite claims by the authors to the contrary, because the newer methods have little proven advantage over the well-established familiar procedures.

Future research is required to outline reagents which will give direct weighing forms for all the noble metals. These reagents should be selective and be applicable at a moderate acidity. Selective reagents for large amounts of the noble metals could also find use in refining and manufacturing processes.

Titrimetric methods are little used. They have found limited application as rapid pass–fail tests for noble metal products of interest to jewellers and other manufacturers.

XIII. SOLUTION SPECTROPHOTOMETRY

Prior to 1960, spectrophotometric or emission spectroscopy methods were the techniques of choice for the analysis of trace levels of the noble metals. With the widespread acceptance of atomic absorption spectroscopy the general use of techniques of this type has declined sharply. Spectrophotometric methods are still very important for the determination of microgram amounts of iridium, osmium, and ruthenium.

In the case of osmium and ruthenium, the procedures using thiourea and related compounds are the methods of choice (1). These reagents are readily applicable to the acid solutions used as receiver solutions in the distillation procedures for these metals.

In spite of the swing to atomic absorption, the number of new spectrophotometric methods for the noble metals proliferates. This is particularly redundant in the case of palladium. The volume of new spectrophotometric methods for palladium exceeds that for all the other noble metals combined, and yet this element is one of the easiest to determine by atomic absorption spectroscopy. Conversely, there has been relatively little research into spectrophotometric methods for iridium, an element which shows rather poor sensitivity by flame atomic absorption spectroscopy.

Complex interference problems abound with spectrophotometric methods. One of the reasons for the popularity of these techniques for osmium and ruthenium determinations is that the distillation separation used produces a receiver solution virtually free from interfering substances. For the other elements, laborious wet chemical separation schemes are essential to produce adequate purity in the solution prior to the color-forming step. This naturally mitigates against use of such methodology.

Because of the complexity of interference problems, it is essential to test a proposed new spectrophotometric procedure extensively. Commonly this is done by adding a pure solution of the interferent to the analyte in a relatively pure solution. These conditions are usually quite different from those encountered when the interferent is carried through the entire procedure. Also it is common to evaluate the effect of one interferent at a time. Again, this approach does not mirror what happens in a real analysis.

Useful field tests for the precious metals have been developed using color-forming reagents. This is still an important area for the spectrophotometric approach.

XIV. STANDARD REFERENCE MATERIALS

Of ultimate importance in any field of analysis is the accuracy of the results. In the noble metal field, until recently, the lack of any generally available standard reference materials has been a continuous source of problems.

The Department of Energy, Mines and Resources, Ottawa, Canada, prepared two standards in 1970, copper–nickel matte (PTM) and an alluvial black sand (PTA). In 1973 a sulfide concentrate (PTC) was also prepared.

In 1975 the National Institute of Metallurgy, Johannesburg, South Africa, distributed a standard produced from samples taken from the Merensky Reef.

Other standard reference materials may exist, particularly in the USSR, but the present authors have few details about such samples.

The scientists and groups responsible for the production and certification of noble metal standard reference materials are to be highly commended. It is to be hoped that more noble metal standards of other sample types will become available in the near future.

XV. ANALYTICAL PROCEDURES

A. Lead Fire Assay (1)

Despite frequent claims to the contrary, a fire assay is still essential for the reliable analysis of noble metals in most geological samples. The classical lead assay remains the most generally acceptable approach. Good additional information on the classical assay can be obtained from books by Beamish (1) and Bugbee (34).

No assay procedure has been developed that is applicable to all sample types. The following method may be used for the collection of platinum, palladium, gold, rhodium, iridium, osmium, and ruthenium from a wide range of materials. If osmium and ruthenium are to be analyzed, the cupellation step must be avoided.

Table 8.1
Suggested Flux Compositions

	Gold ores, flux A (g)	Nickel ores Flux B (g)	Flux C (g)	Flux D (g)
Ore[a]	15	30	15	30
PbO	85.0	275	250	266
SiO_2	10	15	10	60
Na_2CO_3	21.1	40	12	46
CaO	4.5	12	—	12
$Na_2B_4O_7$	—	—	—	—
KNO_3	—	16	9	18
Flour	1–3	—	—	—

[a]For blanks, 15 g of silica replaced 15 g of gold ore and a mixture of 12 g of silica, 3 g of copper sulfide, 3 g of nickel sulfide, and 12 g of iron sulfide replaced 30 g of nickel ore.

For rhodium and iridium a cupellation into a gold rather than a silver bead must be done.

Reagents and Equipment

A furnace capable of reaching 1200°C is required. Cupels, roasting dishes, and assay pots are available from many brick manufacturers.

Flux components must be free from noble metals but otherwise do not need to be of a high grade. Suggested flux compositions are given in Table 8.1.

Procedure: Assay

If the ore contains sulfides, the following roasting procedure must be done. Transfer the weighed ore to a 15 cm porcelain evaporating dish and place it for a few minutes at the front of the furnace with the door open. The initial temperature should be about 600°C. Over a period of about 5 min move the dish to the furnace center, stirring intermittently. Partially close the door and allow to roast for 2 h at 950°C, stirring frequently for the first 30 min to avoid agglomeration of the concentrate. Remove the dish, cool, transfer the contents to a mortar, and grind to a fine powder. Take care to avoid any loss of powder.

From the available analytical data calculate a suitable flux composition. In general, a bisilicate slag is a suitable medium for the collection of platinum metals in oxidizing ores. For samples with high proportions of associated base metals—copper, nickel, etc.—it may be desirable to increase the proportion of litharge, e.g., flux C. Decide on a proper ratio of flux to ore. If a silver bead is to be prepared, add to the flux an amount of silver powder or silver in solution to give a ratio of approximately 20 : 1 of silver to total platinum metals expected. Some suggested fluxes are given in Table 8.1.

Arrange a cellophane or plastic sheet in a suitable position, and pass the roasted ore through a 45 mesh sieve to the center of the sheet. Add a portion of the flux to the sieve to remove any traces of ore. Transfer most of the remaining flux to the mixing sheet and mix thoroughly. Place the mixture in an assay pot of an appropriate size. Add the remaining flux to the sheet, mix, and transfer to the pot, taking care to brush the sheet free of the mixture.

Place the pots in the furnace at about 950°C, and raise the temperature at the maximum rate to 1200°C. This fusion period should be approximately 1 h. Remove the pots, pour the mixture into conical iron molds, and allow to cool. Remove the button, taking care to retain all of the slag. Free the button of slag by gently tapping with a small iron rod. Set the button aside. Transfer the slag to a grinding mill or to a mortar, grind to pass a 45 mesh sieve, and place the sample on the original mixing sheet. Clean the mill, mortar, and screen with sufficient litharge to produce a second button, then transfer to the slag on the sheet. Mix well as before and transfer the mixture to the original pot. Fuse as before and clean the lead button as described above. If necessary, a fusion of the second slag can be made to produce a third button.

When the noble metals are to be concentrated to form a silver bead, clean the button thoroughly by gentle tapping with an iron rod and transfer it to a bone ash cupel that has been preheated at 900°C for at least 10 min. Continue heating the cupelling button at about 1000°C, with a plentiful supply of air. Remove the cupel over a period of a few minutes after the completion of the cupellation process. The cupellation bead may be parted and analyzed as described later for platinum, palladium, and gold. Lead buttons must be used for the analysis of iridium, osmium, and ruthenium.

B. Distillation for the Recovery of Osmium and Ruthenium (35)

Cupellation is negated when osmium and ruthenium are to be analyzed. The following is the accepted procedure for the recovery of these elements from a lead button. Subsequently, atomic absorption or solution spectrophotometric analysis can be done on the receiver solutions.

Procedure

Transfer a lead alloy (20–50 g), obtained before cupellation by the above procedure, to the distillation flask (Fig. 8.2). Add 100 ml of water to the trap, 25 ml of a 3% hydrogen peroxide solution to the first receiver, and 5 ml of the same hydrogen peroxide to each of the other two receivers. Chill the receivers in an ice bath. Pass water through the condenser, and apply suction to produce 2 or 3 bubbles/s. Add 75 ml of 72% perchloric acid to the distillation flask and heat very gently until the lead is completely dissolved and effervescence of hydrogen has ceased. Continue the heating until the white fumes of perchloric acid have disappeared and a colorless liquid is refluxing on the still wall.

Cool to about 60°C and add 8 ml of 36% perchloric acid. Heat again to the

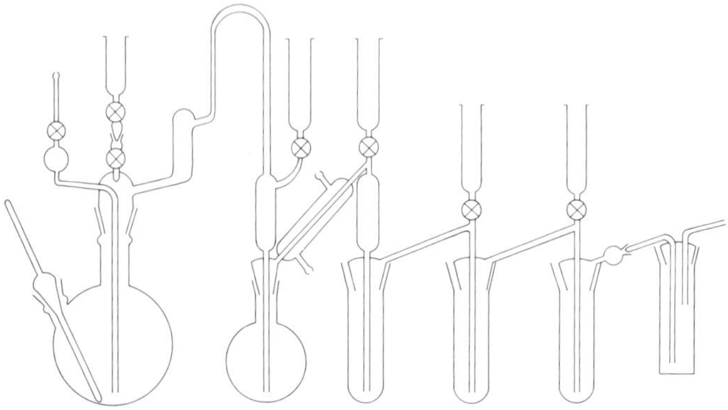

Figure 8.2. Distillation apparatus for osmium and ruthenium. (Reprinted with permission from Ref. 35. Copyright 1954 American Chemical Society.)

removal of brown fumes. Repeat the addition of 8 ml of 36% perchloric acid, intermittently with heating to ensure the complete removal of the osmium and ruthenium tetroxides. A complete distillation requires 0.5 to 1 h. Add 15 ml of 72% perchloric acid to the trap, and boil the solution for 30 min. Transfer the chilled receiving solution to a second chilled distillation flask as quickly as possible to prevent loss of osmium by volatilization, and wash the receivers with cold 3% sulfuric acid. Then wash thoroughly with water to remove the sulfuric acid.

Add 100 ml of water to the trap, 30 ml of a 5% thiourea solution in 1 : 1 hydrochloric acid to the first receiver, and 10 ml of the same solution to each of the other receivers. Chill the receivers in an ice bath. Add 40 ml of 30% hydrogen peroxide and 5 ml of concentrated sulfuric acid to the chilled distillation flask containing the osmium and ruthenium distillates. Boil gently for 30 min. Add 15 ml of perchloric acid to the trap and boil for 25 min. Draw a stream of air slowly through the system (about 3 to 5 bubbles/s). Transfer the contents of the receivers to a volumetric flask and dilute to volume.

To recover the ruthenium in the pot liquid, add 100 ml of water to the trap, 30 ml of 3% hydrogen peroxide and 1 ml of 48% hydrobromic acid to the first receiver, and 10 ml of 3% hydrogen peroxide to each of the other two receivers. Cool the receivers in an ice bath. Add 10 ml of concentrated sulfuric acid to the distillation flask and then add cautiously an excess of a 10% sodium bromate solution (about 20 ml). Apply a gentle suction, and distill cautiously over a low flame for 1 h. Then add 15 ml of the perchloric acid to the trap, and boil for 25 min. Disconnect the receivers from the water condenser, maintaining the connection between the two receivers. Add 8 ml of 48% hydrobromic acid to the first receiver and 4 ml of the acid to the second receiver. Boil the liquid for 10 min. Transfer the contents of the receiver to a volumetric flask, rinse the tubes and the receivers with 10% hydrobromic acid, and dilute to volume.

1. Determination of Osmium in Thiourea Receiver Solutions (36, 37)

Spectrophotometric Method Heat the volumetric flask containing the osmium receiver solutions to 85°C for 5 min. Cool to room temperature. Measure the absorbance of the solution at 480 nm. Prepare standards in 1 : 1 hydrochloric acid : ethanol and treat in a similar manner to the sample above.

Atomic Absorption Method Destroy the osmium thiourea complex of samples and standards by adding 2 ml of 30% hydrogen peroxide for each 5 ml of thiourea solution to a 25 ml volumetric flask. Dilute to volume with distilled water. Stopper and allow to stand at room temperature for 2.5 h to allow the reaction to proceed to completion. Aspirate the osmium solution into the nitrous oxide flame of the atomic absorption unit. Use the 290.9 nm osmium line, a lamp current of 8 mA, 94 kPa acetylene, and 224 kPa nitrous oxide. Adjust the burner height so that the light beam passes directly over the thin white part of the flame.

2. Determination of Ruthenium in Receiver Solutions (38)

Transfer an aliquot portion containing 100–500 μg of ruthenium to a 10 ml volumetric flask and add 0.4 ml of the uranium solution and 0.8 ml of aqua regia. Dilute to the mark with 40% hydrochloric acid. If an aliquot portion of more than 3 ml is required, evaporate the respective aliquot portion to about 3 ml, add 0.4 ml of uranium solution (1 ml = 250 mg uranium) and 0.8 ml of aqua regia, and, after transferring the solution to a 10 ml volumetric flask, dilute to volume with 40% hydrochloric acid.

Using the following instrumental parameters for the atomic absorption spectrometer, measure the absorption of the ruthenium and compare it with that of suitable standards measured at the same time:

Wavelength	349.9 nm
Width of slit	50 μm
Lamp current	10 mA
Air pressure	105 kPa

Acetylene flow is adjusted to give optimum readings. In the same way, measure the absorption of a reagent blank and subtract the value obtained from the values obtained for the samples.

Prepare a 2000 μg/ml standard ruthenium solution from metal or sponge. Dilute this solution so that it contains 100 μg Ru/ml in 40% hydrochloric acid. Take aliquot portions of 1, 2, 3, 4, and 5 ml, add 0.4 ml of uranium solution and 0.8 ml of aqua regia, and dilute to 10 ml in a volumetric flask with 40% hydrochloric acid. This procedure gives standard ruthenium solutions of 10, 20, 30, 40, and 50 μg Ru/ml. Measure the absorption of standards and samples relative to a blank solution containing uranium, aqua regia, and 40% (v/v) hydrochloric acid in a 10 ml volume.

C. Determination of Gold, Platinum, and Palladium in Silver Assay Beads by Flame Atomic Absorption (23)

The method, as originally published, recommends sample solutions containing 6 *M* hydrochloric acid to keep the silver in solution. This high acid content may not be desirable when used with some atomic absorption equipment because of its corrosive properties. It is possible to work with dilute hydrochloric acid solutions as long as the silver chloride is allowed to settle prior to the analysis.

A variety of interference problems were encountered. These were eliminated by using 1% lanthanum as a releasing agent.

Reagents and Equipment

A Perkin-Elmer 303 atomic absorption spectrometer fitted with a Boling burner was used. The following wavelengths, slit widths, and lamp currents were used, respectively: platinum, 265.9 nm, 1.0 nm, 25 mA; palladium, 247.6 nm, 1.0 nm, 30 mA, gold, 242.8 nm, 3.0 nm, 14 mA.

Standard solutions were prepared from pure metals of each material. Other reagents were Fisher certified grade and were checked for the absence of the precious metals.

Procedure

Add 5 ml of nitric acid to a silver bead in a 50 ml beaker and heat to leach the silver. Evaporate the solution to 0.5 ml and add several milliliters of concentrated hydrochloric acid. After all bubbling has ceased, add additional portions of concentrated hydrochloric acid until no further gases are evolved. Add enough lanthanum to make the final concentration in the flask 1%. Wash the mixture into a volumetric flask and dilute to volume with 6 *M* hydrochloric acid. Determine the platinum, palladium, and gold absorbances against standards containing 1% lanthanum and 6 *M* hydrochloric acid by running closely a lower concentration standard, the sample, and a higher concentration sample in quick succession.

Aspirate water frequently to minimize corrosion.

Detection Limit

In these sample solutions, the detection limits (micrograms per milliliter) are: palladium, 1; gold, 1; platinum, 5.

D. Determination of Platinum in Ores by Fire Assay and X-Ray Fluorescence (39, 40)

Comment on the Method

Platinum in ores is concentrated in lead by means of a fire assay. The silver bead obtained by cupellation is flattened to constant thickness for x-ray determination of the platinum content. Results of analyses of two ores were in good agreement with the results obtained by previous analyses. There was no interference from the other noble metals.

Equipment and Chemicals

- Fire assay furnace.
- X-ray fluorescence spectrometer. The operating conditions for the spectrometer are shown in Table 8.2. The x-rays were filtered through 5 mm molybdenum foil and then collimated. The detector was placed immediately below the sample out of the path of the main beam and was maintained at the temperature of liquid nitrogen. The output signals were amplified and sorted by a 1024-channel pulse height analyzer. The data were recorded in digital form on a teletype readout.
- Platinum standard solution, 1 mg/ml: One gram of pure platinum wire was dissolved in aqua regia. Nitrogen oxides were removed by repeated evaporations with hydrochloric acid. The solution was filtered and diluted to 1 liter.

Preparation of Standard Silver–Platinum Beads

Prepare silver beads of approximately 11 mg which contain 0.2–4.2% platinum by salting lead boats. Form the boats of thin lead foil. Carry a blank through the procedure. Heat magnesia cupels at 960°C for 10 min before use. Place the boat on a cupel in the furnace. Close the furnace door for 5 min to melt the lead. Then open the door by placing a 0.6 cm steel plate under the door. These conditions will result in driving the lead at the rate of 1 g/min. When driving has been completed, shut the door for 5 min. Withdraw the cupels slowly and allow them to cool. Flatten the cool beads in two stages. Scrape away any adhering cupel material from the bead. Flatten the bead to a thickness of about 5 mm between two steel blocks. Heat the bead in a porcelain crucible at red heat for 2–3 min. Then flatten the bead in a hydraulic press at 10 tons. The bead should be 0.10–0.12 mm thick and 3.70–4.10 mm in diameter.

Preparation of Calibration Curve

Mount the bead on Mylar foil (450 $\mu g/cm^2$) stretched over aluminum foil and held in position by vacuum grease. Place the sample in the chamber of the x-ray in-

Table 8.2
Instrumental Conditions

Tube	Molybdenum
Voltage	40 kV
Filament current	20 mA
Detector	Kevex Si(Li) with an active area of 30 mm^2 and resolution of 195 eV full width at half-maximum (FWHM) at 5.9 keV
Counting time	8 min

Source: Ref. 39.

strument and evacuate the chamber. Use the conditions in Table 8.2. Count the sample for 8 min. Record the Pt L_α and Ag K_α lines. Subtract a blank for the Pt L_α line. Calculate the intensity ratios of Pt L_α/Ag K_α, and plot these against percentages of platinum. The calibration is linear over the range specified.

Analysis of Ores

Crush the ore to pass 100 and 200 mesh sieves. Place the entire sample (0.5–1 kg) on a large cellophane sheet. Mix it by tumbling (see Chapter 4) for about 25 min. Spread the ore evenly and divide it into 2.5 cm squares with a spatula. Obtain an assay sample by removing small portions from each square. Place the sample in a small plastic bag together with 85 g lead oxide, 21.1 g sodium carbonate, 4.5 g calcium oxide, and 2 g flour. Add some silica to reduce some of the basicity of the mixture. Use 15 g of silica as a blank ore. Add a known quantity of silver as silver nitrate solution. Place the bag in a fire assay pot, and dry the sample in a 70°C oven for several hours. Break up the lumps and shake the mixture for 4–5 min.

Place the pot in a 950°C furnace, and raise the temperature to 1200°C at the maximum rate. After 15 min at the high temperature, remove the pot and pour the contents into a conical iron mold. Separate the lead from the cool slag. Cupel the lead as described above under preparation of standards.

Measure the platinum content of the silver bead following the same procedure as in the preparation of the calibration curve.

Comparison with Other Results

Data for three ores are shown in Table 8.3.

Samples S4 and Float Concentrate (National Institute of Metallurgy, Johannesburg, South Africa) were analyzed previously by means of fire assay–flame atomic absorption. Sample USBM 31 (U.S. Bureau of Mines, Reno, Nevada) was analyzed by eight laboratories using fire assay and various instruments; it is an average value. (There was no satisfactory explanation of the high result.)

Table 8.3
Platinum Contents of Ores

	S4	Float Concentrate	USBM 31
X-ray, ppm	5.31 ± .69	91.0 ± 1.7	6.55 ± 0.75
n (this work)	7	5	8
Previous analysis, ppm	5.23	82.6	4.9

Source: Ref. 39.

E. Determination of Iridium in Gold Assay Beads by Flame Atomic Absorption (25)

Cupellation into silver assay beads is not permissible because of the insolubility of iridium in silver. The following procedure involves a classical fire assay followed by cupellation into a gold bead. A copper–sodium sulfate buffer system is employed to overcome any potential interferences. The procedure was found to give excellent results for iridium on a standard dunite sample. The fire assay is also suitable for rhodium, platinum, and palladium, although no details are given by the authors for the determination of these elements.

Reagents and Equipment

Gold wire, 99.999% pure, was used. For the mixed copper–sodium solution, dissolve 13.75 g of copper sulfate pentahydrate and 4.64 g of sodium sulfate in 50 ml of hydrochloric acid and make up to 100 ml with water.

Standard solutions of iridium prepared from ammonium hexachloroiridate, a stock solution containing 1.000 mg/ml iridium in 2% (v/v) hydrochloric acid, were used. Prepare other solutions by dilution with 2% (v/v) hydrochloric acid.

Cylindrical alumina crucibles, 2 ml capacity, were used for sodium peroxide fusions. Coors AD-999 alumina ceramic crucibles are available from Coors Porcelain Company, Golden, Colorado.

A Perkin-Elmer Intensitron hollow-cathode tube with a Perkin-Elmer model 303 instrument was used in the experiments. The conditions were as follows: wavelength, 264.0 nm; slit, 0.3 nm; hollow-cathode current, 30 mA; fuel acetylene pressure 69 kPa (10 psi), flow meter setting, 8.5; oxidizer air pressure 190 kPa (28 psi), flow meter setting 7.5; flame, oxidizing; burner, standard head; aspiration, adjusted for uptake of 3 ml/min. Conditions for other instruments must be found by trial and error.

Procedure

The fire assay fusion and cupellation procedures used here follow generally accepted practices as described by Bugbee (34). A bisilicate slag composition is to be preferred. For a 20 g sample of dunite, a flux consisting of 50 g of lead(II) oxide, 35 g of sodium carbonate, 159 g of silica, 19 g of sodium tetraborate, and 4 g of flour will yield both a satisfactory lead button and a bisilicate slag. The fusion is made in the presence of 50 mg of added gold for the quantitative collection of iridium, palladium, platinum, and rhodium.

Transfer the gold bead obtained on cupellation to a small beaker. Add 5 ml of aqua regia, cover, and allow the mixture to react at room temperature for approximately 1 h. Heat on a steam bath for several hours more. Add 3 ml of water and a small amount of paper pulp, mix, and filter through a 42.5 mm medium-porosity filter paper. Wash the iridium residue with water. Ignite the residue in a 2 ml alumina crucible at 60°C, starting with a cold furnace. Add 100 ± 10 mg of sodium peroxide by calibrated dipper and carefully fuse the residue. Heat for 5 min more, maintaining in the molten state, and then cool. Add 1.5 ml of water, cover, and allow to stand at room temperature for approximately 15 min. Warm the mixture until the melt disintegrates, and transfer the solution to a 25 ml beaker by several alternating washes with 1 ml portions of concentrated hydrochloric acid, nitric acid, and water. Add 0.15 ml of sulfuric acid 1 : 1 by pipet to convert sodium salts into sulfate and evaporate the solution on a steam bath. Add 2 ml of hydrochloric acid (1 : 1), warm briefly to dissolve salts, and transfer the solution to a 10 ml volumetric flask with water. Dilute to volume with water and mix. Transfer a 5 ml aliquot of the solution to another 10 ml volumetric flask. Add 1 ml of hydrochloric acid (1 : 1) and 1 ml of copper sulfate solution. Adjust to volume with water and mix. Prepare iridium standards and a blank, each containing 1 ml of mixed copper–sodium solution in a 5 ml volume. Determine iridium in all solutions by atomic absorption. Samples containing as low a concentration as 2.5 ppm iridium can be determined.

F. Determination of Rhodium in Chromite Concentrates by Flame Atomic Absorption (26)

Two procedures for the collection of rhodium are given, a tellurium precipitation and a fire assay. In the former, a sodium peroxide fusion of the sample is followed by dissolution and then coprecipitation of the rhodium with tellurium. The fire assay procedure includes a cupellation step yielding a gold bead. Atomic absorption analysis can be performed on sample solutions of either product. To overcome potential interferences, 1% lanthanum sulfate is used.

Reagents and Equipment

- Tellurium solution: Prepare 1 mg/ml in 10% hydrochloric acid. Dissolve tellurium metal in aqua regia and remove nitrate by evaporation with hydrochloric acid.
- Tin(II) chloride solution: Dissolve 20 g of fresh stannous chloride dihydrate in 17 ml of hydrochloric acid. Dilute to 100 ml with water.
- Gold wire for fire assay, 99.999% pure and 0.1 mm in diameter; cut into 2.5 mg segments.
- Lanthanum sulfate solution: Dissolve 14.66 g of lanthanum oxide in 25 ml of hydrochloric acid and then add 15 ml of 1 : 1 sulfuric acid. Evaporate the solution. Dissolve the residue in 125 ml of hydrochloric acid and dilute to 500 ml with water.
- Standard solutions of rhodium: Prepare from the ammonium chloro salt a stock solution containing 1.00 mg/ml rhodium in 2% (v/v) hydrochloric acid. Prepare other solutions by dilution by factors of 10 with 2% (v/v) hydrochloric acid.
- Alumina crucibles for sodium peroxide fusions, Coors Ad-999 alumina ceramic, available from Coors Porcelain Company, Golden, Colorado.

- Instrument parameters and settings. A Perkin-Elmer model 303 instrument was used with the following operating conditions:

Wavelength	343.5 mm
Slit	0.3 nm
Hollow-cathode current	20 mA
Acetylene flow setting	6
Air flow setting	6.8
Flame	Oxidizing
Burner	Standard head
Aspirator	Adjusted for optimum uptake

Procedure

Tellurium precipitation procedure

Fuse over a burner in an alumina crucible 3.0 g of chromite concentrate with 10 g of fresh sodium peroxide. Heat for approximately 15 min after the charge becomes molten. After cooling the melt, place the crucible in approximately 100 ml of water in a beaker and carefully add 60 ml of hydrochloric acid. Detach the melt and remove the crucible.

Heat the solution to approximately 60°C and then, while stirring, very carefully add 5–7 ml of 30% hydrogen peroxide. Heat this solution on the steam bath for 30 min or more to destroy peroxide and then filter the solution through Schleicher and Schüll 589 White Ribbon paper (or equivalent).

Add 2.5 ml of tellurium solution, then 25 ml of tin(II) chloride solution by pipet while stirring. Adjust solution volume to approximately 200 ml. Digest the tellurium precipitate on a steam bath for approximately 2 h, filter it off on a Schleicher and Schüll 589 White Ribbon paper, and then wash with hot 10% (v/v) hydrochloric acid. Discard the filtrate.

Dissolve the precipitate from the paper by slowly adding 50 ml of hot aqua regia (8 volumes of hydrochloric acid, 2 of nitric acid, and 5 of water), collecting the filtrate in a 100 ml beaker. Wash finally with hot 10% (v/v) hydrochloric acid. Pass this filtrate through a 15 ml medium-porosity fritted glass Buchner-type filter funnel to remove paper fibers. Wash the filter with 10% (v/v) hydrochloric acid and then evaporate, treating with 3 ml portions of hydrochloric acid.

Add 2 ml of lanthanum sulfate solution to the residue and warm briefly on the steam bath. Transfer the solution to a 5 ml volumetric flask and adjust to volume with water. Prepare rhodium standards and a blank containing 2 ml of the lanthanum solution in a 5 ml volume. Determine rhodium in all solutions by atomic absorption.

Fire assay procedure

Add 3.0 g of sample to a flux consisting of 35 g of sodiumcarbonate, 11 g of silica, 19 g of anhydrous sodium tetraborate, 50 g of lead oxide, and 4.2 g of flour contained in a fire assay crucible and mix thoroughly.

Place in a furnace at 850°C and gradually raise the temperature to 925°C. Heat for 10 min at this temperature. Total heating time should be approximately 50 min. Pour the melt into an iron mold. Collect the lead button and shape it into a cube. Make a linear indentation on one surface of the cube by tapping a knife edge against this surface. Place a segment of the gold wire in the indentation and then carefully hammer the cube to secure the wire in place. Cupel the lead button at approximately 950°C. Analyze the bead as for iridium.

G. Rapid Fire Assay Method for the Determination of Platinum, Palladium, and Gold in Ores and Concentrates (Atomic Absorption) (41)

A modification of the tin collection method is used to preconcentrate the noble metals platinum, palladium, and gold in ores and concentrates. Tellurium is employed to prevent dissolution of these metals during parting with hydrochloric acid. For this purpose, 15–20 mg of tellurium is sufficient. Although tellurium causes a slight depression in the absorbances of the noble metals, it presents no problems at the concentration used.

The procedure was verified for the determination of platinum, palladium, and gold in four certified reference materials prepared by the Canadian Reference Materials Project and using a standard South African ore.

Apparatus and Reagents

- Furnaces. A 15 kW Globar type with suitable thermocouple and temperature controller, capable of accommodating six 40 g assay crucibles and maintaining their temperature at 1250°C; a Jelrus Handy-Melt portable electric furnace (Jelrus Technical Products Corp., New Hyde Park, New York); or similar. The Jelrus is a small vertical furnace equipped with removable graphite crucibles used in this work for melting tin-based assay buttons before their granulation in water. It is recommended that after 4–6 months of relatively constant use the bottom of the crucibles be examined for small holes.
- Tin(IV) oxide. BDH reagent grade is preferred because it has consistently given a suitable gold blank value of 150–200 mg/g.
- Tellurium powder, reagent grade.
- Standard solutions of platinum, palladium, and gold: Prepare these solutions by dissolving accurately weighed quantities of metal powder or sponge (minimum purity of 99.9%) in aqua regia. Evaporate each solution to dryness; then dissolve the residues in concentrated hydrochloric acid, and evaporate the solutions to dryness again. Repeat several times. Finally, dissolve the salts in, and dilute to volume with, 1 *M* hydrochloric acid. Standardize the gold solution gravimetrically using the hydroquinone precipitation, and standardize the platinum and palladium solutions spectrometrically.

- Mixed cadmium–copper sulfate solution: Prepare by dissolving 98 g of $CuSO_4 \cdot 5H_2O$ and 57 g of $3CdSO_4 \cdot 8H_2O$ in 500 ml of 12 *M* hydrochloric acid and 300 ml of water, followed by dilution to 1 liter with water.
- Flux for fire assay: stannic oxide 40 g, sodium carbonate 50 g, sodium tetraborate 10 g, flour 35 g, tellurium 25 mg, silica 10–20 g according to the amount of silica in the sample. Make enough flux for a sample up to 1 assay ton (29.17 g) in size.

Procedure

Before the crucible fusion procedure, all samples except those of copper–nickel matte are roasted at 750–800°C for approximately 1 h to decompose sulfides and volatilize arsenic and antimony. Place the sample on a shallow fireclay dish and stir it intermittently during the roasting process. In cases where only a few grams of material (particularly sulfides) are to be roasted, place the sample on a bed of silica to prevent possible loss of the resultant calcine to the surface of the dish. (The silica is included as part of that required in the flux above.)

Leaching is performed to remove the bulk of the copper and nickel from the residue of precious metals. Place the sample, weighing up to 2 assay tons, in a 1500 ml beaker and treat it with 25 g of ammonium chloride and 100–200 ml of 12 *M* hydrochloric acid. Heat the sample until the amount of insoluble matter appears not to exceed 2–3 g. With large samples, it may be necessary to treat the residue once or twice more with fresh acid after intervening filtrations.

Dilute the combined sample solution (~100 ml) with an equal volume of water and filter the solution containing most of the nickel and copper through a moderately fast paper. Transfer the solids to the paper with dilute (5%) hydrochloric acid. Dry the washed residue and paper at ~100°C for about 1 h and then mix them with the recommended assay flux for fusion.

Chromite is not completely decomposed during the fusion process, and samples containing appreciable proportions must be subjected to a pretreatment. Decomposition is accomplished by sintering with sodium peroxide. Mix the sample with 1.5 times its weight of sodium peroxide; then place it on a 10 g bed of silica in a roasting dish, and roast it at 700°C for about 1 h. Grind the sinter cake and underlying silica together in a mortar and mix them with the flux for the crucible fusion process. The weights of sodium peroxide and silica are subtracted from the weights of sodium carbonate and silica, respectively, in the flux described above.

For powdered samples, the standard assay practice of blending the samples with the flux on glazed paper and transferring the charge to a 40 g crucible is followed.

When solutions are to be mixed with the flux, place approximately one-third of the flux in the crucible and then press a 30 cm square of thin commerical wrapping film into the crucible to form an envelope. Add the solution carefully to the envelope. Then transfer the remainder of the flux slowly to the crucible to avoid wetting the film or crucible walls. Heat the crucible in a drying oven at 100°C for at least 2 h. After drying, grind the material in the wrapping film in a mortar, mix it well, and place it back in the film in the crucible. It is to be noted that, after drying, the salted portion of the charge is lumpy and difficult to pulverize and mix with the rest of the charge. This could lead to occasional spurious results.

Fusion

Place the crucible in the assay furnace at 1250°C for about 90 min to fuse the charge. At the completion of this fusion period, the melt should not be viscous or lumpy, nor should there be extensive crust formation at the top of the melt.

Pour the melt into a conical steel mold and, when it is cool, separate the tin button from adhering slag by tapping with a small hammer.

Place the button in the crucible of the Jelrus furnace, from which air is purged by nitrogen delivered through a ceramic tube placed directly over the button. Increase the temperature until the button melts (600–1000°C, depending on composition) and then pour the melt into a pail of water to granulate the alloy. Any large pieces are easily reduced in size with metal shears.

Dissolution and analysis

Treat each sample of granulated tin alloy with 150 ml of 12 *M* hydrochloric acid in a covered 600 ml beaker and heat until the excess of tin has dissolved and vigorous evolution of bubbles from the residue has ceased. Add another 15–25 ml portion of acid and boil the sample for approximately 10 min. Add water to give a volume of approximately 400 ml and allow the residue to settle. Decant the supernatant solution through a filter pad. Wash the residue in the beaker several times, by decantation, with 15% (v/v) hydrochloric acid, the washings being passed through the filter pad.

Treat the residue in the beaker with a mixture of 15 ml of 12 *M* hydrochloric acid and 5 ml of 30% hydrogen peroxide, and heat the beaker gently for a few minutes to ensure complete dissolution of the residue. Elute the residue on the filter pad with 20 ml of a 3 : 1 mixture of 8 *M* hydrochloric acid and 30% hydrogen peroxide, and add the washings to the beaker.

Add approximately 50 mg of sodium chloride, and evaporate the sample solution to dryness. When the evolution of fumes has nearly ceased, remove the beaker from the evaporator and wash the sides with 10 ml of a 7 : 2 mixture of hydrochloric and hydrobromic acids. Evaporate the sample again to dryness to volatilize the remaining tin. Cool the beaker, add 10–15 ml of 12 *M* hydrochloric acid, and, while the beaker is being swirled, add 30% hydrogen peroxide cautiously until it is evident that an excess is present. Heat the beaker for a few minutes, and, after cooling, wash the sides with water. Filter the solution through a fast paper into a 400 ml beaker and wash the paper several times with 15% hydrochloric acid. Add approximately 5 ml of aqua regia and evaporate the solution to approximately 1 ml.

To the cooled sample solution, add 5 ml of cadmium–copper sulfate solution and transfer the mixture to a 25 ml volumetric flask. Dilute to volume with water. The platinum, palladium, and gold content of the sample is then determined by atomic absorption using the manufacturer's recommended conditions. Any silver, rhodium, ruthenium, or iridium remaining in the solution will not interfere.

Notes

For milligram amounts of the precious metals, the solution obtained after the volatilization of tin is filtered into a 100–500 ml flask and diluted to volume with 15% hydrochloric acid to prevent hydrolysis. An aliquot is taken and treated by the procedure given above.

Calibration curves for gold and palladium are linear in the ranges 0.2–3 and 0.4–3 ppm, respectively.

Because all batches of stannic oxide tested in this laboratory were found to contain gold, it is deemed necessary to carry a blank through the analytical scheme.

H. Nickel Sulfide Fire Assay for Platinum Group Elements and Gold in Rocks and Ore (Instrumental Neutron Activation) (31)

A large sample size (up to 50 g) can be used with this procedure. The platinum group metals and gold are collected in a nickel sulfide button. Hydrochloric acid is added and the button is dissolved. The residue (which contains the platinum group metals and gold) is collected by filtration. Instrumental neutron activation analysis is used to analyze the residue. Losses are less than 6% for all elements. This method is relatively rapid and gives very good detection limits. The detection limits are presented in Table 8.4.

Equipment

For the fusion procedure a large home-made muffle furnace capable of reaching 1000°C was used. For the filtration procedure a Gelman Instrument Company vacuum filter apparatus (catalog No. xx1004720) was used.

Table 8.4
Detection Limits (2σ)

Element	Detection limit (ppb)	Time counted (s)
Rh	1	200
Pd	50	200
	5	500–2500
Pt	5	700–5000
Ir	0.1	700–5000
Ru	3	700–5000
Os	2	700–5000
Au	0.1	700–5000

Source: Ref. 31.

Irradiations were performed (SLOWPOKE II reactor, University of Toronto) with a thermal neutron flux of 1.0×10^{11} n/cm$^2 \cdot$ s. Palladium and rhodium were counted on a horizontally mounted Ge(Li) detector which has a resolution of 1.93 keV (full width at half-maximum), a peak-to-compton ratio of 32 : 1 for the 1332 keV peak of ^{60}Co, and a relative efficiency of 6.7%. The detector was connected to a Canberra Instruments 8180 4096-channel analyzer. The Pd, Pt, Ir, Os, Ru, and Au radioisotopes were counted with a Princeton Gamma-Tech Ge(Li) detector with an active crystal volume of 65 cm^3, connected to a Tracor Northern TN-1700 4096-channel analyzer. This system has a resolution of 1.75 keV (FWHM), a peak-to-compton ratio of 43.1 : 1 for the 1332 keV peak of ^{60}Co, and a relative efficiency of 12.4%.

Reagents

All reagents must be tested to make sure they are free of noble metals. Commercially available nickel products may contain substantial amounts of noble metals. Blank fire assay charges were analyzed whenever new reagent batches were introduced to ensure absence of noble metals. The following reagents were used in the fusion procedure: fused, ground sodium borate (Fisher Scientific S-252); calcined dry, purified sodium carbonate (Fisher Scientific S-261); silica floated powder, ~240 mesh (Fisher Scientific S-153); sublimed sulfur (J. T. Baker Chemical Co. 5-4088); nickel carbonyl powder (Inco Metals Co.); 30 g fire assay crucible (Canlab 8522-30C). During the dissolution stage 12 *M* HCl (J. T. Baker Chemical Co. 3-9535) and Metricel GA-1 (pore size 5.0 μm, 47 mm) filter papers (Gelman Sciences Inc., No. 60003) were used.

Sample Preparation

Fusion

Nickel sulfide fire assay buttons were prepared from 30–50 g samples according to the fusion procedure of Robert *et al.* (33) with slight modifications. Instead of mixing the reagents and sample on glazed paper, it was more satisfactory and convenient to mix them directly in the crucible; nickel powder, prepared by the nickel carbonyl process, was used instead of nickel oxide; after fusion the crucible contents were allowed to solidify *in situ* instead of pouring the crucible contents into an iron mold.

Dissolution and filtering

The button was crushed in a hardened steel piston-type device to obtain a particle size of ~1 mm. The crushed pellet was heated with 400 ml of 12 *M* hydrochloric acid in a covered beaker on a hot plate until the sample dissolved completely. When cooled, the solution was filtered under vacuum and washed with ~400 ml of distilled water. The black residue on the filter paper was folded into a triangular shape and sealed in plastic ready for the irradiation and counting procedure.

Instrumental Neutron Activation Analysis Measurements

The noble metal content was determined by a procedure involving two to three countings. Rhodium and palladium were determined during the first count, palladium if required (low levels, below ~50 ppb, are not detected in the first count) during a second count, and the balance of the noble metals in a third count.

Rhodium and palladium

The samples were irradiated serially for 5 min and then allowed to decay for 60 s so that very short-lived radioisotopes, e.g., ^{77m}Se ($t_{1/2}$ = 17.5 s), could decay. Samples and standards were counted for 200 s for the ^{109m}Pd and ^{104m}Rh x-rays. Nuclear data for the radioisotopes are given in Table 8.5 (42). Peak areas, corrected for background, were compared for samples and standards.

Palladium, platinum, osmium, ruthenium, iridium, and gold

Up to 40 samples, internal standard, and flux monitors were sealed in a 27 cm^3 plastic irradiation container, which was irradiated for 16 h. Samples were allowed to decay for ~4 h from the end of the irradiation and were counted for ^{109}Pd for 500–2500 s, depending on the palladium content.

The samples were allowed to decay for a further 7–9 days, depending on the rhenium content of the sample. The 155 keV ^{185}Re γ peak interfered with the 158 keV ^{199}Au photopeak. The samples were recounted for ^{199}Au (for platinum content), ^{191}Os, ^{103}Ru, ^{192}Ir, and ^{198}Au. Integrated peak areas, corrected for background, were entered into a computer program, which took into account sample weights, yield through the crushing procedure, irradiation time, counting time, and duration

Table 8.5
Nuclear Data[a]

Element	Induced nuclear reaction	Isotope counted	% Isotopic abundance target isotope (θ)	Cross section (barn)	Half-life of nuclide produced ($t_{1/2}$)	γ ray used (keV)
Rhodium	^{103}Rh(n, γ) ^{104m}Rh	^{104m}Rh	100.0	800	4.41 min	51
Palladium	^{108}Pd(n, γ) ^{109}Pd	^{109}Pd	26.7	12	13.5 h	88
	^{108}Pd(n, γ) ^{109m}Pd	^{109m}Pd	26.7	0.2	4.69 min	186
Platinum	^{198}Pt(n, γ) ^{199m}Pd (β^-)^{199}Au	^{199}Au	7.2	4	3.15 d	158
Iridium	^{191}Ir(n, γ) ^{192}Ir	^{192}Ir	38.5	750	74.2 d	317 468
Osmium	^{190}Os(n, γ) ^{191}Os	^{191}Os	26.4	3.9	14.6 d	129
Ruthenium	^{102}Ru(n, γ) ^{103}Ru	^{103}Ru	31.5	1.4	38.9 d	497
Gold	^{197}Au(n, γ) ^{198}Au	^{198}Au	100.0	98.8	64.8 h	412

Source: Ref. 31.
[a]Data from Bureau of Radiological Health and the Training Institute Environmental Control Administration (42).

of count. Radioisotopes were corrected for decay and the activity was compared with standards of known concentration.

I. Determination of Platinum Group Metals and Gold in Ores by ICP–AES (28)

Comment on the Method

The platinum group metals and gold are concentrated in nickel sulfide buttons from fire assays. Nickel is dissolved in a solution of hydrochloric acid and ammonium chloride, leaving a residue of precious metals. The aqua regia solution of the residue is used for analysis. Results were compared with recommended values and values obtained by atomic absorption spectrometry. The agreements were satisfactory. Osmium cannot be determined by this procedure because it is lost during the dissolution of nickel sulfide buttons.

Equipment

- Spectrometer, ARL model 33000 Sequentometer (1 m). The grating was ruled at 1920 lines/mm and had a reciprocal linear dispersion (first order) of 0.5 nm/mm. The primary slit width was 20 μm and secondary slit widths were 50 or 75 μm.
- Computer, Digital Equipment Corp. PDP1105 with 8K memory.
- Nebulizer, Meinhard design of glass pneumatic, concentric nebulizer connected to a dual spray chamber. Solutions were fed to the nebulizer at a constant rate by a peristaltic pump. The carrier gas was humidified to prevent blockage of the nebulizer with salts.

The operating conditions are shown in Table 8.6.

Table 8.6
Operating Conditions for Plasma

Forward rf power	1.25 kW
Reflected power	<10 W
Argon coolant/support gas flow rate	10 liters/min
Argon flow rate for aerosol transport	1 liter/min (stabilized)
Height of observation above load coil	26 mm
Sample solution uptake rate	1.2 ml/min
Integration time	10 s

Reprinted with permission from Ref. 28. Copyright 1978 American Chemical Society.

Standard Solutions

Prepare standard solutions of Pt, Pd, Ir, Rh, Ru, and Au from pure metals and salts at concentrations of 1000 μg/ml. Dilute solutions of the standards should contain 10% (vol) nitric acid.

Interference Effects

The mutual platinum group metal interference effects as well as the interference effects of the major extraneous base metals (copper, nickel, and iron) are shown in Table 8.7. The analyte emission lines used were chosen on the basis of lowest detection limit, lowest background equivalent concentration, and absence of interfering lines from other platinum metals and the three major base metals. The interference effects were negligible for solutions containing low and comparable concentrations of platinum metals and of the base metals.

Detection Limits

The background equivalent concentrations and detection limits for this determination are shown in Table 8.8.

Calibration

Prepare calibration curves for the platinum metals and gold using the conditions described above. The calibration ranges are shown in Table 8.9. Use three composite standard solutions and a blank. Analyze each standard solution in duplicate, and calculate the averages.

Procedure for Low-Grade Samples

Choose a weight of sample (50–100 g) such that the anticipated concentrations of the metals in the final solutions fall within the calibration ranges. Mix the sample with a premixed flux containing 32 g nickel carbonate, 45 g sodium carbonate, 90 g sodium borate, and 12.5 g sulfur. If the sample is low in silica, add 20 g silica.

Table 8.7
Additive Interferences Expressed in Terms of Micrograms per Milliliter Analyte

Analyte emission line (nm)	Interferents (1000 μg/ml)									
	Pt	Pd	Rh	Ir	Ru	Os	Au	Cu	Ni	Fe
Pt 265.9	—	0	0.8	7.3	12.4	1.4	0	0	0	0
Pd 340.5	0	—	0	0	0	0	0	0	0	0.4
Rh 343.5	0	0	—	3.7	0.3	0	0	0	0.2	0
Ir 322.1	0	0	3.7	—	0	0	0	0.5	0	0
Ru 349.9	0	0	6.6	0.2	—	0	0	0	0.2	0
Os 290.9	0	0	0	0	0.5	—	0	0	0	0
Au 267.6	0	0	0.3	0	0.2	0	—	0	0	0

Table 8.8
Background Equivalent Concentrations and Detection Limits for Selected Platinum Group Emission Lines

Element	BEC (μg/ml)	DL in solution (μg/ml)
Pt	9.8	0.05
Pd	1.8	0.04
Rh	4.2	0.05
Ir	14.5	0.2
Ru	4.7	0.05
Os	2.2	Not determined
Au	2.6	0.01

Reprinted with permission from Ref. 28. Copyright 1978 American Chemical Society.

Fuse the mixture at 1100°C for 2 h in a fireclay crucible. Pour the melt into a mold. When the sample has cooled, separate the nickel sulfide button from the slag.

Powder the slag, and mix it with a premixed flux containing 7 g nickel carbonate, 10 g sodium carbonate, 30 g sodium borate, and 3 g sulfur. Fuse the mixture as described previously, and separate the second button. This treatment removes any residual metals from the first slag.

Crush the two nickel sulfide buttons, and transfer the solids to a large beaker. Add 300 ml of hydrochloric acid (37%) and 100 g of ammonium chloride. Boil the sample gently for 2 h to dissolve the base metals. Collect the residue of precious metals on a Millipore filter. Dissolve the metals in 20 ml of aqua regia. Make up the solution to a volume of 100 ml or less, depending on the anticipated concentrations.

Measure the concentrations using the instrumental conditions described above.

Table 8.9
Calibration Ranges

Analytical line (nm)	Calibration range (μg/ml)
Pt 265.9	0.5–200
Pd 340.5	0.4–100
Rh 343.5	0.5–20
Ir 322.1	2–5
Ru 349.9	0.5–50
Au 267.6	0.1–10

Reprinted with permission from Ref. 28. Copyright 1978 American Chemical Society.

Table 8.10
Comparison of Results Obtained by ICP with Recommended and Atomic Absorption Values (μg/g)

Sample No.	Pt R[a]	Pt ICP	Pt AA	Pd R	Pd ICP	Pd AA	Rh R	Rh ICP	Rh AA	Ir R	Ir ICP	Ir AA	Ru R	Ru ICP	Ru AA	Au R	Au ICP	Au AA
1A		3.7	3.6		1.3	1.5		0.3	0.2		0.1	<0.1		0.4	0.6		0.2	0.2
1B		4.0	3.7		1.6	1.6		0.3	0.3		0.05	<0.1		0.4	0.6		0.2	0.3
Av.	3.6	3.9	3.7	1.6	1.5	1.6	0.2	0.3	0.3	0.1	0.1	<0.1	0.4	0.4	0.6	0.3	0.2	0.3
2A		66.0	67.6		38.4	41.6		6.4	6.8		0.4	<1		9.8	12.6		3.2	3.8
2B		64.4	58.2		37.0	39.6		6.4	6.0		<0.4	<1		9.8	11.6		2.8	3.8
Av.	71.4	65.2	62.9	42.7	37.7	40.6	6.5	6.4	6.4	1.9	?	<1	12.7	9.8	12.1	4.0	3.0	3.8
3A		0.7	0.8		0.3	0.3		0.1	0.1		<0.04	<0.1		0.2	0.3		0.1	0.1
3B		0.6	0.8		0.3	0.3		0.1	0.1		<0.04	<0.4		0.2	0.3		0.1	0.1
Av.	0.7	0.7	0.8	0.3	0.3	0.3	0.1	0.1	0.1	0.04	<0.04	<0.1	0.2	0.2	0.3	0.3	0.1	0.1

[a]R; Recommended values.

Results for Standard Materials

The results of analyses of three low-grade standard samples are shown in Table 8.10. The values are compared with the recommended values and also with values obtained by means of atomic absorption. The weights of samples and final dilutions were: sample 1, 100 g, 20 ml; sample 2, 50 g, 100 ml; sample 3, 100 g, 20 ml.

J. Determination of Platinum, Palladium, Rhodium, Ruthenium, and Iridium in Ore by Electrothermal Atomic Absorption (43)

Comment on the Method

Low concentrations of the five platinum metals are concentrated in nickel sulfide in a fire assay. The precious metals are separated from base metals, and a hydrochloric acid solution is used for electrothermal atomization. The limits of the determinations ranged from 0.002 to 0.03 mg/liter for the various elements. There are no interelement interferences at ratios of 10 : 1 of interferent to analyte element. The results of analysis of a reference material (SARM 7) were satisfactory.

Equipment

- Pye Unicam SP9 spectrometer equipped with an SP9 video furnace programmer and an SP9 computer
- Pyrolytically coated graphite tubes
- CR 6525 two-pen chart recorder (JJ instruments)

The instrumental conditions are listed in Table 8.11.

Standard Solutions

Prepare standard solutions of the five elements from pure metals and sponges. Working standards should contain 10% (v/v) hydrochloric acid.

Procedure

Fire assay using nickel sulfide as a collector (33)

Use a 50 g sample for platinum group metal concentrations of less than 50 ppm. Transfer the finely powdered sample, 60.0 g of fused borax, 30.0 g of soda ash, 32.0 g of nickel carbonate, and 12.5 g of powdered sulfur to a glazed paper. For chromite samples add 90.0 g of fused borax and 45.0 g of soda ash. At least 10 g of silica should be present for a good fusion. This is usually contributed by the sample. If the sulfur content of the sample is larger than 0.5%, reduce the amount of powdered sulfur added accordingly. If the sulfur content is not known, roast the sample. If there is a large excess of sulfur during fusion, the button will be much larger than 25 g and may disintegrate when it is cooled.

Since nickel compounds can contribute platinum metals, a blank determination should be made with each new batch of nickel carbonate.

Roll the mixture from corner to corner of the paper until it is thoroughly mixed, and transfer it to a No. 1 fireclay crucible. Wipe the surface of the glazed paper with a piece of tissue paper, and place this on top of the mixture in the crucible. With the aid of furnace tongs, transfer the crucible to a furnace at a temperature of 1000°C. Leave the mixture to fuse for 1¼ h.

Remove the crucible and pour the contents into an iron mold. Allow the mold to cool for 30 min, and remove the button and the slag.

Table 8.11
Instrumental Conditions

Element	Wavelength (nm)	Spectral bandpass (nm)	Background correction	Maximum ash temperature (°C)	Atomization temperature (°C)
Platinum	265.9	0.2	Used	1700	2800
Palladium	244.8	0.2	Used	1200	2400–3000
Rhodium	343.5	0.5	Not used	1600	2800
Ruthenium	349.9	0.2	Not used	1700	2800
Iridium	208.9	0.2	Used	>2000	2800

Source: Ref. 43.

Weigh the button and record the weight. An ideal weight for the button is between 25 and 30 g. Break up the button by pressing it in a hydraulic press. Transfer the pieces to the small bowl of a Siebtechnik mill and grind for 2 min.

Brush the sample into a 600 ml squat beaker, add 400 ml of concentrated hydrochloric acid, and cover the beaker with a watch glass. Leave the beaker on a steam bath set at a low temperature for approximately 16 h. The operation is usually carried out overnight. Remove the beaker from the steam bath, wash down the lid and the sides of the beaker with water, and stir the contents with a glass rod. Allow the noble metal sulfides to settle and cool for about 1 h.

Filter the solution under vacuum, using a No. 542 11-cm filter paper in a Fisher funnel. Wash out the beaker with cold 50% hydrochloric acid, and wash the paper at least five times with water to remove all traces of nickel. Discard the filtrate. With a fine jet of water, wash the residue that can be readily removed from the filter paper into the original beaker. Pour 10 ml of concentrated hydrochloric acid and 10 ml of 30% hydrogen peroxide onto the filter paper contained in the funnel, and cover it immediately with a watch glass.

As the noble metals dissolve, collect them in the original 600 ml squat beaker. Wash the paper three times with water. Retain the filter paper together with any undissolved residue. Transfer the covered beaker to a hot plate, and boil until the solution is reduced to a volume of about 20 ml. Add 2 ml of hydrogen peroxide and boil for 10 min.

Filter the solution into a 150 ml squat beaker using a No. 540 filter paper, and wash with 10% hydrochloric acid.

Ignite the filter paper together with the above washed paper in a 15 ml porcelain crucible. Add 10 ml of aqua regia to the crucible, cover it with a watch glass, and heat strongly until the residue is dissolved. Add this solution to the original solution in the 150 ml beaker and evaporate to near dryness. Add 3 ml of hydrochloric acid and again evaporate to near dryness. Add 5 ml of 40% hydrochloric acid and warm to dissolve. Dilute the solution to a suitable volume for atomic absorption measurements. The final solution should contain 10% hydrochloric acid.

Atomic absorption measurements

Use the instrument conditions shown in Table 8.11 for analyses of standards and samples. The linear calibration ranges, detection and determination limits, and relative standard deviations for 10 determinations are listed in Table 8.12.

Interference

Since one fusion with nickel sulfide as the collector effectively separates the noble metals from the base metals and other constituents in the sample material, the resulting solutions contain only the platinum group metals, gold, silver, and a small amount (<10 mg) of occluded nickel. Only mutual interferences between the platinum group metals, gold, silver, and nickel were therefore tested. There were no interferences at a ratio of 10 : 1 interferent to analyte.

Table 8.12
Eletrothermal Measurements of the Platinum Metals

Element	Linear calibration range (mg/liter)	Detection limit[a] in solution (mg/liter)	Determination limit[b] in solution (mg/liter)	s^c
Platinum	0–0.4	0.008	0.016	0.05 (0.30)
Palladium	0–0.06	0.001	0.002	0.08 (0.02)
Rhodium	0–0.05	0.002	0.005	0.04 (0.05)
Ruthenium	0–0.1	0.015	0.03	0.05 (0.03)
Iridium	0–0.3	0.012	0.025	0.08 (0.10)

Source: Ref. 43
[a]Twice the average blank value of a used graphite furnace.
[b]Four times the average blank value of a used graphite furnace.
[c]Relative standard deviation for 10 determinations. Figures in parentheses represent the concentrations (milligrams per liter) at which *s* was calculated.

Accuracy

The accuracy of the method is illustrated by the results of analysis of the reference material SARM 7 in Table 8.13.

K. Chlorination–Atomic Spectrometry Method (10)

This method does not involve a fire assay. It can often be used in the place of a fire assay for difficult to dissolve samples such as concentrates, alloys, precious metal coatings, and precious metal-containing ceramic products. The approach, though simple and very effective, is seldom used because of perceived dangers in using chlorine at high temperatures. However, experience in the present authors' laboratories suggests that the method can be employed safely in a laboratory with a properly functioning fume hood.

The method involves mixing the sample with NaCl and then treating the mixture (held in a porcelain boat) in a stream of dry chlorine at 500–600°C for 1–2 h. The chlorinated product is dissolved in hydrochloric acid and then can be analyzed

Table 8.13
Results of Analysis of Reference Material (SARM 7)

Element	Analytical results (mg/liter)	Recommended values (mg/liter)	Number of analyses
Pt	3.75 ± 0.12	3.74 ± 0.045	8
Pd	1.46 ± 0.05	1.53 ± 0.032	10
Rh	0.22 ± 0.02	0.24 ± 0.013	10
Ru	0.45 ± 0.05	0.43 ± 0.057	10
Ir	0.061 ± 0.004	0.074 ± 0.012	10

Source: Ref. 43.

by flame atomic absorption or plasma atomic emission spectrometry. The elements covered are gold, platinum, palladium, and rhodium. Results were checked by fire assay and satisfactory agreement was obtained. Detection limits are in the 100 μg/g range.

Equipment

A tube furnace, Lindberg single-zone furnace, 1000°C maximum temperature, (Lindberg, Watertown, Wisconsin), which holds a Vycor tube of 31 mm o.d. and 100 cm long, was used. The Vycor tube is fitted at each end with a rubber stopper through which is inserted a short length of glass tubing of 5 mm o.d. At one end the glass tubing is connected by Tygon tubing to a chlorine gas cylinder. At the other end the glass tubing is connected to a water bubbler flask which has an exit up the rear of a fume hood. The complete apparatus is placed in a fume hood. Silica boats used to hold the sample mixture were 2 cm wide by 1 cm high by 10 cm long.

Platinum, palladium, gold, and rhodium were determined by a dc plasma emission spectrometer (model III, SpectraMetrics Co., Andover, Massachusetts). A model 4000 atomic absorption spectrophotometer (Perkin-Elmer Corp., Norwalk, Connecticut) was used for the determination of palladium and gold. The operating parameters of these instruments are given in Table 8.14.

Reagents

All reagents were of ACS reagent grade purity. Standard solutions of precious metals were prepared by dissolving the pure metals or pure metal chlorides in hydrochloric acid and adjusting the final solution acidity to 10% hydrochloric acid.

Calibration standards were prepared from these solutions. These should contain sodium chloride and any matrix ingredient which appears in an amount greater than about 0.5% in the final sample solutions. The final acidity of calibration standards was 10% hydrochloric acid.

Table 8.14
Instrument Operating Parameters for dc Plasma Emission (DCP)[a] and Atomic Absorption (AA)

Element	Wavelength (nm)		Slit		Air flow rate (AA) (liters/min)	Acetylene flow rate (AA) (liters/min)
	DCP	AA	DCP (μm)	AA (spectral) bandwidth) (nm)		
Au	267.6	242.8	300 × 50	0.7	2.0	15
Pt	256.9	—	300 × 50	—	—	—
Pd	363.5	247.6	300 × 50	0.2	2.0	15
Rh	339.7	—	300 × 50	—	—	—

Source: Ref. 10.

[a]DCP conditions: nebulizer argon pressure, 140 kPa; electrode argon pressure, 350 kPa; signal an average of two 8 s integrations; peristaltic pump used to introduce sample at a flow rate of 2.8 ml/min.

Procedure

Weigh up to a 1 g sample and mix well with 1 g of sodium chloride powder. Transfer the mixture to a silica boat. Place the boat in the center of a Vycor chlorination tube. A length of glass rod can be used for this purpose. Insert the rubber stoppers into the ends of the Vycor tube. Adjust the chlorine gas flow to about 2 bubbles per second. Inspect all lines to ensure that the fittings are secure. Allow 2–3 min of chlorine gas flow to be certain that all air has been displaced from the system before placing the tube in the furnace, and begin heating. The chlorination temperature should be 500°C for alloys and coatings and 600°C for concentrates. When this temperature has been reached, the heating is continued for 1.5 h. The chlorine gas flow should be checked from time to time and adjusted if necessary. When the chlorination is finished, remove the Vycor tube from the furnace and place on a refractory brick to cool. The chlorine gas is left flowing during this period. After cooling, turn off the chlorine and displace the remaining chlorine gas with air. Remove the rubber stoppers from the ends of the Vycor tube. Carefully slide the silica boat out of the tube and into a beaker containing 100 ml of hot 20% hydrochloric acid. Rinse the interior of the Vycor tube with hot 1% hydrochloric acid solution into the beaker. Then cover with a watch glass and boil for ½ h to dissolve chlorinated sample. The solution in the beaker can be evaporated at this stage if a concentration step is necessary. Cool the solution and transfer to an appropriate-sized volumetric flask and dilute to volume. The final solution acidity should be 10% in hydrochloric acid.

Run the samples and standards on a plasma emission or atomic absorption spectrometer. The instrument operation parameters are listed in Table 8.14.

REFERENCES

1. F. E. Beamish, "The Analytical Chemistry of the Noble Metals." Pergamon, Oxford, 1966.
2. T. W. Steel, *Rep.—Natl. Inst. Metall. (S. Afr.)*, **1696** (1975).
3. T. Wilm, *Ber. Dtsch. Chem. Ges.* **18,** 2536 (1855).
4. F. E. Lathe, *Can. J. Res., Sect. B* **18,** 333 (1940).
5. S. G. Wall and A. Chow, *Anal. Chim. Acta* **70,** 425 (1974).
6. A. Diamantatos, *Anal. Chim. Acta* **94,** 49 (1977).
7. M. E. V. Plummer and F. E. Beamish, *Anal. Chem.* **31,** 1141 (1959).
8. G. H. Faye and P. E. Moloughney, *Talanta* **23,** 377 (1976).
9. R. V. D. Robert, E. Van Wyk, and R. Palmer, *Rep.—Natl. Inst. Metall. (S. Afr.)* **1341** and **1487** (1971).
10. J. C. Van Loon, M. S. Szeto, W. W. Howson, and I. A. Levin, *At. Spectrosc.* **5,** 43 (1984).
11. F. W. E. Strelow, E. C. Feast, P. M. Matthews, C. J. C. Bothma, and C. R. Van Zyl, *Anal. Chem.* **38,** 115 (1966).
12. R. Gilchrist, *J. Am. Chem. Soc.* **57,** 2565 (1931).
13. F. E. Beamish and J. J. Russell, *Ind. Eng. Chem., Anal. Ed.* **8,** 141 (1936).
14. H. Zachariasen and F. E. Beamish, *Anal. Chem.* **34,** 964 (1962).
15. Y. Samiullah, *J. Geochem. Explor.* **23,** 193 (1985).
16. A. Walsh, *Spectrochim. Acta* **7,** 108 (1955).
17. T. W. Steele, *Rep.—Natl. Inst. Metall. (S. Afr.)* **1696** (1975).
18. R. C. Mallet, D. C. G. Pearton, and E. J. Ring, *Rep.—Natl. Inst. Metall. (S. Afr.)* **1086** (1970).

19. R. C. Mallet and R. L. Breckenridge, *Rep.—Natl. Inst. Metall. (S. Afr.)* **1318** (1971).
20. J. C. Van Loon, *At. Absorpt. Newsl.* **8,** 6 (1969).
21. A. E. Pitts, J. C. Van Loon, and F. E. Beamish, *Anal. Chim. Acta* **50,** 195 (1970).
22. W. B. Rowston and J. M. Ottaway, *Anal. Lett.* **3,** 411 (1970).
23. J. C. Van Loon, *Z. Anal. Chem.* **246,** 122 (1969).
24. A. E. Pitts and F. E. Beamish, *Anal. Chim. Acta* **52,** 405 (1970).
25. F. S. Grimaldi and M. M. Schnepfe, *Talanta* **17,** 617 (1970).
26. M. M. Schnepfe and F. S. Grimaldi, *Talanta* **16,** 1461 (1969).
27. M. E. Farago and P. J. Parsons, *Analyst (London)* **107,** 1218 (1982).
28. R. B. Wemyss and R. H. Scott, *Anal. Chem.* **50,** 1694 (1978).
29. P. Pille, P. R. Lowe, and L. R. P. Butler, "Analytical Chemistry in the Exploration, Mining and Processing of Materials," Abstr. No. 92, p. 533. IUPAC, Johannesburg, 1976.
30. R. Gijbels and J. Hoste, *Symp. Nucl. Anal. Tech.*, p. 81 (1971).
31. E. L. Hoffman, A. J. Naldrett, R. G. V. Hancock, A. Manson, and J. C. Van Loon, *Anal. Chim. Acta* **102,** 157 (1978).
32. A. Chow and F. E. Beamish, *Talanta* **13,** 539 (1966).
33. R. V. D. Robert, E. van Wyk, R. Palmer, and T. W. Steele, *Rep.—Natl. Inst. Metall. (S. Afr.)* **1371** (1971).
34. E. E. Bugbee, "A Textbook of Fire Assaying," 3rd ed. Wiley, New York, 1940.
35. A. D. Westland and F. E. Beamish, *Anal. Chem.* **26,** 739 (1954).
36. T. W. Oslinsky and N. H. Knight, *Appl. Spectrosc.* **22,** 532 (1968).
37. E. B. Sandell, "Colorimetric Determination of Traces of Metals," 3rd ed. Wiley (Interscience), New York, 1959.
38. M. M. Kruger and R. V. D. Robert, *Rep.—Natl. Inst. Metall. (S. Afr.)* **1432** (1972).
39. R. J. Coombes, A. Chow, and R. H. Flint, *Anal. Chim. Acta* **91,** 273 (1977).
40. R. J. Coombes and A. Chow, *Talanta* **26,** 991 (1979).
41. P. E. Moloughney and G. H. Faye, *Talanta* **23,** 377 (1976).
42. Bureau of Radiological Health and the Training Institute Environmental Control Administration, "Radiological Health Handbook," p. 522. U.S. Public Health Serv., Washington, D.C., 1970.
43. J. Haines and R. D. V. Robert, *S. Afr. J. Chem.* **37,** 121 (1984).

9

Biogeochemical Analysis

I. INTRODUCTION

Two biological approaches to prospecting for ores are biogeochemical and geobotanical methods. Geobotanical prospecting involves using the relationship of plant morphology and the distribution of species of plants for finding ores and minerals. (This subject will not be covered in this book.) Biogeochemical prospecting involves the chemical analysis of plant tissue for the purpose of finding ore bodies. In the latter case there is a rapidly growing literature which shows the great importance of biogeochemistry used together with other approaches to the location of ores.

Goldschmidt in the 1930s speculated that plant tissue analysis would be a useful approach to prospecting. Beginning in the 1950s and accelerating during the 1970s and 1980s, there have been many published reports on biogeochemical analysis methodology. Most of the very early work was on copper and zinc applications.

Several of the so-called micronutrient elements, e.g., copper, manganese, molybdenum, and zinc, are important ore-forming elements. Because these elements are required for plant health it was assumed that they would show useful biogeochemical activity. Hence, much of the early work was concentrated on these elements. More recently it has been shown that anomalous levels of elements such as gold, silver, arsenic, selenium, lead, nickel, and mercury can be found in elevated amounts in plants grown on soils containing these constituents. Thus, there has been much recent work on the application of such elements to biogeochemistry. In many cases, these (so-called) toxic elements have been better for outlining anomalies than the essential micronutrients.

The toxic elements frequently become lodged in the plant root system. However, sufficient quantities travel to other plant tissue to make their use quite feasible. Micronutrient elements travel readily to the tissues in which they are essential. For most elements it has been found that twigs of the previous year's growth are excellent for biogeochemical surveys. Evergreen needles have been very satisfactory for biogeochemical purposes. Deciduous tree leaves and leaves of lower plants have also proved to be useful in such work. However, bark and augered samples of wood are often unreliable in biogeochemical work. Generally, deep-rooted vegetation has been found to be superior to shallow-rooted material in biogeochemistry. It is important to use only a single sample species in a biogeochemical survey.

A. Choice of Methodology

It may be necessary to begin the sample preparation in the field. For example, if deciduous tree leaves are collected, they must be field-dried. If this is not done the leaves wilt and tend to become mold-covered in the sample bags and thus to spoil. Drying should be done by spreading out the sample on a noncontaminating support material in the sun. The same treatment should be used for leaves of lower plants and for needles. Sample containers should be noncontaminating.

The moisture content of plant material varies greatly. Generally, it is advisable to take a wet weight of at least 25 g to ensure an ash of more than 1 g. If it is desirable to report results on a wet weight basis (seldom done), several samples can be weighed while wet and then oven-dried and weighed once again. An average of these results can be used to make the conversion.

If dry ashing is used, twigs and stems should be cut into small pieces prior to treatment. Segments between 1.2 and 2.5 cm in length are satisfactory. (Care must be taken not to do this with a metal tool, which might cause contamination). Wet ashing requires that sample pieces not exceed 1 mm in extent, with 30 mesh being better. Obviously, therefore, hard tissue samples are seldom treated directly by wet ashing.

Leaf tissue can easily be reduced to a size suitable for direct wet ashing by freeze-drying, followed by grinding. The grinding must be done in a device which does not contaminate the sample with metals. Freeze-dried samples thus produced can be stored in appropriately sized wide-mouth plastic bottles (e.g., 25 to 100 ml capacity).

The reader is referred to sections in Chapter 4 on the virtues of wet versus dry ashing. In summary, the following can be stated. With dry ashing there is the possibility that volatile metals may be lost and/or that metals will become trapped in insoluble residues or on container walls. Even at temperatures below 500°C several volatile metals may be lost (e.g., arsenic, selenium, mercury). With wet ashing, the sample particle size must be small and the technique is comparatively labor-intensive. Unless perchloric acid is used (together with excess nitric acid for safety), metals may not be completely released from tissues. Except in the case of the very volatile metals, dry ashing at about 450°C is used in the authors' laboratory. For mercury, arsenic, and selenium a highly oxidizing wet ashing treatment is used to retain these very volatile elements.

The more recent literature suggests low-temperature oxygen ashing as a method to ensure a useful sample ash without loss of metals. In the authors' view this approach is not practical on the basis of time (too long) or size of sample (too small).

Even if dry ashing is employed, an acid treatment of the sample ash is essential if atomic absorption or plasma emission is used. For this purpose a mixture of nitric acid (added first) and hydrochloric acid (added after about 30 min of digestion) can often be employed. It is important to point out that such a procedure does not, in most cases, result in total metal recovery. However, in the majority of cases recoveries are greater than 80%, which is adequate in biogeochemical analysis. If a total digestion is essential, mineral acid mixtures together with hydrofluoric acid must be employed.

Fusions are seldom used in biogeochemistry because of their high labor intensity and the high salt contents of the sample solutions. An important exception to this rule is the use of a sodium (or potassium) hydroxide fusion of plant material for the analysis of arsenic and selenium. If a fusion is used the sample solution should be analyzed by hydride generation atomic absorption or plasma emission.

Sample solutions for biogeochemistry surveys are usually analyzed by atomic absorption or plasma emission spectrometry. If both instruments are available the choice is usually dependent on the number of elements which must be determined per sample. In this regard, the authors recommend that when more than three elements are to be determined per sample plasma emission spectrometry should be employed.

Some of the less abundant elements such as molybdenum may be present at levels which are below flame atomic absorption or plasma emission detection limits. In such cases furnace atomic absorption may be necessary. However, this technique is very time-consuming and should not be used unless a definite need is demonstrated.

II. DETERMINATION OF METALS IN VEGETATION BY FLAME ATOMIC ABSORPTION SPECTROSCOPY (1)

Comments on the Method

Wet or dry ashing can be done as a first step in the decomposition of organic samples. Dry ashing is generally less labor-intensive and is preferred where applicable. However, there is danger of loss of some of the more volatile metals at the elevated temperatures required.

Dry Ashing Procedures

The following procedure is commonly used for plant materials including bark, wood, leaves, stems, seeds, fruit, roots, and highly organic soils. Dry ashing is done at 450°C. At this temperature mercury will be lost. Arsenic and the other chalconides will most likely be lost in varying amounts. Other volatile metals such as lead, zinc, and cadmium are normally retained. When appreciable siliceous matter is present, e.g., in leaves and stems of many plants, hydrofluoric acid must be included in the acid digestion mixture.

The main advantages of a dry ashing procedure are that large sample sizes can be used and perchloric acid need not be employed. As emphasized elsewhere, this acid requires the use of a special fume hood and has been known to cause violent explosions when handled improperly.

The sample should be broken into small pieces. Although 30 mesh is best, up to 1 mm lengths are permissible. Place the desired sample weight in a borosilicate beaker of appropriate size. Place in a large cool oven. Begin heating at a very low rate. The temperature should rise from room temperature to 450°C in 6–8 h. Heat at 450°C for 2 h or until ashing is complete. Then use the aqua regia method described in the next section. If no hydrofluoric acid is necessary, decomposition can be done in the original beaker. If hydrofluoric acid is required, use Teflon dishes.

Wet Ashing Procedure

This procedure is not applicable to all sample types and all metals at all concentrations. It is designed to be a good procedure for a wide range of metals in many of the plant materials but not in twigs.

Two basic approaches to decomposition, both involving nitric acid, are given below. In one perchloric acid is employed, in the other aqua regia. For samples containing appreciable silica, hydrofluoric acid can be included in either mixture.

Methods of dissolution involving perchloric acid are usually best. However, many laboratories are not equipped with proper fume hoods to allow use of this acid.

A worker contemplating use of perchloric acid should consult the supplier on precautions to be observed. The following is a short list of the most important rules.

1. Nitric acid should always be present in excess with perchloric acid at the beginning of a decomposition. Perchloric acid should never be used without nitric acid.
2. When very easily oxidizable matter is present, a digestion with nitric acid alone should precede the addition of perchloric acid.
3. A stainless steel fume hood with facilities for washing down the exhaust vents and the complete hood interior should be used. There have been many reports of workers using perchloric acid for years in ordinary fume hoods without incident. Then, suddenly, a serious explosion occurred due to the accumulation of perchlorates in the venting mechanism.

The aqua regia (1 : 3 nitric acid : hydrochloric acid) method does not give as strongly oxidizing conditions as fuming perchloric acid. However, in many instances, this mixture leaches over 95% of the heavy metals from an organic matrix.

The sample sizes cited below may be altered. Samples up to 0.5 g can be handled in 100 ml beakers or Teflon dishes. Acid volumes should be increased but not proportionately; e.g., a 1 g sample requires 25 ml nitric acid and 10 ml perchloric acid.

Dissolutions can often be done in test tubes. Evaporations are difficult in these tubes and, hence, the strongly oxidizing conditions obtained by fuming perchloric acid are not obtained. However, for easily dissolved samples, test tubes are satisfactory. Heating of test tubes can be done in a sand bath, aluminum block, or boiling-water bath. The last is the least effective.

The following metals can be analyzed after the above decompositions: copper, manganese, iron, cobalt, nickel, chromium, zinc, cadmium, lead, bismuth, vanadium, molybdenum, antimony, and silver.

In the case of silver, chloride can invalidate the results. As a rule, silver contents of solutions up to 0.5 ppm can be analyzed without problem in the presence of chloride. When silver is to be measured at parts per million and higher levels, hydrochloric acid must be avoided. Lead, unless in very high concentration, is not precipitated by chloride. Sulfuric acid is avoided in the procedures to prevent formation of lead sulfate.

Agemian and Chau (2) report loss of chromium as the volatile chromyl chloride when perchloric acid is fumed. Hence, aqua regia is preferred for this element.

For most metals, the following method is good only down to the microgram per gram level. It is of great importance to employ background correction, particularly when working near the detection limit.

A lean air/acetylene flame must be employed to avoid interferences in the determination of chromium by several base metals. This, unfortunately, also degrades the detection limit. A nitrous oxide/acetylene flame obviates the interference problem due to base metals.

Several of the base metals interfere significantly with nickel. Again, a lean flame minimizes this problem.

A number of elements interfere in a molybdenum analysis. Preparation of samples and standards containing 1000 ppm aluminum overcomes this problem.

Aluminum interferes in the determination of vanadium. It is best to prepare samples and standards to contain 1000 ppm aluminum. A fuel-rich nitrous oxide/acetylene flame must be used.

The procedure has been tested on a variety of botanical standard reference materials. Percentage recoveries depend on the element and the sample type. In general, the aqua regia method gives 80–100% recoveries, compared to better than 95% by the perchloric method.

Reagents and Equipment

Perkin-Elmer 303, 503, and 603 atomic absorption spectrometers are used in the authors' laboratory. The experimental parameters given in Table 9.1 are related to this equipment.

Reagent grade concentrated acids are generally adequate for decompositions prior to flame atomic absorption work.

Table 9.1
Instrumental Parameters[a]

Element	Analytical line (nm)	Slit width (nm)	Flame type[b]	Lamp source[c]
Cr	357.9	0.7	A/A (lean)	HCL
Mn	279.5	0.2	A/A	HCL
Fe	248.3	0.2	A/A	HCL
Co	240.7	0.2	A/A	HCL
Ni	232.0	0.2	A/A	HCL
Cu	324.7	0.7	A/A	HCL
Zn	213.9	0.7	A/A	HCL
Cd	228.8	0.7	A/A	HCL
Pb	283.3	0.7	A/A	HCL
Bi	223.1	0.2	A/A	HCL
V	318.4	0.7	N/A (rich)	HCL
Mo	313.3	0.7	N/A (rich)	HCL
Sb	217.6	0.2	A/A	HCL
Ag	328.1	0.7	A/A	HCL

[a]All elements not indicated otherwise are run with lean flames. In the case of chromium this condition gives poorer detection limits but minimizes interferences with air/acetylene flames. Nitrous oxide/acetylene flames are often used for chromium.
[b]A/A, Air/acetylene; N/A, nitrous oxide/acetylene.
[c]HCL, Hollow-cathode lamp.

Calibration standards are made by diluting 1000 μg/ml stock solutions prepared from pure metal or metal salts dissolved in a minimum amount of the appropriate acid. Stock and calibration standards must contain 1% acid. Multielement standards may be used for most elements.

Procedure

Method 1

Perchloric, nitric, and hydrofluoric acids. Weigh a 0.2 g sample into a 100 ml beaker. Add 10 ml nitric acid. Heat on a hot plate set at medium heat (just below boiling) for 30 min. Top up with nitric acid to 10 ml and add 4 ml perchloric acid. Using medium heat, evaporate to dryness in a perchloric acid fume hood. Add 1 ml of nitric acid and warm. Add 5 ml of water and filter into a 25 ml flask. Cool and dilute to volume. Make appropriate dilutions, as required, keeping the acid content to 1%.

If the sample contains siliceous material, the following step should be inserted after addition of 1 ml of nitric acid.

Wash the material from the beaker with a minimum of water into a 100 ml Teflon dish. Scrub the beaker walls with a plastic stirring rod and rinse. Evaporate to dryness. Add 2 ml hydrofluoric acid, 1 ml nitric acid, and evaporate to dryness. Add 1 ml nitric acid and warm. Proceed as above.

Method 2

Nitric, hydrochloric, and hydrofluoric acids. Weigh a 0.2 g sample into a 100 ml beaker. Add 5 ml nitric acid and 15 ml hydrochloric acid. Place a watch glass over the beaker and digest at medium heat for 60 min. Evaporate to dryness. Add 1 ml nitric acid and warm. Add 5 ml of water and warm and filter into a 25 ml flask. Cool and dilute to volume. Make appropriate dilutions as required, maintaining the acid content at 1%. If the sample contains appreciable siliceous material, do the above dissolution in Teflon vessels. Insert the hydrofluoric acid procedure above in place of the perchloric acid procedure.

Analysis of Solutions

Run samples and standards using the experimental conditions listed in Table 9.1. Standards should be repeated every 10–20 samples.

It is well to analyze a standard reference sample, e.g., NBS Orchard Leaves, with each sample set. The reference sample is run together with samples and calibrating standards.

Detection Limits

In general, detection limits below 0.2 ppm (solution) are not obtainable.

III. DETERMINATION OF NICKEL AND ZINC IN PLANT MATERIAL BY FLAME ATOMIC ABSORPTION (3)

Comments on the Method

This procedure was used for the determination of nickel and zinc in European species of *Thlaspi* L. (Cruciferae). Levels of nickel and zinc expected were above 1000 μg/g dry weight.

A dry ashing at 500°C is employed. This is followed by dissolution in 2 ml of 2 *M* hydrochloric acid. Thus it is not likely that all the zinc and nickel are extracted. Determinations are made by flame atomic absorption spectrometry using background correction. No precision or accuracy assessment was made. Detection limits were 2–6 μg/g dry ash for nickel.

Reagents

The 2 *M* hydrochloric acid was prepared by dilution of redistilled analytical grade concentrated hydrochloric acid (no instructions for preparation of standards were given).

Equipment

A Varian Techtron AA5 flame atomic absorption spectrometer was used. The instrument was equipped with a background correction system.

Procedure

Ash enough sample at 500°C to give an ash of at least 20 mg. Accurately weigh from 5 to 20 mg into appropriate size borosilicate test tubes (e.g., 25 ml capacity). Add 2 ml of 2 *M* hydrochloric acid. Place the test tubes in a heating apparatus (e.g., an aluminum block or a water bath) and heat gently for 30 min. Filter the solutions into 10 ml volumetric flasks. Wash with 2 *M* hydrochloric acid and dilute to volume with 2 *M* hydrochloric acid. Analyze the solutions for zinc and nickel by flame atomic absorption using experimental conditions recommended by the instrument manufacturer. Background correction should be employed. Use an appropriate concentration range of nickel and zinc standards prepared in 2 *M* hydrochloric acid.

IV. DETERMINATION OF ARSENIC AND SELENIUM IN PLANTS BY HYDRIDE GENERATION (ATOMIC ABSORPTION) (4, 5)

Comments on the Method

Samples are digested in a ternary acid solution of nitric, perchloric, and sulfuric acids. The hydrides are generated by means of sodium borohydride. The hydrides are carried directly into a shielded, nitrogen-diluted hydrogen diffusion flame. Several NBS standard reference materials (orchard and tomato leaves, spinach, and pine needles) were analyzed, and there were good agreements with accepted values. The same procedure was used for analyses of animal tissues.

Apparatus

A Perkin-Elmer model 403 atomic absorption spectrometer equipped with three 3-slot, 10 cm Boling-type burner head was used. Electrodeless discharge lamps were used for arsenic and selenium. The operating conditions for the instrument are listed in Table 9.2.

Table 9.2
Operating Parameters

Element	As	Se
Source	EDL[a]	EDL
Source power or current, W	8	6
Slit width, mm	3	3
Spectral bandwidth, nm	2.0	2.0
Wavelength, nm	197.2[b]	196.0

Reprinted with permission from Ref. 4. Copyright 1976 American Chemical Society.

[a]EDL, Electrodeless discharge lamp.

[b]The 193.7 nm resonance line of As is about twice as sensitive as the 197.2 nm line and can be employed with some sacrifice of linear range.

The flame is protected from room air currents by the shield illustrated in Fig. 9.1. The shield improved precision approximately twofold.

The hydride generator is shown in Fig. 9.2. The reaction vessel is a 20 cm heavy-wall test tube. It is connected to a reaction head made of acrylic plastic. The head incorporates two Teflon metering valves, a vent tube, and a rubber stopper. A four-cam, 90 s/rev motor-driven timer is connected to two Teflon-bodied, relay-actuated solenoid valves. Nitrogen is controlled by a single-stage low-pressure gas regulator. Reagents are contained in two 1 liter reservoirs.

The gaseous reaction products from the generator are delivered to the burner via latex rubber tubing and into the burner mixing chamber. The nebulizer is replaced by a single glass adapter.

Reagents

- Sodium borohydride ($NaBH_4$) solution, 4%: Dissolve 4 g of powder or pellets in 50 ml of 10% (w/v) sodium hydroxide. Dilute to 100 ml with 10% sodium hydroxide.

Figure 9.1. Flame shield for Boling burner head. (Reprinted with permission from Ref. 4. Copyright 1976 American Chemical Society.)

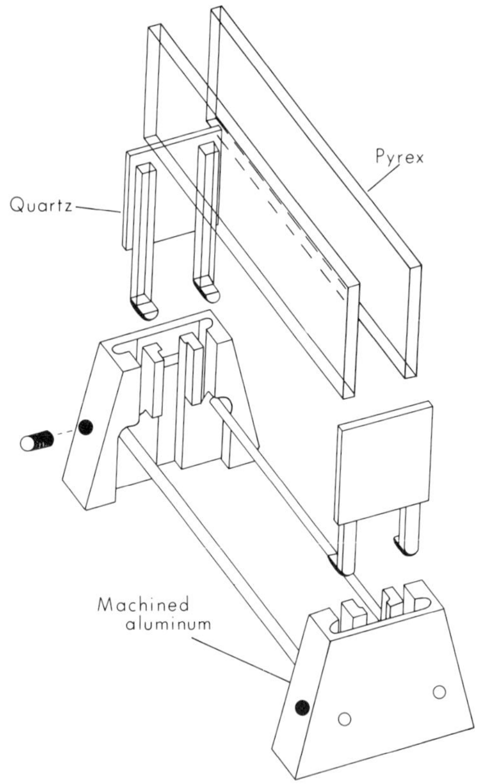

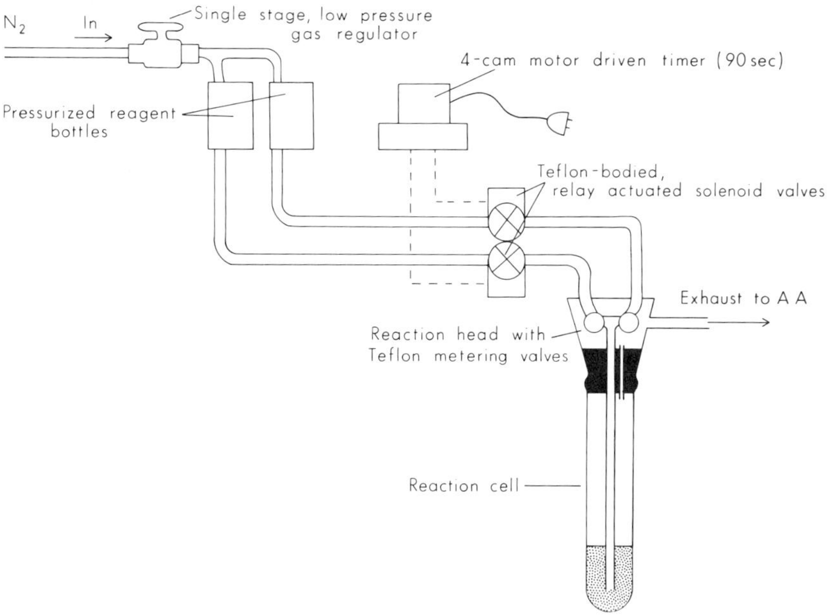

Figure 9.2. Hydride generator. (Reprinted with permission from Ref. 4. Copyright 1976 American Chemical Society.)

- Sodium iodide solution, 10% (w/v).
- Acid diluent: Add 50 ml of sulfuric acid to 300 ml of water. Then add 300 ml of hydrochloric acid and dilute to 1 liter with water. (The water must be distilled and deionized.)

Procedure

Instrumental conditions

Stopper the burner drain tube. Adjust the hydrogen flow rate to 2 liter/min (69 kPa) and the nitrogen flow rate to 6 liter/min (200 kPa). Connect the nitrogen to the auxiliary oxidant port. Light the flame and adjust the position of the burner head to a height slightly below the optical path.

Typical flow rates for $NaBH_4$ solution and distilled water wash are 24–32 ml/min and 20–40 ml/min, respectively. Adjust the flow rates by means of the needle valves in the reaction head and the pressure on the reservoirs (28 kPa). Use the timer to open the solenoid valve on the $NaBH_4$ line for 15 s and that on the water line for 5–15 s. The volume of $NaBH_4$ solution (6–8 ml) yields well-defined and reproducible peaks for peak height measurements. No background correction is required.

Sample preparation

Transfer 1–3 g dry weight of sample into a 100 ml Kjeldahl flask. Add 30 ml of ternary acid solution. This consists of nitric, sulfuric, and perchloric acids in the ratios 4 : 1 : 1. Prepare a blank with each set of 10–12 samples. Carry out the oxidation with a micro-Kjeldahl digestion rack equipped with a glass manifold, water-cooled condenser, receiving flask, and 10% (w/v) sodium hydroxide scrubbing tower. Heat the sample very gently until the foaming subsides. Then raise the temperature so that steady boiling results. Avoid charring the sample. If charring occurs, add 1 ml volumes of nitric acid, or cool the sample rapidly to avoid losses. Boil off the perchloric acid, and heat to strong fumes of sulfuric acid. All organic material must be oxidized so that quantitative release of the hydrides can be obtained. The digestion requires 1.5–2 h. Cool the digest, and transfer it to a 100 ml volumetric flask. Add 30 ml of hydrochloric acid, and dilute to 100 ml with water.

Determination of arsenic

Transfer an aliquot (maximum 20 ml) of the diluted digest to a reaction test tube. If necessary, dilute to 20 ml with the acid diluent. Add 0.5 ml of sodium iodide reductant solution. Wait 1 min. Attach the test tube to the reaction head, and start the timer. Record the transient signal.

Prepare standards containing 0.05–0.07 μg arsenic from a 10 μg/ml stock solution prepared each day in 30% (v/v) hydrochloric acid. Dilute each to 20 ml with the acid reductant. Follow the same procedure as for the samples.

Determination of selenium

Follow the same procedure as described for arsenic. Do not add sodium iodide solution.

Table 9.3
Analysis of NBS Reference Materials

Material	ppm (dry weight)	
	Hydride	NBS (or literature value)
Orchard Leaves, SRM 1571		
As	11 ± 1	11 ± 2
Se	0.055 ± 0.009	0.08 ± 0.01
Tomato Leaves, SRM 1573		
As	0.29 ± 0.02	0.27 ± 0.05
Se	<0.1	0.06 ± 0.02
Spinach, SRM 1570		
As	0.16 ± 0.01	0.15 ± 0.05
Se	<0.1	0.039 ± 0.015
Pine Needles, SRM 1575		
As	0.19 ± 0.03	0.21 ± 0.04
Se	<0.1	0.049 ± 0.004

Source: Ref. 5.

Accuracy

The calibration curves are linear to about 0.5 μg of arsenic and antimony. The results of analyses of several NBS standards are shown in Table 9.3. The detection limits based on 1 g of digested sample are 10–20 ng/g for each element. At levels of 0.10 μg, the measurement precision is 1–2% relative standard deviation.

V. DETERMINATION OF MOLYBDENUM IN BOTANICAL MATERIAL BY GRAPHITE ROD ATOMIC ABSORPTION (6)

Comments on the Method

A sample is digested in nitric acid. The filtered, diluted solution is analyzed by electrothermal atomic absorption spectrometry. Recovery of molybdenum from NBS Orchard Leaves was excellent. The relative standard deviation was 10% for a concentration of 0.50 μg per gram of molybdenum in plant material.

Apparatus

A Varian-Techtron model AA6 atomic absorption spectrometer equipped with a model 90 carbon rod atomizer was used. The instrument settings were:

Spectral bandpass: 0.5 nm

Wavelength: 313.3 nm

Dry: 90°C/30 s

Ash: 700°C/30 s

Atomize: 2750°C/3 s

A Varian A-25 recorder with a range of 2 mV for measurements of peak heights was also used.

Reagents

Prepare a solution containing 1000 mg/liter of molybdenum from ammonium molybdate. Use this solution to prepare standards containing 0, 10, 20, 30, and 40 ng/ml molybdenum.

Procedure

Dry the plant material, and grind it in a stainless steel Willey mill to pass a 40 mesh screen. Transfer a 2.000 g sample to a 250 ml Erlenmeyer flask. Add 20 ml of 7 *M* nitric acid (Baker Ultrex). Heat the mixture to a gentle boil. After 20 min, cool the sample and filter it through a Whatman No. 42 filter paper. Collect the filtrate in a 100 ml flask. Dilute to volume with water.

Use 5 ml aliquots of solution for analysis according to the conditions listed above.

Treat the standards in the same manner as the samples. The calibration curve is linear over this range of concentrations.

Accuracy and Precision

The molybdenum content of NBS Orchard Leaves determined by multiple additions was 0.39 ± 0.03 μg/g. A value of 0.36 ± 0.04 μg/g was obtained from standards. The relative standard deviation was 10.4% over a period of 1 year (for this standard). (The certified value is 0.3 ± 0.1 μg/g.)

VI. DETERMINATION OF MERCURY IN PLANT SAMPLES BY COLD VAPOR ATOMIC ABSORPTION (7)

Comments on the Method

The following procedure was used to determine mercury in a variety of oven-dried (80°C) plants. These included *Abies lasiocarpa, Pinus contorta, Picea glauca, Populus tremuloides, Populus trichocarpa, Salix* sp., *Achillea millefolium, Castilleja miniata, Epibolium auqustifolium, Shepherdia canadensis,* and *Taraxacum officinale.* A variety of plant organs from each species were analyzed (i.e., roots, flowers, stems, needles, or leaves). Variations in mercury contents between the species and organs were relatively small. The biggest variation was noted between plants growing in mineralized and nonmineralized areas. Thus, the authors conclude that local vegetation can be analyzed without much need for worry about species chosen.

The samples are oven-dried at no greater than 80°C. Dissolution is accomplished in capped vials using concentrated nitric acid alone. Sodium borohydride is used as the reducing agent to produce elemental Hg vapor.

Reagents

A 3% solution of sodium borohydride ($NaBH_4$) was prepared in 1% sodium hydroxide. Solutions of 5% potassium permanganate, 10% Antifoam B, and 1.5% nitric acid were also prepared for use in the reaction vessel. (No details were given on the preparation of mercury stock or calibrating solutions.)

Equipment

A Perkin-Elmer model 603 spectrometer was used. It was equipped with a Perkin-Elmer NHS-10 cold vapor device, a Perkin-Elmer model 56 recorder, and a PRS-10 printer sequencer.

Procedure

Dry samples at 80°C or below in a well-ventilated oven. Grind the dried product to −40 mesh in a steel mill. Wash all root samples thoroughly in an ultrasonic bath prior to drying.

Weigh 0.100 g sample into a scintillation vial equipped with a plastic-lined lid. Add 1 ml concentrated nitric acid and close the vial. Allow it to stand overnight at room temperature to predigest most of the organic matter. Wash down the walls with an additional 2 ml of concentrated nitric acid. Heat the samples in the closed vials at 80°C for 15 min. Further dilutions of samples containing more than 10 μg/g mercury are required.

Add sample solutions (10–100 ml) to 10 ml 1.5% nitric acid in the reaction vessel. Add 5 drops of 10% Antifoam B solution and 1–10 drops of nitric acid to ensure nonfoaming of the oxidized sample solution. Add sufficient 3% $NaBH_4$ solution to fully reduce the mercury in the sample (triplicate readings were made which showed a variation of 10%).

VII. DETERMINATION OF GOLD IN VEGETATION BY GRAPHITE ROD ATOMIC ABSORPTION (8)

Comments on the Method

A sample is decomposed in fuming nitric acid. The gold is extracted from a hydrochloric acid solution into 4-methylpentan-2-one. Iron is removed, and gold in the organic layer is measured by electrothermal atomic absorption spectroscopy. The lower limit of detection was 0.2 ng of gold in a sample weight of 1 g. There was reasonable agreement in the gold content of an NBS standard vegetation sample.

Equipment

- Varian-Techtron AA-5 spectrometer equipped with a deuterium lamp background corrector.
- Varian-Techtron model 63 carbon rod atomizer. The operating conditions are shown in Table 9.4.

Procedure

Use a finely ground sample of plant material (leaves, stalks, twigs, etc.). Transfer 1 g to a 50 ml borosilicate beaker. Add 10 ml of fuming nitric acid. Warm the sample on a water bath. Dissolution takes place in about 5 min. Then add 8 ml of hydrochloric acid and a little bromine vapor. Evaporate the sample to a volume of about 3 ml on the water bath. Transfer the sample to a 20 ml glass centrifuge

Table 9.4
Operating Conditions for Carbon Rod Atomizer

	Temperature (°C)	Time (s)
Dry	120	30
Ash	400	15
Atomize	2000	2

Source: Ref. 8

tube, and add 6 *M* hydrochloric acid to a final volume of 5 ml. Then dilute to 15 ml with water to reduce the acid concentration to about 2 *M*.

Add 1 ml of 4-methylpentan-2-one (methyl isobutyl ketone, MIBK). Stopper and shake the tube for 2 min. Allow the layers to separate. Remove all but 4 ml of the aqueous phase by means of a pipet. Add distilled water to restore the volume to 15 ml. Shake the sample for 2 min. Centrifuge the sample for 1 min to clarify the MIBK. The organic layer will be about 0.55 ml.

Measure the gold contents of 10 μl aliquots of the MIBK layer according to the instrument conditions in Table 9.4. Carry standards and a blank through the same procedure.

The limit of detection is 0.2 ng of gold in a sample weight of 1 g.

Precision and Accuracy

Gold content of 1 g sample	RSD
170 ng	10.1%
5 ng	25%

The gold content of NBS standard Orchard Leaves was 4.0 and 4.5 ± 1 ng/g as compared with a value of 1.8 ± 1 ng/g obtained by neutron activation.

VIII. ANALYSIS OF PLANT MATERIALS BY ICP–AES (9)

Elements Determined

Mn, Zn, Cu, Mo, Ni, Pb, Co, Cr, Cd

Comments on the Method

The authors have been unable to find a suitable ICP–AES procedure meant specifically for the analysis of plant materials for exploration geochemistry. The following procedure developed in the authors' laboratory can be used for this purpose but was developed to allow total decomposition of plants. The decomposition is carried out in a Teflon bomb. For geochemical prospecting purposes, a Teflon open dish would suffice. Two decomposition approaches are used (methods 1 and 2), both involving hydrofluoric, perchloric, and nitric acids. Hydrofluoric acid was found essential for most plant tissue because of the presence of relatively high levels of silicon. Method 2 is more rapid than method 1, involving no evaporation, and could probably be used for most geochemical prospecting surveys.

The method was tested on NBS standard reference Orchard Leaves and NIES (Japan) standard reference *Chlorella* and Pepper Bush. Acceptable agreement with the certified values was obtained.

Reagents

Reagent grade (Baker Analyzed) chemicals were found to be pure enough for the trace elements covered in the proposed procedures. Stock 1000 ppm standard solutions were prepared from metals or metal salts. A 1 : 1 mixture of perchloric : nitric acids was added to stock and working solutions to give a 6% final acid content.

Equipment

An Applied Research Laboratories model 34000 was employed. Wavelengths and operating parameters used are given in Tables 9.5 and 9.6, respectively. A commercially available pressure cooker suitable for home cooking was employed. Teflon bombs were made at the University of Toronto from Teflon available from Canplas Industries Ltd. (Ontario, Canada).

Figure 9.3 is a photograph of the Teflon digestion vessel used. These are cleaned using a concentrated nitric acid leach followed by potassium permanganate treatment, which in turn is followed by hydrochloric acid cleaning to remove residual manganese oxide.

Procedures

Method 1

Weigh about 0.5 g of dried or 0.8–1 g of wet plant tissue and transfer it to a Teflon vessel. Add 4 ml of aqua regia (1 ml of concentrated nitric acid and 3 ml of concentrated hydrochloric acid). Keep at room temperature in the closed vessel for about 0.5 h. Heat in the oven at 100°C for 2–3 h or in a pressure cooker for 1 h. Evaporate the digest on a hot plate to about 1 ml. Cool and add 3 ml of concentrated

Table 9.5
Spectral Lines Used

Element	Line (nm)	Order
Al	308.22	2
Ag	328.07	2
As	189.04	3
Ca	317.93	2
Cd	226.50	3
Co	228.62	3
Cr	267.72	2
Cu	324.75	2
Fe	259.94	2
Mg	279.08	2
Mn	257.61	3
Mo	202.03	3
Ni	231.60	3
P	178.28	4
Pb	220.35	3
Sb	217.59	3
Se	196.03	3
Zn	213.86	3

Source: Ref. 9.

Table 9.6
Operating Parameters

Plasma and nebulizer premix chamber
Argon flow rates, liter/min
Coolant gas, 12
Plasma gas, 0.8
Carrier gas, 1.0
Nebulizer
Concentric pneumatic with automatic tip wash
Sample uptake rate, 2.5 ml/min
Viewing height of plasma above coil, 15 mm
RF generator
Frequency 27.12 MHz
Operating power, 1.2 kW
Reflected power, <3 W
Instrument
ARL model ICPQ 34000
Grating spectrometer that has 32 fixed channels and can be evacuated
Wavelength-scanning accessory consisting of a moving primary source
Variable channel comprising a 0.25 Spex monochromator also available
Optics
Laser-ruled tripartite concave Al-coated on SiO_2 blank
1080 lines/mm
Metal–dielectric–metal narrow bandpass filters for order sorting
Blaze angle, 600 nm (first order)
Resolving power, 43,200 (first order)
Reciprocal linear dispersion, 0.926 nm/mm (first order)
Four orders used
Primary slit, 20 μm
Secondary slits, 35–50 μm

Source: Ref. 9.

nitric acid and 1 ml of concentrated perchloric acid. Heat again in the oven or in the pressure cooker for 0.5 h and evaporate the digest to white fumes. After cooling, add 1 ml of concentrated hydrofluoric acid and evaporate the sample to near dryness. (In case of a high silica content the volume of hydrofluoric acid must be increased.) Dilute using 10 ml of a 6% (v/v) mixture of 1 : 1 nitric acid : perchloric acid and heat on the hot plate for about 5–10 min. Transfer the solutions to 25 ml volumetric flasks and adjust the volume with the 6% acid solution. Store samples in polyethylene bottles. Run three blanks with each series of samples. Prepare calibration standards to have the same acid composition as the samples.

Method 2

Weigh 0.3–0.4 g of wet sample into a Teflon vessel. Add 4 ml of concentrated nitric acid, 1 ml of concentrated perchloric acid, and 0.5 ml of concentrated hydrofluoric acid. Allow the mixture to react in the closed vessel at room temperature for about 0.5 h. Heat in the oven at 140°C for about 2–2.5 h or in the pressure

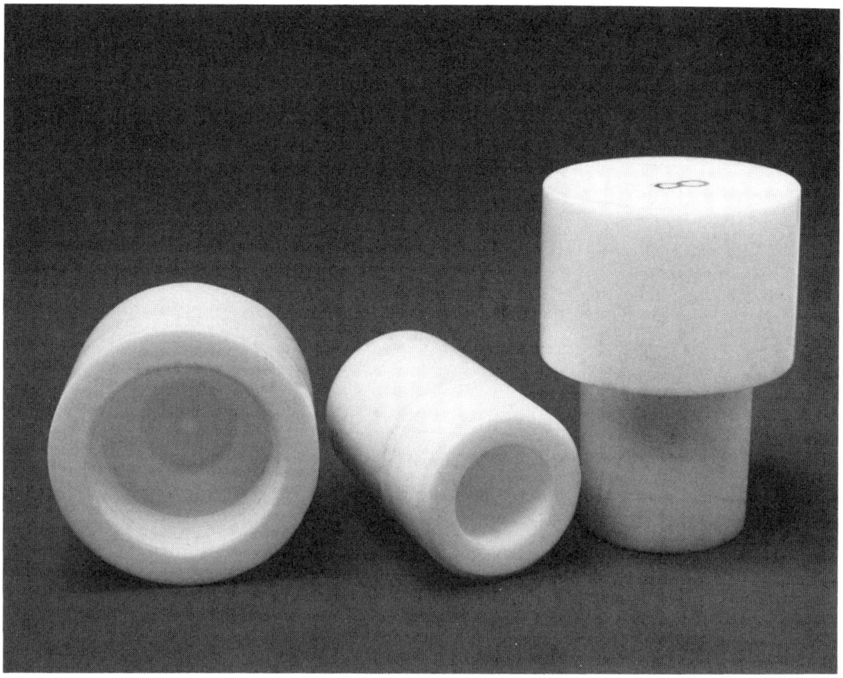

Figure 9.3. Teflon closed-tube digestion vessel (9).

cooker for 1.5 h. After cooling, transfer the sample to a 100 ml volumetric flask and adjust to volume with distilled water. Store the samples in polyethylene bottles. Run three blanks with each sample set. Prepare calibration standards to have the same acid composition as the samples.

Note that hydrofluoric acid is necessary for complete dissolution of plant material because of the presence of silica. The accuracy and precision were assessed using the samples listed in Table 9.7.

Method 1 is much more time-consuming than method 2 and is subject to contamination problems. The latter should be used when applicable. It is only when very low levels are encountered (e.g., nickel in these samples) that the greater dilution which is required in method 2 invalidates its use.

IX. X-RAY FLUORESCENCE ANALYSIS OF VEGETATION SAMPLES

A. Introduction

Aspects of wet and dry ashing of vegetation samples which influence the recovery of minor and trace elements have been presented in the previous sections. Oxidation of samples may not be necessary if the elements are determined by x-ray

Table 9.7
Results of Analysis of Plant Tissue (Micrograms per Gram of Dry Matter)

Element	Orchard leaves, NBS 1571			Chlorella No. 3, NIES[a], Japan			Pepperbush No. 1, NIES, Japan		
	NBS value	Method 1	Method 2	NIES value	Method 1	Method 2	NIES value	Method 1	Method 2
Ca	20,900 ± 300	20,680 ± 500	19,980 ± 200	4200–5560	5000 ± 200	4600 ± 150	13,800 ± 700	12,000 ± 500	12,900 ± 500
Mg	6200 ± 200	6300 ± 220	6150 ± 150	2900–3800	3200 ± 300	3200 ± 200	4000 ± 200	3200 ± 400	3850 ± 150
P	2100 ± 100	1990 ± 120	1980 ± 50	14,500–18,200	1600 ± 400	15,600 ± 400	1100	1060 ± 25	1085 ± 25
Al	(409)[b]	408 ± 26	356 ± 10	98–210	324 ± 30	192 ± 26	—	419 ± 40	376 ± 35
Fe	300 ± 20	460 ± 45	285 ± 12	1500–2100	1950 ± 150	1700 ± 150	205 ± 17	275 ± 35	220 ± 28
Mn	71 ± 4	89.6 ± 5	89 ± 2	54.1–76	63.6 ± 5	62 ± 4	2030 ± 170	1685 ± 200	1780 ± 180
Zn	23 ± 4.5	25 ± 3	26± 2.5	6.3 ± 26.6	20.0 ± 3	19.5 ± 2.5	340 ± 20	355 ± 25	336 ± 20
Cu	12 ± 1	12 ± 1.5	13.0 ± 0.8	3.4–4.9	2.0 ± 0.6	2.2 ± 0.5	12 ± 1	9.7 ± 1.5	10.8 ± 1.2
Mo	0.3 ± 0.1	1.0 ± 0.4	0.3 ± 0.2	—	1.6 ± 0.4	1.2 ± 0.4	—	1.6 ± 0.4	1.4 ± 0.3
Ni	1.3 ± 0.2	1.3 ± 0.2	ND[c]	0.4–0.84	0.8 ± 0.3	ND	8.7 ± 0.6	17.4 ± 2.5	1.6 ± 1.2
Pb	45 ± 3	42 ± 3.8	48 ± 3	—	1.5 ± 0.8	1.3 ± 0.6	(5.5)	2.2 ± 1	3.5 ± 1.1
Co	(0.2)	0.2 ± 0.1	0.1 ± 0.1	0.65–1.2	1.2 ± 0.5	1.1 ± 0.4	23 ± 3	17.4 ± 4	19.8 ± 4
Cr	2.6 ± 0.3	2.6 ± 0.4	1.8 ± 0.5	—	1.7 ± 0.5	1.6 ± 0.4	(1.3)	1.2 ± 1	1.2 ± 0.8
Cd	0.11	0.1 ± 0.1	ND	0.025–0.055	0.1 ± 0.1	ND	6.7 ± 0.5	5.3 ± 1.2	6.0 ± 0.7
As	10 ± 2	15.3 ± 4	11 ± 4	0.11 ± 0.5	1.8 ± 0.8	ND	—	16 ± 1.2	10 ± 1.4
Se	(0.008)	ND	ND	—	—	—	—	—	—
Sb	2.9 ± 3	4.5 ± 3.2	3.5 ± 2.5	0.017–0.8	ND	ND	—	1.5 ± 1.6	1.6 ± 1.4
Ag	—	ND	ND	—	1.2 ± 0.7	1.0 ± 0.6	—	ND	ND

Source: Ref. 9.

[a]National Institute for Environmental Studies, Tokyo, Japan.

[b]Numbers in parentheses are not certified, given for information.

[c]ND, Not detected.

spectroscopy. Many workers have shown that the instrumental methods have sufficient sensitivity that analyses can be performed on the dried materials. The dried vegetation is ground to a fine powder and pressed to form small pellets. This usually means a saving in sample preparation time that can be significant when large numbers of samples must be analyzed. Another advantage is that standards can be prepared by adding measured amounts of solution standards to pure cellulose powder.

A detailed study of nine elements of low atomic number (less than 20) and four of high atomic number has been made by Norrish and Hutton (10, 11). An example of an application to mineral exploration was reported by Smith and Ball (12), who showed that determinations of calcium, barium, and lead in vegetation delineated a barite/fluorite/galena/sphalerite ore body.

B. Determination of Low Atomic Number Elements (10)

Elements Determined

Ca, K, Cl, S, P, Si, Al, Mg, Na

Comments on the Method

Pellets are formed from the dried, ground samples. Matrix and background corrections are measured and applied. Calibrations are made using previously analyzed plants or synthetic cellulose samples. Results of analyses of 13 agricultural samples were compared with average chemical values obtained in five laboratories. Agreements were good.

Instrument and Equipment

- The instrument used was a Phillips model PW1540 all-vacuum spectrometer. The instrument was equipped with 1 kW x-ray tubes, a sign θ amplifier, and a dead-time corrector. Details of instrument conditions and counting times are contained in Table 9.8.
- A ring and puck mill equipped with 10 cm^3 or 100 cm^3 vials for grinding samples
- Hydraulic press

Sample Preparation

Grind subsamples of the dried vegetation in a mill. Use 0.5–1 g of material in a 10 cm^3 vial or 2.5 g in a 100 cm^3 vial. A time of 2 min is sufficient for many samples, but an 8 min grinding time may be necessary for harder materials.

About 0.1 g sample is sufficient for a 2.5 cm disk. The pressure applied is 10–20 tonne. A spongy pellet can be the result of insufficient grinding or pressure.

Calibration standards are prepared from previously analyzed vegetation, or from cellulose powder. After solutions are added to the dry, weighed cellulose powder, the standard is dried and ground up again.

Table 9.8
Instrumental Conditions for Determination of Low Atomic Number Elements[a]

Element	Crystal		Counting time[b] (s)	To give (counts/%)
Ca	Ge	(111)	2.4	10^5
K	Ge	(111)	4	10^5
Cl	Ge	(111)	7.4	4×10^4
S	Ge	(111)	8.7	4×10^4
P	Ge	(111)	7.8	4×10^4
Si	PE	(002)	13	2×10^4
Al	PE	(002)	30	4×10^4
Mg[c]	ADP	(101)	21	2×10^3
Mg	TlAP	(001)	30	10^4
Na[d]	KAP	(001)	73	10^3
Na	RbAP	(001)	30	10^3
Na	TlAP	(001)	22	2×10^3

Source: Ref. 10. Reproduced by permission of John Wiley and Sons Limited.

[a]Cr x-ray tube, 44 kV, 20 mA; 480 μm collimator. Flow-proportional counter with 1 μm polypropylene window; pulse height analysis used with 0.75–1.5 V channel. All *K* lines; measurements made in vacuum.

[b]The scalar-timer equipment has inbuilt facilities to display total counts divided by 4.2 or direct, and to adjust counting times automatically to compensate for losses due to dead time.

[c]For Mg *K*α using ADP (101) crystal the amplifier gain was increased 20% above the symmetrical setting to decrease the background arising from fluorescence of the crystal.

[d]For Na *K*α using KAP (001) crystal the gas in the flow-proportional counter was changed from 90% argon–10% methane to pure methane.

Counting

Table 9.8 shows information on counting for this determination.

Matrix Corrections

Variations in composition of the vegetation are sufficient that matrix corrections must be used. Linear calibration procedures are not recommended. Concentration is expressed as $C = K(R_p - R_b)\mu$, where C is the concentration of the element, $(R_p - R_b)$ is the net intensity of the line, μ is the mass absorption coefficient of the sample, and K is a constant. The authors (10) state that for samples that consist mainly of low atomic number elements μ can be replaced by the mass absorption coefficient of the sample for the fluorescent radiation. All elements which make a significant contribution to variations in absorption must be measured, and the matrix coefficients are used to correct for interelement absorption effects.

Measure the matrix coefficients of the elements using aqueous solutions or prepared cellulose standards. The values obtained by the authors are shown in Table

Table 9.9
Coefficients Used to Correct Nominal Concentration (Generally Applicable)

Analytical line	Cel[b]	Ca	K	Cl	S	P	Si	Al	Mg	Na
Ca $K\alpha$	0.721	0.55	10.15	7.48	6.27	5.17	4.14	3.23	2.44	1.71
K $K\alpha$	0.850	0.65	0.42	8.85	7.42	6.13	4.89	3.82	2.82	2.02
Cl $K\alpha$	0.911	0.66	0.42	0.00	7.78	6.44	5.14	4.04	3.07	2.14
S $K\alpha$	0.934	0.65	0.41	0.00	−0.18	6.55	5.21	4.09	3.12	2.18
P $K\alpha$	0.951	0.65	0.40	−0.02	−0.19	−0.33	5.27	4.15	3.17	2.21
Si $K\alpha$	0.975	0.65	0.39	−0.03	−0.20	−0.36	−0.49	4.20	3.20	2.24
Al $K\alpha$	0.975	0.61	0.37	−0.05	−0.21	−0.36	−0.50	−0.60	3.19	2.22
Mg $K\alpha$	0.922	0.60	0.35	−0.06	−0.24	−0.38	−0.52	−0.63	−0.72	2.24
Na $K\alpha$	1.000	0.57	0.33	−0.08	−0.25	−0.40	−0.52	−0.64	−0.73	−0.79

Source: Ref. 10. Reproduced by permission of John Wiley and Sons Limited.
[a]Matrix coefficients for cellulose (Cel) and elements (element coefficients are for unit weight fraction).
[b]A carbon, oxygen, nitrogen, and hydrogen matrix assumed to have the same absorption coefficient as cellulose and to be constant.

9.9. They have been normalized so that the matrix correction for each element will be 1.0 for a plant of average chemical composition.

For example, a matrix correction for calcium is made as follows. Let the true percentage concentrations of the elements be Ca, K, etc. and the nominal concentrations be Ca*, K*, etc. (The nominal concentrations are obtained from the calibration graphs.)

$$\mathrm{Ca} = \mathrm{Ca}^*[0.721 + 0.01(0.55\mathrm{Ca} + 10.15\mathrm{K} + 7.48\mathrm{Cl} + \cdots + 1.71\mathrm{Na})]$$

In the first calculation, the nominal values for the elements are used in place of the true values. Then successive iterations are used until a constant result for each element is obtained. Usually, two iterations are sufficient.

(The factor 0.01 converts percentage concentration to weight fraction.)

Background Corrections

Background intensities are low. Table 9.10 contains corrections measured by the authors (10) for the instrumental conditions in Table 9.8. As an example, a correction for magnesium is obtained as:

$$\text{Background \%} = 0.01(0.93 \times \mathrm{Ca}^* + 0.50 \times \mathrm{K}^* + \cdots + 0.16 \times \mathrm{S}^*) + 0.016$$

Subtract background corrections from nominal concentrations prior to matrix correction.

Nominal Concentrations

Nominal concentrations are obtained by using calibration standards as explained earlier.

Table 9.10
Background Coefficients (Applicable Only to Conditions Set Out in Table 9.8)

Analytical line	Constant background (%)	Additional background (% due to unit weight fraction of element)					
		Ca	K	Cl	S	P	Si
Ca $K\alpha$	0.001						
K $K\alpha$	0.001	0.002					
Cl $K\alpha$	0.001	0.013	0.018				
S $K\alpha$	0.001	0.017	0.017	0.037			
P $K\alpha$	0.001	0.043	0.035	0.025	0.036		
Si $K\alpha$	0.002	0.032	0.021	0.017	0.027	0.037	
Al $K\alpha$	0.0008	0.031	0.022	0.015	0.014	0.010	0.030
Mg[a] $K\alpha$	0.016	0.93	0.50	0.20	0.16		
Mg[b] $K\alpha$	0.013	0.60	0.18	0.085	0.065	0.050	
Na[c] $K\alpha$	0.060	2.50	0.65	0.30	0.25	0.15	
Na[d] $K\alpha$	0.060	1.50	0.95	0.70	0.45	0.25	
Na[b] $K\alpha$	0.030	0.40	0.12	0.095	0.075	0.080	

Source: Ref. 10. Reproduced by permission of John Wiley and Sons Limited.
[a]ADP crystal.
[b]TlAP crystal.
[c]KAP crystal and CH_4 gas.
[d]RbAP crystal.

Obtain intensity readings from a series of standards using the instrumental conditions in Table 9.8. Calculate nominal concentrations from the matrix corrections in Table 9.9. For example,

$$\text{Ca}^* = \text{Ca}/[0.721 + 0.01(0.55\text{Ca} + 10.15\text{K} + \cdots + 1.71\text{Na})]$$

Add the background correction.

Plot the resulting values for each element against intensities to obtain the calibrations, and use these for subsequent analyses of vegetation samples.

C. Determination of Elements with Atomic Number Greater than 20 (11)

Elements Determined

Mn, Fe, Cu, Zn

Comments on the Method

Pellets are formed from the dried, ground samples. Calibrations are carried out by means of synthetic cellulose compositions. Matrix and background corrections are applied. Four elements in a series of 13 previously analyzed samples were determined. The agreement was satisfactory even at very low concentrations.

Instruments and Equipment

These were the same as in the previous method for low atomic number elements. The instrumental conditions are listed below.

X-ray tube: Au target. Window reduced with a 10 mm diameter circular lead port.

Analyzing crystal: LiF (200).

Collimator: 480 μm.

Pulse height selection: 0.75–1.50 V.

Sample Preparation

The sample is dried and ground as described in the previous method. Press a 3.2 cm diameter disk from at least a 2.5 g sample, or a 4.0 cm diameter disk from 4.0 g sample.

Prepare calibration standards from previously analyzed vegetation. Alternatively, prepare standards containing 250, 500, and 750 ppm of the analytes by adding measured volumes of standard solutions to cellulose powder. Then add a few drops of ammonium hydroxide. Dry, grind, and press the standards in the same manner as the samples.

Mass Absorption Coefficient

The concentration of an element in a plant sample is expressed by the equation $C = K(R_p - R_b)\mu$, where C is the concentration of the element, $(R_p - R_b)$ is the net intensity of the line, μ is the mass absorption coefficient of the sample, and K is a constant.

In this work with high atomic number elements, μ is related linearly to μ_2, the mass absorption coefficient of the sample for the fluorescent radiation of the element being measured. The authors (11) describe three methods of measuring μ_2. The direct method is described here. The method requires the preparation of 12.5 cm sample disks that can be inserted in front of the counter. The fluorescent radiation is obtained by placing a piece of metal foil or metal powder in the sample compartment. Measurements are made with and without the sample disk in the beam.

$$\mu_2 = \frac{\pi r^2}{w} \, log \, \frac{T_x}{T_s}$$

where w is the weight of the disk (grams), r the radius of the disk (centimeters), T_x the time to accumulate N counts with sample in beam (seconds), and T_s the time to accumulate N counts without sample in beam.

Background Correction

This is an important measurement. The time of the background count must be as long as the peak count. The background is measured in an off-peak position. This value must be multiplied by a factor which relates off-peak background to true background under the peak. The authors recommend using pure ashless cellulose for both measurements (11). Backgrounds must be measured for each element.

X. DETERMINATION OF URANIUM IN PLANTS BY LASER FLUOROMETRY (13)

Comments on the Method

Ward and Price (14) critically reviewed the major approaches for uranium exploration. The most commonly used methods are "conventional" fluorometry (i.e., with other than a laser source) and delayed neutron counting. Conventional fluorometry gives sensitivities of 0.1–0.2 ppm (solids). Sensitivities are less than 1 ppm (solids) and 0.02 ppb (water—uranium trapped on resin) by neutron activation analysis. Laser-induced fluorescence gives 0.02–0.05 ppb sensitivity for waters but is subject to disadvantages because of the considerable length of time required for sample preparation. Nuclear fission track techniques are capable of sensitivities down to 0.01 ppb with water.

This method requires a laser fluorometer (e.g., Scintrex model UA-3). Levels of uranium typically found in plants are usually below the detection limit of conventional fluorometry. The determination limit for the proposed method is 0.05 ppm, which was found to be satisfactory for detecting uranium in all but one of the 74 plant samples analyzed.

The sample is dry ashed at 450°C followed by dissolution of the ash in 2.5 *M* nitric acid saturated with aluminum nitrate. The uranium is extracted into ethyl acetate. Ethyl acetate is then destroyed by ignition and the residue is redissolved in 0.005% nitric acid.

Apparatus

A model UA-3 Scintrex (Concord, Ontario, Canada) uranium analyzer was used together with polyethylene bottles and syringes.

Reagents

- Nitric acid, 2.5 *M:* Mix 160 ml concentrated acid with 840 ml water.
- Nitric acid, 5%: Dilute 7.2 ml concentrated acid to 100 ml.
- Nitric acid, 0.005%: Dilute 70 μl concentrated acid to 1 liter with water.
- Nitric acid–aluminum nitrate reagent: Dissolve 950 g of $Al(NO_3)_3 \cdot 9H_2O$ in 600 ml of 2.5 *M* nitric acid. This is a saturated solution. The amount of aluminum nitrate required to achieve saturation will vary depending on the water content of the nitrate; if the water content is very high, the salt will dissolve in its own water of hydration and saturation is not achievable. However, the solution can still be used. The reagent may have to be washed with ethyl acetate just prior to use to remove traces of uranium. To do this, place the reagent in a separatory funnel, add 10–15 ml of ethyl acetate, and shake for 2 min. Allow the phases to separate before using the acid.
- Ethyl acetate, reagent grade.
- Sodium hexametaphosphate reagent: Dissolve 7.4 g $(NaPO_3)_6$ in water and dilute to 100 ml.

- Uranium standard solution, 1000 ppm: Dissolve 0.211 g $UO_2(NO_3)_2 \cdot 6H_2O$ in 100 ml of 5% nitric acid.
- Uranium standard solution, 100 ppm: Dilute 10 ml of the 1000 ppm standard solution with water to 100 ml.
- Uranium standard solution, 10 ppm: Dilute 10 ml of the 100 ppm standard solution with water to 100 ml.
- Uranium standard solution, 100 ppb: Dilute 5 ml of the 10 ppm standard solution with water to 500 ml.
- Uranium standard solution, 2 ppb: Dilute 5 ml of the 100 ppb standard solution with water to 250 ml.

Procedure

Dry and grind the plant tissue to produce a homogeneous sample. Ignite a sufficient amount of the ground sample at 450°C to produce at least 100 mg of ash.

Weigh 100 mg of ash into a 25 × 150 mm screw-cap culture tube. Add 8 ml of 2.5 *M* nitric acid previously saturated with aluminum nitrate and heat in a boiling-water bath for 1 h. Cool, add 10 ml ethyl acetate, screw the cap onto the tube, and shake the tube and contents for 2 min. Allow the phases to separate—centrifuge if necessary—and pipet a 5 ml aliquot of the ethyl acetate into a clean, dry platinum dish. Ignite the organic solvent and allow it to burn off.

After ignition, dissolve the residue in 7 ml of 0.005% nitric acid. Transfer the dilute nitric acid sample solution to the UA-3 sample holder and read the fluorescence of the solution following the procedure given in the UA-3 manual. In the absence of a proprietary reagent, a dilute solution of sodium hexametaphosphate can be substituted to establish optimum conditions for uranium fluorescence. The uranium content of the sample is calculated by comparison of the meter reading of the sample with that of a set of standards (containing 0, 0.005, 0.01, 0.02, and 0.05 μg uranium) prepared in a similar manner.

REFERENCES

1. J. C. Van Loon, internal laboratory method, unpublished.
2. H. Agemian and A. S. Y. Chau, *Anal. Chim. Acta* **80,** 61 (1975).
3. R. D. Reeves and R. R. Brooks, *J. Geochem. Explor.* **18,** 275 (1983).
4. J. A. Fiorino, J. W. Jones, and S. G. Caspar, *Anal. Chem.* **48,** 120 (1976).
5. J. W. Jones, S. G. Caspar, and T. C. O'Haver, *Analyst (London)* **107,** 353 (1982).
6. D. R. Neuman and F. F. Munshower, *Anal. Chim. Acta* **123,** 325 (1981).
7. H. V. Warren, S. J. Horsky, A. Kruckeberg, G. H. N. Towers, and J. E. Armstrong, *J. Geochem. Explor.* **18,** 169 (1983).
8. R. R. Brooks and S. D. Naidu, *Anal. Chim. Acta* **170,** 325 (1985).
9. A. Brzezinska, A. Balicki, and J. C. Van Loon, *Water, Air, Soil Pollut.* **21,** 323 (1984).
10. K. Norrish and J. T. Hutton, *X-Ray Spectrom.* **6,** 6 (1977).
11. J. T. Hutton and K. Norrish, *X-Ray Spectrom.* **6,** 12 (1977).
12. T. K. Smith and T. K. Ball, *Adv. X-Ray Anal.* **26,** 409 (1982).
13. T. F. Harms, F. N. Ward, and J. A. Erdman, *J. Geochem. Explor.* **15,** 617 (1981).
14. F. N. Ward and V. Price, *J. Geochem. Explor.* **13,** 97 (1980).

10

ICP Source Mass Spectrometry

I. INTRODUCTION

One of the most exciting developments in elemental analysis is the use of an inductively coupled plasma as a source for mass spectrometry. The application of this technique to geochemical analysis should be particularly rewarding. In addition to its capability for trace element analysis at sub-parts per billion levels this approach, for the first time, opens up the possibility of determining natural isotopic ratios routinely. Isotopic ratio sample throughputs should approach those obtained for ordinary multielement trace element analysis. Also of particular relevance to geochemical analysis is the possibility of doing isotope dilution mass spectrometry routinely. Isotope dilution mass spectrometry is the technique by which most other approaches are measured.

The use of plasmas as sources for mass spectrometry dates back to the 1960s to work by O'Halloran and Flügge (1). But it was the work by Gray (2), using a direct-current plasma, and Houk *et al.* (3), using an inductively coupled plasma, which resulted in an abrupt upswing in activity in the plasma mass spectrometry field. The result of this new interest was the recent introduction of commercial offerings by VG Instruments Ltd., Winsford, Cheshire, and Sciex Co. Ltd., Thornhill, Ontario, Canada. Much of the pioneering work was done in the laboratories of Gray (4–6), Fassel (7–9), and Douglas (10–12).

Douglas and Houk (13) and Date (14) have written reviews of inductively coupled plasma–mass spectrometry (ICP–MS). These publications are of great use to those interested in acquiring a basic understanding of this technique.

ICP–MS equipment consists of an inductively coupled plasma source, an interface, and a quadrupole mass spectrometer. A typical arrangement is shown in Fig. 10.1 (after the Sciex instrument).

The inductively coupled plasma (in contrast to ICP–optical equipment) is in a horizontal position. The ions generated in the plasma are sampled by a conical

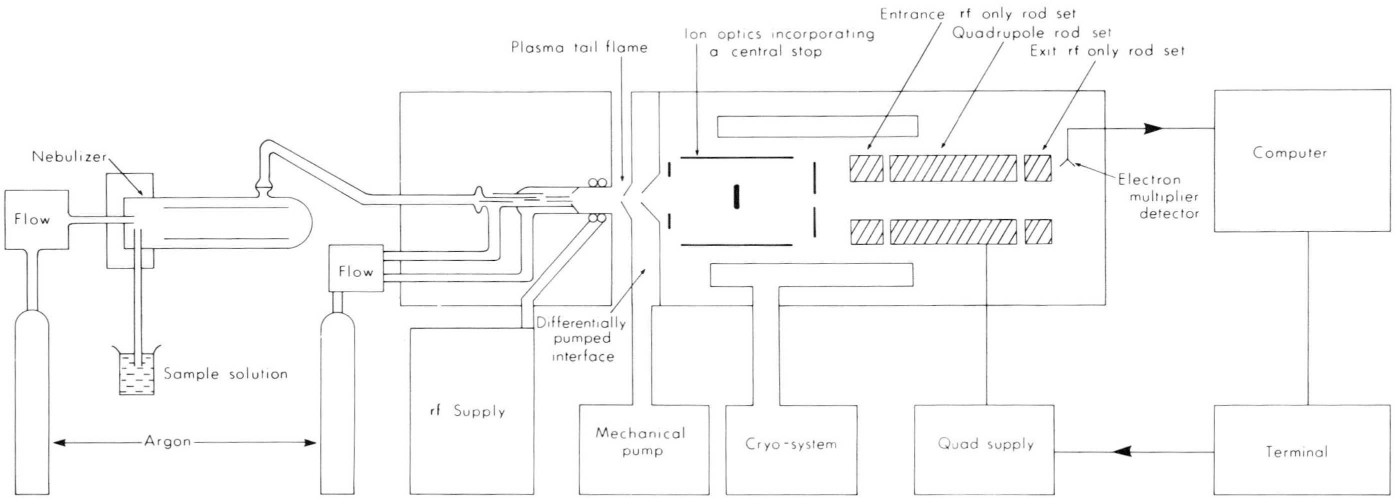

Figure 10.1. Schematic diagram of an ICP–MS system. (Reprinted with permission from Ref. 13. Copyright 1985, Pergamon Press, Ltd.

sampler containing an orifice between 0.3 and 1.0 mm in diameter. Ions which enter the sample orifice pass into a region which is mechanically pumped down to about 1 torr. A skimmer, which has a larger orifice (~3×) than the sampler, is the entry point of ions into a region cryogenically pumped down to about 10^{-6} torr.

The ion optics, which help focus ions (while rejecting photons) into the quadrupole, consist of Einsel lenses and a Bessel box, Fig. 10.1 (Sciex system). A photon stop exists behind the skimmer and another stop is placed in the center of the Bessel box. An electron multiplier is the detector, and this device is kept out of the axis of the ions in case stray photons still exist. A deflector is used to direct the ions onto the detector.

The quadrupole mass spectrometer has at its entrance and exit rf-only quadrupole lenses as focusing elements to overcome defocusing caused by fringing fields at the ends of the quadrupole. The quadrupole, as the name suggests, consists of four rods. These are connected so that opposite pairs are electrically interconnected. An rf signal of about 1 Hz is superimposed over a dc signal on the rod pairs. The device acts as a mass filter so that only ions of a particular mass/charge ratio reach the detector for a given electrical condition of the quadrupole. A mass scan is done by sweeping the dc and rf potentials, keeping the ratio between them constant.

A mass spectrum of a solution of a number of elements is shown in Fig. 10.2. At the present state of development of ICP–MS systems the most serious problems are related to mass spectral overlap at mass 80 and below and ionization suppression. One of us (JCVL) obtains up to 10% drift per hour in routine operation. The drift is worst during early minutes of operation and often becomes more stable with time. The drift can be +*ve* or −*ve* (mainly negative) as shown in Fig. 10.3.

Figure 10.2. Mass spectrum of a solution of transition metals at 10 μl/ml.

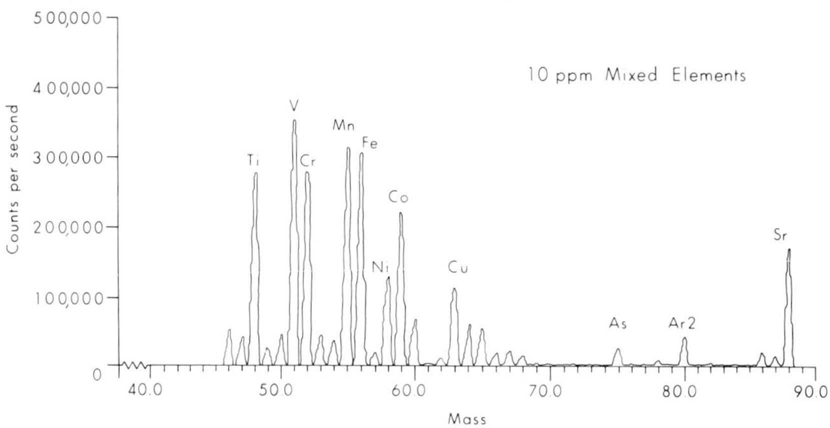

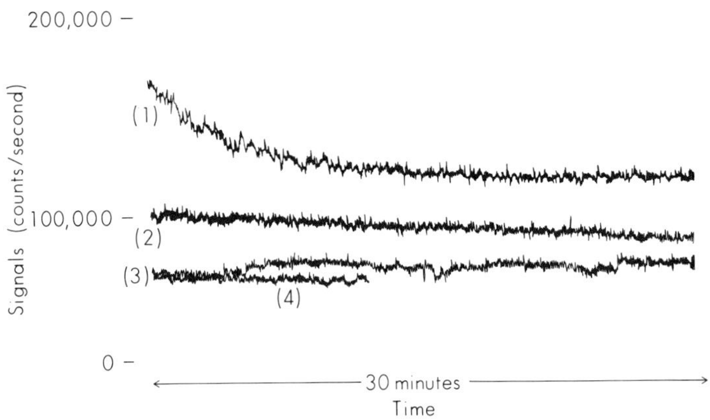

Figure 10.3. Drift of signal from ICP–MS.

II. INTERFERENCES

The worst interference problem is ionization suppression. At typical ICP temperatures in the region sampled, of 5000–7000 K, and electron densities of 5×10^{14} cm^{-3} most elements of the periodic table are predicted to be efficiently ionized (most singly ionized). Thus at a high matrix concentration (i.e., >1000 ppm) almost any element (certainly all those with lower ionization energies than argon—8 eV) could cause ionization suppression of analyte ions. At these levels, which can handle up to 10,000 ppm salts without problem, the matrix ions typically become the plasma ions (i.e., the plasma is now a matrix ion plasma). Ionization interference can be corrected for as follows:

1. Matrix matching of samples and standards
2. Internal standardization (but the internal standard must have its ionization suppressed to a similar degree as the analyte)
3. Isotope dilution

Mass spectral overlap can occur. This type of interference is very much less than spectral interference in ICP emission. Mass spectral interference results from background peaks (due to plasma gas etc.), oxide ions, isobaric singly charged ions, doubly charged ions, and molecular ions (due to use of acids etc.).

Background peaks and possible interferences are listed in Table 10.1.

Oxide ions of analyte compete for the singly charged positive analyte ion population. Thus, in the case of refractory elements a small proportion of total analyte ion may be tied up as the oxide ions. This effect is present only when samples are aspirated into the plasma. Other techniques for sample introduction such as electrothermal volatilization and laser ablation do not suffer from this problem.

Table 10.1
Background Peaks Observed in ICP–MS

Mass	Assignments	Interferes With
12	C^+	$^{12}C^+$
13	$^{13}C^+$, CH	$^{13}C^+$
14	N^+	
16	O^+	
17	OH^+	
18	H_2O^+	
19	H_3O^+	
20	$H_2{}^{18}O^+$, H_2DO^+	
21	$H_3{}^{18}O^+$	
28	$N_2{}^+$, CO^+	$^{28}Si^+$
29	N_2H^+ (?)	$^{29}Si^+$
30	NO^+	$^{30}Si^+$
31	NOH^+ (?)	$^{31}P^+$
32	$O_2{}^+$	
33	O_2H^+	
34	$^{18}O^{16}O^+$	
35	$^{16}O^{18}OH^+$	$^{35}Cl^+$
36	$^{36}Ar^+$	
37	$^{36}ArH^+$	$^{37}Cl^+$
38	$^{38}Ar^+$	
39	$^{39}ArH^+$	$^{39}K^+$
40	Ar^+	$^{40}Ca^+$
41	$^{40}ArH^+$	$^{41}K^+$
42	$^{40}ArD^+$	
43	?	
44	$CO_2{}^+$	$^{44}Ca^+$
45	?	
46	$NO_2{}^+$	
54	ArN^+	$^{54}Fe^+$
56	ArO^+	$^{56}Fe^+$
76	$^{36}Ar^{40}Ar^+$	$^{76}Se^+$
78	$^{38}Ar^{40}Ar^+$	$^{78}Se^+$
80	$^{40}Ar^{40}Ar^+$	$^{80}Se^+$
81	$^{40}Ar^{40}ArH^+$	$^{81}Br^+$

III. SAMPLE PREPARATION

To avoid a high salt content, as much as possible, fusion techniques are seldom used. High salt contents are more serious problems in ICP–MS than in ICP emission. Thus acid decomposition of samples is commonly used in sample preparation. Positively charged ions resulting from the required use of relatively high levels of acids can present spectral overlap problems. The following is a comment on each of the common acids for ICP/MS work.

Nitric acid: Best acid (gives only ArN^+-type ions—one for each isotope).

Hydrochloric acid: Bad (gives ClO^+-type and Cl_2^+ and all isotopes of these).

Sulfuric acid: Bad (gives SO^+, SOH^+, SO_2^+, SO_2H^+, etc. and all isotopes of each).

Phosphoric acid: Do not use as it erodes the sampler.

If organic solvents are a part of the sample preparation they can be introduced into the plasma with little problem. In this regard the most severe limitation is carbon buildup on the sample orifice. Addition of 10% oxygen to the plasma gas helps minimize this effect. Ion species which may cause potential spectral overlap resulting from organic solvent use are, for example, C^+, C_2^+, CO_2^+, and ArC^+.

Buildup of scale on the sampler orifice can occur due to high concentrations of a variety of matrix materials. In this regard it may be necessary to scrape the scale off the orifice from time to time.

IV. DETECTION LIMITS

Table 10.2 (13, 15) is a list of currently obtainable detection limits. These are 3σ values for pure solutions. Work with real samples results in a degradation. These values are in many cases close to furnace atomic absorption spectrometry detection limits.

V. NOVEL SAMPLE INTRODUCTION METHODS

In most cases a nebulizer is used for sample introduction in ICP–MS, the pneumatic nebulizer being most commonly employed. There are a few reports of the use of ultrasonic nebulization. In these cases, signal enhancements of up to a factor of 10 have been recorded.

Other types of sample introduction include laser ablation (16), use of an electrothermal atomizer, and a direct insertion device. These approaches all avoid problems related to oxygen-containing molecular ions. In addition, signal enhancements of severalfold have been recorded. If the electrothermal vaporizer is chosen, potentially interfering substances in the matrix can be ashed way prior to the atomization/ionization step.

VI. LINEAR WORKING RANGE

A linear working range of up to 5 orders of magnitude is achievable with ICP–MS. Thus, this approach is capable of practical multielement analysis of geological samples for a wide range of elements.

Table 10.2
Comparison of Detection Limits[a]

Element	Detection limits (ng/ml)		
	ICP–MS[b,c]	ICP–AES[b]	F–AAS[b]
Li	0.1	2.8	3
B	0.4	4.8	1500
Be	0.03		
Mg	0.7	0.15	0.3
Si	10		
Ti	0.1		
Al	0.2	22	30
Cr	0.3	6.1	4.5
Mn	0.1	1.3	4.5
Co	0.01	6	7.5
Zn	0.2	1.8	0.9
Ge	0.02	1.5	75
As	0.04	52	150
Se	0.8	45	150
Sr	0.04		
Zr	0.06		
Nb	0.03		
Ag	0.03	7	3
Cd	0.06	2.5	1.5
Sn	0.06		
In	0.07	63	45
Te	0.09	40	105
La	0.05	10	2400
Ce	0.04	48	
Pr	0.02		
Nd	0.03		
W	0.05	30	750
Au	0.06	16	15
Hg	0.02	25	300
Pb	0.05	42	
Th	0.02	64	
U	0.03	255	10,500

[a]A signal which is greater than three standard deviations of the noise while running the blank.
[b]Ref. 15.
[c]Ref. 13.

VII. TYPICAL MASS SPECTRA

A typical mass spectrum of a plasma into which distilled water is being sprayed is given in Fig. 10.4. A mass spectrum of a lead solution showing the predominant Pb isotopes is shown in Fig. 10.5. From the latter spectrum it can be seen how readily isotopic ratio work can be done. Of particular importance to geologists would be the use of such data in determining the origins of magmas.

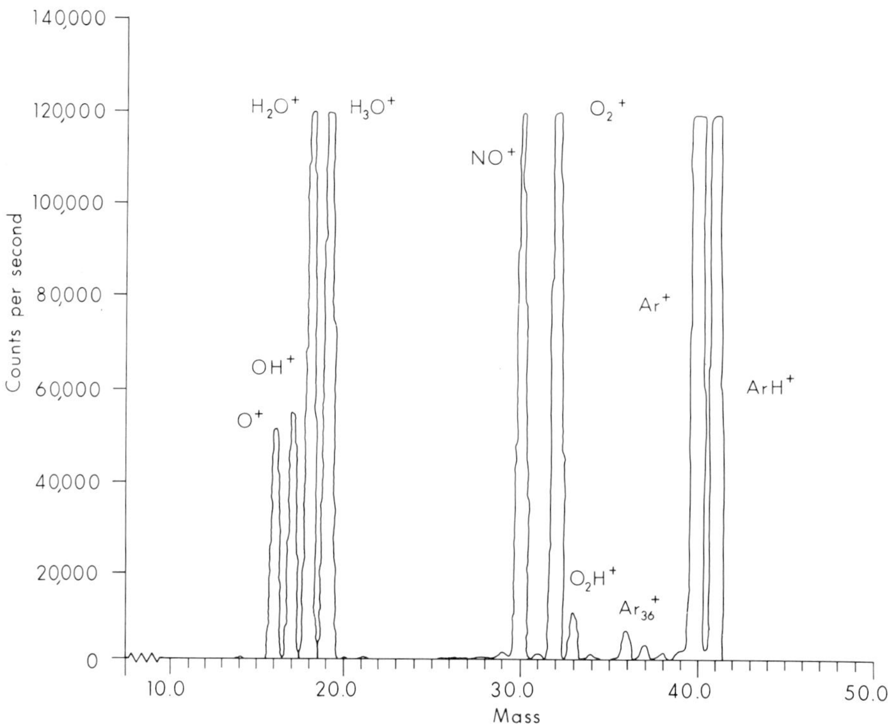

Figure 10.4. Background spectra with a distilled water sample, ×10. The ion optics were detuned to keep the major peaks on scale. (Reprinted with permission from Ref. 13. Copyright 1985, Pergamon Press, Ltd.)

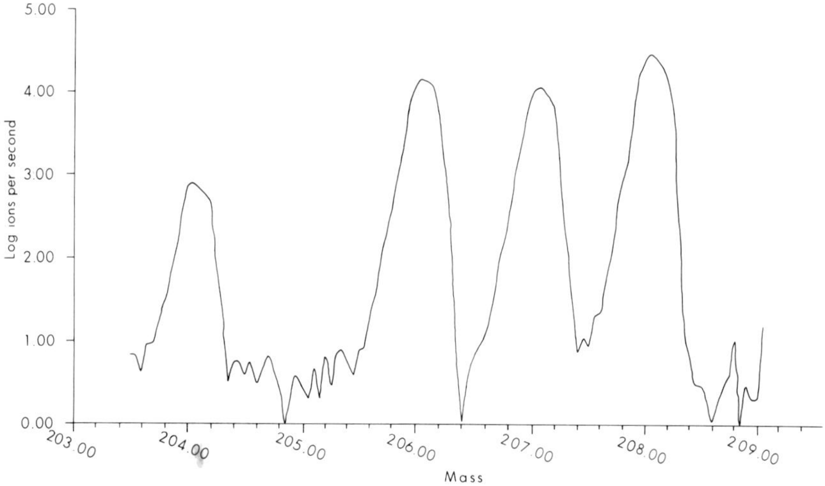

Figure 10.5. Mass spectrum of lead isotopes on a logarithmic scale showing four orders of magnitude drop in signal between peaks. (Reprinted with permission from Ref. 13. Copyright 1985, Pergamon Press, Ltd.)

VIII. ADVANTAGES OVER ICP–OPTICAL EMISSION SPECTROMETRY

There are several important advantages to be expected using ICP–mass spectrometry compared to ICP–optical emission spectrometry. These include the following.

1. Elemental response throughout the periodic table does not vary as greatly with the mass spectrometry approach. (Background noise is fairly uniform throughout in mass range and most elements are nearly 100% ionized.)
2. Detection limits already achieved by the mass spectrometry technique are often 2 to 3 orders of magnitude superior to optical values.
3. The ratios of natural isotopes can be determined.
4. Isotope dilution mass spectrometry can be done routinely.

IX. ISOTOPE DILUTION (ID) MASS SPECTROMETRY BY ICP–MS

Isotope dilution mass spectrometry is the technique by which other methods of analysis are evaluated. ICP–MS provides a rapid method of doing ID–MS.

ID–MS is done as follows:

1. Measure the *natural isotopic abundance* of two stable isotopes of an element in a sample.
2. Add a spike of a solution (containing the element of interest) which has a well-known isotopic ratio of the same two stable isotopes. (Often the spike is made to contain 100% of one of the two isotopes.)
3. Measure the *altered* isotopic *ratio*.
4. Compute the original concentration of the element as follows:

$$[\mathrm{M}] = \frac{(\mathrm{Ask} - R \times \mathrm{Bsk}) \times \mathrm{Wsk}}{(R \times \mathrm{Bsa} - \mathrm{Asa}) \times \mathrm{Wsa}}$$

where sk denotes the spike (Ask = isotope A in spike, Bsk = isotope B in spike, Wsk = weight of analyte in spike in micrograms), sa the sample (Asa = natural isotope A abundance, Bsa = natural isotope B abundance, Wsa = sample weighting), and R the altered isotopic ratio (after the spike has been added to the sample).

X. DETERMINATION OF RARE EARTH ELEMENTS BY ICP–MS (17)

Comments on the Method

This procedure is for the direct determination of rare earth elements (REE) in rock sample solutions. This is possible with ICP–MS (in contrast to optical emission) because of the relatively few spectral interferences and the high sensitivity of the technique.

A hydrofluoric : hydrochloric : perchloric (10 : 1 : 1) acid mixture is used for the primary digestion, and this is followed by a treatment using a 15% hydrochloric : perchloric (15 : 1) acid mixture. The residue is dissolved in hydrochloric acid and the sample is diluted to the mark in a volumetric flask. In this way 0.2 g of rock is dissolved and diluted to 100 ml. The final solutions of the sample and standards are made 10% in nitric acid.

The procedure was tested on a variety of standard reference rocks available from the U.S. Geological Survey (USGS), the Canada Centre for Mineral and Energy Technology, and the National Institute of Metallurgy in South Africa. Good agreement with the accepted values was generally obtained.

Equipment

All determinations were performed on an ELAN model 250 ICP–MS spectrometer system (Sciex). Additional data manipulations were carried out on an Apple IIe personal computer. ICP–MS operating conditions for the determinations are listed in Table 10.3.

It should be noted that in establishing optimum parameters for the determination of the REE, a reference point for the position of the ICP torch with respect to the sampler interface was established while aspirating a solution containing 1000 μg/g yttrium. During the nebulization, two distinct features can be observed in the plasma: (a) a slender red inner core corresponding to the presence of YO^+ and defined by Koirtyohann *et al.* (18) as the initial radiation zone (IRZ) and (b) a blue plume corresponding to the Y^+ and neutral Y species, defined as the normal analytical zone (NAZ). Since the relative positions of the IRZ and NAZ with respect to any chosen reference point (sampler orifice, tip of torch, load coil) depend on several operating parameters, including nebulizer gas flow, auxiliary gas flow, and plasma power, it is felt by many workers that the distance between the sampler orifice and the tip of the yttrium IRZ provides a more suitable reference point for torch positioning than does the parameter "height above the load coil" commonly used in ICP optical emission work. For example, in the REE determinations, the torch position was established so that the tip of the yttrium IRZ was approximately 7 mm from the sampler orifice; at this distance the boundary layer between the plasma and the sampler exhibited minimal visible (red) YO^+.

Table 10.3
ICP–MS Operating and Measuring Conditions

Plasma conditions	
Plasma gas flow, 14 liters/min	
Auxiliary gas flow, 2.0 liters/min	
Nebulizer back pressure,[a] 1660 kPw	
Forward power, 1350 W	
Reflected power, 5 W	
Sample uptake	
Sample flow (peristalic pump), 1.4 ml/min	
C2 Meinhard nebulizer	
Sciex long torch	
Distance sampler-reference point,[b] 7mm	
Lens settings	
BBA, 60–47[c]	
BBC, 67	
BBB, 75–62[c]	
Ring lens, 10	
Acquisition parameters	
One point per peak at 0.5 s/point	
Eight integrations	

[a]Mass flow controller not yet installed, therefore actual flow rates not measures; reading of 0.45 liter/min on nebulizer gas flow gauge on instrument.

[b]Reference point defined as the distance between the sampler orifice and the tip of the initial radiation zone (IRZ) visible during the nebulization of 1000 ppm Y solution (see text).

[c]Range of lens settings used due to the need to compensate for instrument drift between sample batches.

Final adjustment of operating conditions specifically for the rare earths was done while aspirating appropriate REE solutions.

It should also be noted that the ion lens settings in Table 10.3 (Bessel box lens settings on the ELAN) cover a range of values. This was required because of an observable instrument drift and the need for frequent readjustment of lens settings.

Reagents

High-purity individual-element spectroscopic solutions (Y, La, Ce, Nd, Sm, Eu, Gd, Dy, Yb) at concentrations of 1000 μg/ml were obtained from Spex Industries (Metuchen, New Jersey). High-purity (99.9% pure) solid rare earth element oxides were obtained from Alfa Products (Danvers, Massachusetts). The ion-exchange resin used was Amberlite CG-120 (Na) (type 1, 100–200 mesh) (BDH Chemicals). Analytical grade acids and deionized water were used throughout for the preparation of samples and reference solutions.

Individual-element standard solutions (1000 μg/ml) were prepared for Pr, Tb, Ho, Er, Tm, and Lu by dissolving the appropriate oxides in 10% (v/v) nitric acid.

Two calibration standard solutions were prepared. Solution A, containing Y,

La, Ce, Pr, Nd, Sm, Eu, Yb, and Lu at concentrations of 100 ng/ml, was prepared by combining aliquots of the appropriate individual-element standard solutions and serial dilution with 10% (v/v) nitric acid.

Solution B, containing 100 ng/ml Gd, Tb, Dy, Ho, Er, and Tm, was prepared in a similar fashion.

Procedure

To digest rock powders, place 0.2 g of the powder in a 50 ml PTFE beaker, and add 12 ml of a mixture of 10 ml 48% hydrofluoric, 1 ml hydrochloric (concentrated), and 1 ml perchloric (concentrated) acids. Allow the mixture to digest and evaporate to dryness. Then add 10 ml of a solution containing 15% (v/v) hydrochloric and perchloric (concentrated) acids in a ratio of 15 : 1 to the residue and evaporate the solution again to dryness. While the residue is still hot, add 1 ml of nitric acid (concentrated) and 10 drops of hydrochloric acid (concentrated). Cover the beaker with a watch glass, and boil the solution for a few minutes. After cooling, wash the watch glass and the sides of the beaker with water. Add additional water to the beaker to a total of approximately 12–15 ml, and heat the solution until clear. Transfer the contents of the beaker to a 100 ml volumetric flask and dilute to the mark with 10% (v/v) nitric acid. The final dilution factor is 1 : 500.

Analyze the solutions using the experimental conditions given in Table 10.3. Possible interferences are given in Tables 10.4 and 10.5. Determination limit data, compared with REE abundances in chrondite, are given in Table 10.6.

Table 10.4
ICP–MS Determination of Rare Earth Elements: Mass Number Selection and Isotope Interferences

Element	Mass	Abundance (%)	Interference
Yttrium	89	100	
Lanthanum	139	99.9	
Cerium	140	88.5	
Praseodymium	141	100	
Neodymium	142	27.1	Ce (11.1%)[a]
Samarium	147	15	
Europium	153	52.2	
Gadolinium	158	24.7	Dy (0.1%)
Terbium	159	100	
Dysprosium	164	28.2	Er (1.6%)
Holmium	165	100	
Erbium	166	33.4	
Thulium	169	100	
Ytterbium	174	31.8	Hf (0.2%)
Lutetium	175	97.4	

Source: Ref. 17.
[a]Abundance of isotope causing interference.

Table 10.5
ICP–MS Determination of Rare Earth Elements: Oxide Interferences at Selected Masses

Element	Mass	Oxide[a]
Europium	153	BaO (11.2% Ba)[b]
Gadolinium	158	CeO (11.1% Ce)
		NdO (27.1% Nd)
Terbium	159	NdO (12.2% Nd)
Dysprosium	164	NdO (5.7% Nd)
		SmO (11.2 Sm)
Holmium	165	SmO (13% Sm)
Erbium	166	NdO (5.6% Nd)
		SmO (7.4% Sm)
Ytterbium	174	GdO (24.7% Gd)
Lutetium	175	TbO (100% Tb)

Source: Ref. 17.
[a]Although a number of oxide interferences are listed, the only important correction required in most rock samples is for CeO and NdO at ^{158}Gd.
[b]This interference becomes important at very high levels of Ba; e.g., the Eu value in NIM-S was observed to be 1.2 ppm (0.3 ppm accepted) because of 2400 ppm Ba.

Table 10.6
ICP–MS Determination of Rare Earth Elements: Determination Limits Compared with Chondrite Abundances[a]

Element	Chondrite (ppm)	Determination limit (ppm)
Y	1.96	0.02
La	0.330	0.05
Ce	0.88	0.05
Pr	0.11	0.05
Nd	0.60	0.18
Sm	0.181	0.15
Eu	0.069	0.07
Gd	0.249	0.14
Tb	0.047	0.03
Dy	0.325	0.13
Ho	0.070	0.03
Er	0.200	0.10
Tm	0.030	0.03
Yb	0.200	0.11
Lu	0.034	0.04

Source: Ref. 17.
[a]Based on 4σ of blank.

XI. DETERMINATION OF LEAD ISOTOPES BY PLASMA SOURCE MASS SPECTROMETRY (12)

Comments on the Method

Lead isotopes are determined by the following method and then are used to help classify gold deposits. This procedure is one of the first very useful applications of plasma source mass spectrometry. The authors qualify the paper with the phrase "initial results." However, the methodology is of sufficient importance that it should be included here.

The instrumentation employed is a prototype of an ICP–MS now being commercially offered by Sciex Inc. in Canada. (A competitive instrument is also manufactured by VG Instruments in Great Britain.)

The procedure was tested on NBS 1981. The ratios of Pb isotopes 206, 207, and 208 to 204 and of 207 and 208 to 206 were determined. The results by the proposed procedure varied by 0.2% for 206/204, 2% for 207/204, −0.25% for 208/204, 1.8% for 207/206, and −0.5% for 208/206. Standard deviations on replicate runs ranged from ±0.013 for 207/206 to ±0.86 for 208/204. Gold samples supplied by the Geological Survey of Canada with "known" lead isotopic ratios were also run. Agreements were moderate to good. Because the instrument was a prototype, some difficulties were encountered which do not exist in the commercial model. For example, sensitivities are now improved by a factor of 10 compared to those obtainable by these authors.

Equipment

A prototype model of the newly developed ELAN 250 inductively coupled plasma–mass spectrometer (Sciex) was used throughout this study. A schematic of the instrument is shown in Fig. 10.1. A more detailed instrument description can be found elsewhere (10). This instrument combines an inductively coupled argon plasma (ICP), as the ion source, with a quadrupole mass spectrometer. A sample solution is nebulized into the ICP and the metal ions are extracted from the plasma gas into the mass spectrometer. The interfacing arrangement between the plasma and the mass spectrometer is similar to that reported by Gray and Date (19) but differs in some important details. The sampling orifice lifetime is many months [cf. 30 h reported by Gray and Date (19)]; doubly charged ions, which are a potential source of spectral interference, are present at much lower levels; and no ions characteristic of the orifice material (Ni) appear in the mass spectrum. This configuration immediately provides the analyst with the capability of determining isotopic concentrations with the same ease as elemental analyses. In fact, any elemental analyses done on this instrument are in essence isotopic analyses. Low detection limits (0.01–1 ng/ml) and a wide dynamic range (5–6 orders of magnitude) are possible, as might be expected of any instrument using an ion counter for spectral measurement.

Table 10.7
Lead Isotope Ratios of Lead-Bearing Samples by Plasma–Mass Spectrometry

Sample	Sample location	$\frac{206}{204}$	$\frac{207}{204}$	$\frac{208}{204}$	$\frac{207}{206}$	$\frac{208}{206}$
NBS 1981	National Bureau of Standards,	16.98 ± 0.60[a]	15.81 ± 0.38	36.63 ± 0.86	0.9311 ± 0.013	2.157 ± 0.032
	United States	16.94 (0.2%)[b]	15.49 (2.0%)	36.73 (−0.25%)	0.9146 (1.8%)	2.168 (−0.5%)
SP-1392[c]	Lake George, Canada	16.90 ± 0.56[a]	15.79 ± 0.78	35.43 ± 1.65	0.9343 ± 0.014	2.096 ± 0.031
		16.51[d]	15.53	36.06	0.9409	2.184
TQ65-14[c]	Echo Bay, Canada	20.19 ± 0.36[a]	15.62 ± 0.37	37.01 ± 1.09	0.7737 ± 0.012	1.830 ± 0.027
		20.35[d]	15.85	39.74	0.7791	1.940
SP-706[c]	Brunswick No. 6, Canada	19.05 ± 0.88[a]	16.32 ± 0.57	39.95 ± 1.54	0.8567 ± 0.013	2.097 ± 0.032
		18.28[d]	15.66	38.22	0.8593	2.091
RFA64-417[c]	Kerr Addison, Canada	14.64 ± 0.75[a]	15.95 ± 0.72	36.20 ± 1.85	1.089 ± 0.016	2.473 ± 0.037
		13.69[d]	14.69	33.53	1.073	2.449
TQ70-150[c]	Detroit Algoma, Canada	25.36 ± 0.71[a]	17.30 ± 0.51	38.33 ± 1.68	0.6822 ± 0.010	1.511 ± 0.015
		25.82[d]	17.32	40.31	0.6710	1.561
M22498[e]	Sandon, Caucasus, USSR	18.63 ± 0.26[a]	15.74 ± 0.31	38.34 ± 0.77	0.8450 ± 0.011	2.058 ± 0.030
M27901[e]	Nigadoo Mines	18.34 ± 0.27[a]	15.59 ± 0.31	37.85 ± 0.76	0.8501 ± 0.009	2.064 ± 0.030
	Bathurst, New Brunswick, Canada	18.20[f]	15.65	38.11	0.8600	2.094
M14679[e]	Worthington Mine,	26.37 ± 0.37[a]	16.89 ± 0.35	51.63 ± 1.03	0.6405 ± 0.008	1.958 ± 0.029
	Sudbury, Canada					
M14093[e]	Secondary Mine	16.19 ± 0.23[a]	15.56 ± 0.28	35.99 ± 0.68	0.9610 ± 0.013	2.223 ± 0.033
	Broken Hill, Australia	15.98[g]	15.37	35.56	0.9616	2.225
M32209[e]	Pine Point Mine,	18.87 ± 0.28[a]	16.31 ± 0.31	40.50 ± 0.80	0.8645 ± 0.015	2.146 ± 0.032
	Northwest Territories, Canada					

[a]Values obtained with the ELAN 250 (ICP–MS).
[b]NBS certified values; ( ) percent difference.
[c]Samples courtesy of Geological Survey of Canada, Ottawa, Ontario, Canada.
[d]Values supplied by Geological Survey of Canada.
[e]Samples courtesy of Royal Ontario Museum, Toronto, Ontario, Canada.
[f]Ref. 20.
[g]Ref. 21.

Procedure

Weigh about 20 mg of sample into a beaker. Add 10 ml of reagent grade hydrochloric acid and reflux the sample for ½ h. Add 5 ml of nitric acid and heat the sample to complete dissolution. Cool and dilute to 100 ml with distilled water. Dilute an aliquot of solution (20 ml) by a factor of 5 for measurement of the lead isotopes. The undiluted solution is used to determine trace element spectra.

The lead isotope data were produced by repeated peak hopping between ^{204}Pb, ^{206}Pb, ^{207}Pb, and ^{208}Pb using a dwell time of 2 s per isotope and a total measurement time of 10 min per sample. A total of 300 isotope measurements were obtained on each sample. After collecting the data on disk, each set of measurements was divided into seven blocks of 40 measurements each. Within each block the average count rate of each isotope was calculated and the isotope ratios determined. The individual ratios of each block were then averaged and the standard deviation calculated. The uncertainties of Table 10.7 (20, 21) are 1 standard deviation (not the standard deviation of the mean). Approximately 1–2 mg of lead was required per sample.

XII. ANALYSIS OF LEAF SAMPLES USING ICP–MS WITH ELECTROTHERMAL SAMPLE INTRODUCTION (22)

Comments on the Method

The following procedure has been proved for the determination of As, Cu, Mn, Pb, Rb, V, and Ag in organic samples and hence is well suited for biogeochemical work. Preliminary work on samples containing high levels of inorganic material suggests that the method can be used on such samples in a semiquantitative way. More specifically, it is suitable for solids containing a good deal of silicate matter if the relative amounts of elements are of interest.

Because the level of the above metals is very low in many types of organic samples (e.g., leaves), a method for sample introduction which would result in an improvement in detection limits was sought. Electrothermal vaporizers (ETVs) have been used in ICP–optical emission spectrometry for this purpose and hence this approach was adapted to the mass spectrometric technique. The detection limits achieved were 1–3 orders of magnitude better than those in ICP–optical emission spectrometry and 1 order of magnitude better than those in ICP–mass spectrometry when conventional nebulization was being used. Detection limits are given in Table 10.8. The method was tested using NBS Orchard Leaves and Oyster Tissue. Values obtained for As, Cu, Mn, Rb, V, and Zn were in acceptable agreement with the certified values. In the case of Pb and Ag the certified values are near the detection limit for this procedure. The results obtained for these elements by the proposed procedure were about 1.2–5 times too high. For best results standardization should be done by the method of standard additions.

Table 10.8
Detection Limits (μg/g)
(Leaves)

As	0.32
Cu	0.07
Mn	0.1
Pb	1.1
Zn	0.2
V	0.21
Rb	0.04

Equipment

The ICP/MS apparatus used was designed by Douglas and French (23) at the Institute for Aerospace Studies, University of Toronto. Further development by Sciex has resulted in a commercial product, the ELAN.

The electrothermal vaporizer was developed for the purposes of this investigation and its design and optimum operation are described in detail in a previous publication (24). A cross-sectional drawing of the ETV is shown in Fig. 10.6. A summary of its characteristics is given in Table 10.9. The filament consists of 0.05 mm thick rhenium metal ribbon (available from REMBAR Laboratory, New York) of sample capacity 5 μl, which can be resistively heated using a crude home-made power supply. This is constructed from a 1 kW transformer and laboratory variac.

Figure 10.6. Cross section of electrothermal vaporizer.

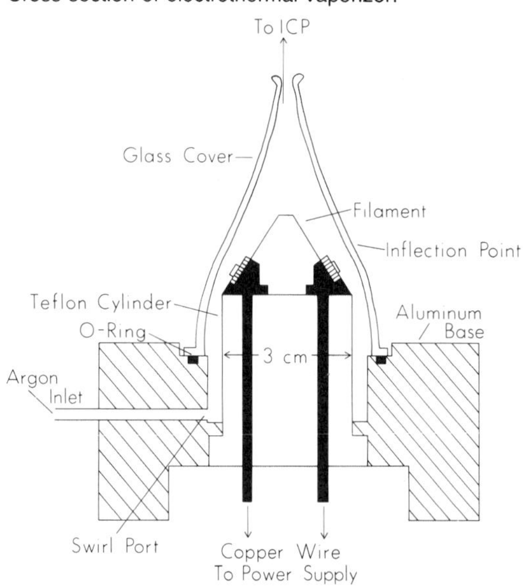

Table 10.9
Summary of ETV Characteristics

Volume of glass dome above filament	5 ml
Inert gas introduction	Tangential at base (2 liters/min), laminar flow (Reynolds number of system is 150)
Filament	Re metal (central sample dimple 1mm deep by 5 mm diameter—5 μl capacity)
Maximum temperature	1800°C (above this severe plasma loading occurs due to Re loss)
Particle size of aerosol	0.5–1.0 μm (Zn)
Transport efficiency[a] at 2 liters/min (%)	Cd (85), PB (83), Cu (84) Fe (85), Ni (83).

[a]Greatly improved compared to previously listed devices.

Maximum output capability is 2.5 V and 30 A. The filament is encased in a glass dome, which is sealed to a metallic base by an O-ring. Argon flows through the device to carry the vaporized sample, via the Tygon tubing, into the ICP.

The ICP was obtained from Henry Radio (model 2500). Output powers up to 2 kW are achievable. This device must be operated and adjusted for the most part manually. A standard cross-flow nebulizer for solution sample introduction, when required, was used for initial optimization of ion optical and MS parameters for each element.

Mass selection is achieved with a quadrupole mass spectrometer. The quadrupole is particularly well suited for mass analysis using a plasma source. The operating pressures of 10^{-4} to 10^{-5} torr tolerated by quadrupoles reduce the pumping speed requirement and allow the use of relatively large orifices. Fringing field effects, which are encountered at the entrance and exit of quadrupoles, are minimized by the use of short sections of ac-only quadrupole rod sets. This provides improved resolution capability.

The interface of the plasma to the ion optical system consists of a sample orifice approximately 0.4 mm in diameter. The orifice is a nickel conical structure mounted on a water-cooled copper flange. Immediately behind the sample, and separated by a region (12 × the diameter of the orifice) which is pumped mechanically to about 1 torr, is a skimmer orifice 1 mm in diameter.

A cryogenic pumping system, supplied by CTI-Cryogenics Ltd., is capable of pumping the mass spectrometer down to a pressure of about 2×10^{-5} torr during operation.

A model 9176B *x–y* recorder, available from Varian, was employed for display of spectra. Operating conditions for the equipment are given in Table 10.10.

Table 10.10
Operating Conditions

Forward power	1000 W
Reflected power	0 W
Coolant flow	14 liters/min
Auxiliary flow	0.7 liter/min
Carrier flow	2 liters/min
Drying	100°C for 20 s
Ashing	(400–800°C)[a]
Vaporization	(up to 1800°C)[a]

[a]Depends on type of sample and element.

Reagents

All metal salts, solvents, and acids used were reagent grade quality or better. Atomic absorption standard solutions of metals (As, Cu, Mn, Pb, Rb, V, Zn, and Ag), 1 mg/ml in concentration, were obtained from Fisher Scientific.

Calibrating solutions (micrograms per milliliter) were prepared by dilution of the 1000 μg/ml standards. For preparation of nanogram per milliliter calibrating solutions the 10 or 1 μg/ml calibration solutions were diluted. All calibration solutions were for single elements and were made to contain 1% nitric acid on final dilution. The concentration range covered was 1 ng/ml to 10 μg/ml.

National Bureau of Standards (Washington, D.C.) standard reference samples were used to test the methodology. Samples analyzed were Orchard Leaves (SRM 1571) and Oyster Tissue (SRM 1566).

Procedure

Pump the system down to 1 torr in the interface region and 2×10^{-7} torr in the mass spectrometer. (This takes approximately 2.5 h from a cold start.) Turn on all electronic systems and ignite the plasma. Aspirate a 1 μg/ml solution of the element to be determined. Optimize the ion optical components so that the maximum analyte signal is obtained. The instrument must then be left operating until any analyte signal drift has stopped, a minimum of 1 h. At this point the system is changed from nebulizer to ETV operation.

Orchard Leaves (SRM 1571) and Oyster Tissue (SRM 1566) samples are dissolved in closed Teflon digestion vessels by the method of Brzezinska *et al.* (25). A mixture of nitric, perchloric, and hydrofluoric acids is used. The acid content of the solutions after dilution to final volume is 1% in nitric acid.

Sample and standard solutions are analyzed as follows. Remove the glass dome from the ETV. Pipet a 2 μl aliquot of sample or standard solution onto the rhenium filament. (The auxiliary argon flow, 2 liters/min of argon to the plasma, bypasses the ETV during this operation.) Replace the glass dome and clamp it tightly onto the ETV base using the toggle clamps. Divert the ETV argon flow of 2 liters/min by means of the three-way stopcock through to a fume hood vent. Dry the solution at about 100°C until the drop of liquid has been completely evaporated (visual observation). Char the standard or sample, if necessary, at 400–800°C for 30–60 s. (This char step is suitable for the determination of Cu, Mn, Pb, Rb, V, Zn, and Ag.) (If As is to be analyzed add a 2 μl drop of 1% nickel nitrate solution after drying the sample or standard solution and repeat the drying step.)

Using a three-way stopcock, switch the ETV argon flow (at 2 liters/min) to pass into the plasma. Then turn off the auxiliary argon flow. Vaporize the sample at up to 1800°C. Record the transient signal obtained. Signal integration must be used in computing elemental concentration in Orchard Leaves and Oyster Tissue solutions.

Calibration is done by the method of standard additions as follows. Run the sample alone, as outlined above. Then add 2 μl of a suitable concentration of calibrating solution to the rhenium filament and dry as indicated above. Remove the glass envelope and add 2 μl of the sample solution and dry again as detailed above. Then vaporize the material on the rhenium filament according to the above procedure. This procedure is then repeated using a 2 μl spike of calibrating solution which is twice the concentration of the first spike. Again, it is important to emphasize that signal integration must be employed in computing the signal values for these runs.

REFERENCES

1. G. J. O'Halloran and R. A. Flügge, *Star Rep.* **NG4–21711,** 1931 (1964).
2. A. L. Gray, *Analyst (London)* **100,** 289 (1975).
3. R. S. Houk, V. A. Fassel, G. D. Flesch, H. J. Suec, A. L. Gray, and C. E. Taylor, *Anal. Chem.* **52,** 2283 (1980).
4. A. L. Gray and A. R. Date, *Analyst (London)* **106,** 1255 (1981).
5. A. R. Date and A. L. Gray, *Analyst (London)* **108,** 159 (1983).
6. A. R. Date and A. L. Gray, *Spectrochim. Acta, Part B* **38B,** 29 (1983).
7. R. S. Houk, V. A. Fassel, and H. J. Svec, *Dyn. Mass Spectrom.* **6,** 234 (1981).
8. R. S. Houk, A. Montaser, and V. A. Fassel, *Appl. Spectrosc.* **37,** 425 (1983).
9. R. S. Houk and J. J. Thompson, *Mineralogist* **67,** 238 (1982).
10. D. J. Douglas, E. S. K. Quan, and R. G. Smith, *Spectrochim. Acta, Part B* **38B,** 39 (1983).
11. D. J. Douglas, *Can. Res.* **16,** 55 (1983).
12. R. G. Smith, E. J. Brooker, D. J. Douglas, and E. S. K. Quan, *J. Geochem. Explor.* **21,** 385 (1983).
13. D. J. Douglas and R. S. Houk, *Prog. Anal. At. Spectrosc.* **8,** 1 (1985).
14. A. R. Date, *Trends Anal. Chem.* **2,** 225 (1983).
15. A. R. Date and A. L. Gray, *Spectrochim. Acta, Part B* **40B,** 115 (1985).
16. J. W. Carr and G. Horlick, *Spectrochim. Acta, Part B* **37B,** 1 (1982).
17. W. Doherty and A. VanderVoet, *Can. J. Spectrosc.* **30,** 135 (1985).
18. S. R. Koirtyohann, J. S. Jones, C. P. Jester, and D. A. Yates, *Spectrochim. Acta, Part B* **36B,** 49 (1981).
19. A. L. Gray and A. R. Date, *Analyst (London)* **108,** 1033 (1983).
20. B. R. Doe and J. Stacey, *Econ. Geol.* **69,** 757 (1974).

21. B. L. Gulson and K. J. Mizon, *J. Geochem. Explor.* **11,** 299 (1979).
22. C. J. Park, J. C. Van Loon, P. Arrowsmith, and J. B. French, internal laboratory method, unpublished.
23. D. J. Douglas and J. B. French, *Anal. Chem.* **53,** 37 (1981).
24. C. J. Park, J. C. Van Loon, P. Arrowsmith, and J. B. French, *Can. J. Spectrosc.* **32,** 29 (1987).
25. A. Brzezinska, A. M. Balicki, and J. C. Van Loon, *Water, Air, Soil Pollut.* **21,** 323 (1984).

Index